Quantum Electrodynamics at High Energies

Translation Series

Quantum Electrodynamics at High Energies

P. S. Isaev

Translated by
David Parsons

Originally published as *Kvantovaya electrodinamika v oblasti vysokih energiy.*

Library of Congress Cataloging-in-Publication Data

Isaev, P. S. (Petr Stepanovich)
Quantum electrodynamics at high energies.

Translation of: Kvantovaia ėlektrodinamika v oblasti vysokikh energiĭ. 1984.
Bibliography: p.
Includes index.
1. Nuclear reactions. 2. Quantum electrodynamics. 3. Particles (Nuclear physics) I. Title. II. Series.
QC794.8.H5I8313 1989 539.7′5 88-34983
ISBN 978-0-88318-602-2

Contents

Chapter 1. Tests of quantum electrodynamics at small distances

Chapter 2. Electromagnetic form factors of elementary particles

Chapter 3. Deep inelastic scattering of leptons by nucleons

Chapter 4. One-photon e^+e^- annihilation

Chapter 5. The $\gamma\gamma$ interaction

Preface

Many important new results have been obtained over the past 10 or 15 years in the field of quantum electrodynamics (QED). For example, scaling has been observed in deep inelastic scattering of leptons by nucleons; the J/ψ and Υ families of particles and charmed particles have been discovered; new limits have been established on the applicability of QED; and new data have been obtained on the form factors of nucleons, π mesons, K mesons, and η mesons. A new direction has formed in elementary-particle physics: the physics of the $\gamma\gamma$ interaction.

Quantum electrodynamics has served as a solid foundation for the construction of quantum chromodynamics (QCD). It is fair to say that such questions as the quark-gluon structure of hadrons, the spins of quarks and gluons, the point-like nature of the quark-gluon interaction, the discovery of new quark species, the properties of the J/ψ and Υ particles, the properties of charmed mesons, the jet nature of the interaction of particles at high energies, and many other topics could not have been studied as thoroughly or as rapidly without QED. All the successes of QED and most of the successes of QCD at high energies have been based on experiments carried out either at the Stanford Linear Accelerator Center (SLAC, at Stanford, California) or at accelerators with colliding e^+e^- beams.

Over the past decade, research in QED and QCD has been developing at a remarkable pace and in impressive depth. So many new results have been obtained that writing a scientific monograph to generalize the recent achievements is at once highly desirable and extremely complicated. Difficulties arise not only in terms of selecting material but also because of the frequent change in the points of view regarding specific physical phenomena.

The factors which led me to take up the extremely nontrivial task of attempting to select and generalize certain advances in QED at high energies (making partial use of QCD) were the absence of monographs on current problems in QED at high energies; my own research on the electromagnetic form factors of elementary particles, deep inelastic scattering of leptons by nucleons, and the $\gamma\gamma$ interaction; and my reading of lectures on these problems to young scientists at the Joint Institute for Nuclear Research.

The development of those branches of QED and QCD which are included in this book can be credited to many institutes and laboratories around the world, many collectives and groups of scientists, and, finally, individual scientists. Accordingly, we will be citing for the most part reviews which have been published in the proceedings of conferences and symposia on elementary-particle physics over the past decade. In those reviews the reader can find comprehensive information on the establishment of this field of physics and the further references which he needs.

The contents of this book are accessible to anyone who knows the quantum field theory in the now-classic monograph *Introduction to the Theory of*

Quantized Fields, by N. N. Bogolyubov and D. V. Shirkov (Nauka, Moscow, 1973; English translation published by Wiley-Interscience, New York, 1980). The derivations of several of the equations given in the text can be found in the well-known monograph *Quantum Electrodynamics*, by A. I. Akhiezer and V. B. Berestetskiĭ (Nauka, Moscow, 1969; an English translation of an earlier edition was published by Wiley, New York, in 1965), and in the book *Lectures in the Physics of Neutrino and Lepton-Nucleon Processes*, by S. M. Bilen'kiĭ (Energoizdat, Moscow, 1981). The latter volume also gives a detailed exposition of the Weinberg-Salam theory, which we will be citing with regard to the theoretical interpretation of certain physical phenomena.

I am deeply indebted to S. B. Gerasimov and A. V. Efremov for reading chapters of this book and for several valuable comments.

P.S.I.

Chapter 1
Tests of quantum electrodynamics at small distances

1. The meaning of tests of quantum electrodynamics at small distances

Calculations for high energies in QED are usually carried out in lowest-order perturbation theory. The constant $\alpha = e^2/\hbar c = 1/137$, which is used in the perturbation-theory series, is small, and the higher orders in α, in which the infinities which arise are eliminated through a renormalization of the divergent expressions, introduce small corrections. In some cases the discrepancies between experimental data and theoretical predictions cannot be explained by contributions of higher-order QED perturbation theories. In such cases it becomes necessary to introduce the strong interactions: It is assumed that QED is correct, and differences between the QED predictions and experimental data are attributed to strong interactions. Since the calculations on the strong-interaction components to QED processes are inaccurate, there is the question of whether it is at all possible to carry out a fundamental test of QED at small distances.

As the energy is increased, weak interactions become comparable in importance to strong interactions in the unitary limit. The overall picture of the interaction of elementary particles becomes extremely complex. At such energies both strong and weak interactions must be taken into consideration in order to analyze discrepancies between QED predictions and experimental data. At the energies attainable at accelerators today, as a rule, weak interactions are generally negligible in QED processes.

The meaning of a test of QED at small distances can be seen in the example of testing the classical Coulomb's law:

$$F = f q_1 q_2 / r^2$$

This law is known to hold for point charges or for charges which are separated by a distance large enough that the shape and dimensions of the charges can be ignored. In general, when nonpoint charges close on each other, the force F will eventually cease to satisfy Coulomb's law. When a deviation from Coulomb's law is observed, it is difficult to determine whether the law is breaking down or whether the charges are not point charges. There is a corresponding meaning in the concept of testing QED at small distances.

If a deviation from QED is observed, we will assume that it is caused by the finite sizes of the particles. This assumption does not touch upon the foundations of QED.

Testing QED at small distances means arranging experiments with a large momentum transfer. It follows from the uncertainty relation

$$\Delta x \Delta p \gtrsim \hbar$$

that as the 3-momentum transfer $|\Delta \mathbf{p}|$ increases, there will be a decrease in the distance involved in the experimental test. At ordinary electron accelerators, even those with large enough energies, the 3-momentum transfer is small: $|\Delta \mathbf{p}|$ increases in proportion to the square root of the energy of the accelerated particle. The Erevan electron synchrotron, for example, provides a beam energy of 6 GeV. In this case the maximum possible momentum transfer is

$$|\Delta \mathbf{p}| \lesssim \sqrt{2mE} \approx 80 \text{ MeV}/c, \tag{1.1}$$

which leads to value $\Delta x_E \sim 2f = 2\times 10^{-15}$ m $= 2$ fm. This distance is comparable to the dimensions of the π-meson "cloud" of a nucleon, which is determined by the Compton wavelength of the π meson ($\lambda \sim \hbar/m_\pi c \sim 10^{-15}$ m), so that experiments with such a momentum transfer are of little help for a thorough study of the electromagnetic structure of the proton.

A large momentum transfer can be achieved at a colliding-beam accelerator. In this case we have

$$\Delta p \lesssim 2E; \tag{1.2}$$

i.e., the momentum transfer increases in proportion to the energy of the accelerated particles. If we construct an accelerator with an energy of 2×6 GeV (i.e., if each of the colliding beams has an energy of 6 GeV), then the distances down to which QED can be tested will be $\Delta x \sim \hbar/\Delta p \sim 10^{-17}$ m, or two orders of magnitude smaller than Δx_E.

Constructing colliding-beam accelerators requires overcoming some serious technical difficulties. Such accelerators are also rather expensive. Even further, they suffer from a low luminosity (Sec. 1 of Chap. 4), and this is their main drawback. Nevertheless, the list of such accelerators in operation today is fairly long (several have now been shut down; see Table 1.1).[1]

The energy of the PETRA accelerator is to be increased to 2×23 GeV. There are also plans for constructing new e^+e^- colliders:*

(1) The LEP (large electron-positron) accelerator with an energy of 2×50 GeV (oppositely directed e^+ and e^- beams, each with an energy of 50 GeV). Construction is to be completed in 1987.

(2) A Cornell accelerator with an energy of 2×50 GeV (construction began in 1983 and was to be completed by 1986).

(3) A Stanford accelerator (a single-pass collider) with an energy of 2×50 GeV (the energy is subsequently to be increased to 2×70 GeV; this accelerator is to be constructed over three years).

Table 1.1. Characteristics of colliding-beam accelerators.

Accelerator	Location	Accelerated particles	E_{max} (GeV)	Started up
Ada	Frascati	e^+e^-	0.25	1960
Princeton-Stanford	Stanford	e^-e^-	0.55	1962
ACO	Orsay	e^+e^-	0.55	1966
VÉPP-2	Novosibirsk	e^+e^-	0.55	1966
ADONE	Frascati	e^+e^-	1.55	1969
BYPASS	Cambridge	e^+e^-	3.5	1971
SPEAR II	Stanford	e^+e^-	4.2	1972
DORIS	Hamburg	e^+e^-,e^-e^-	4.5	1974
VÉPP-2M	Novosibirsk	e^+e^-	0.7	1975
DCI	Orsay	e^+e^-	1.7	1976
VÉPP-4	Novosibirsk	e^+e^-	5.5	1979
PETRA	Hamburg	e^+e^-	19	1978
PEP	Stanford	e^+e^-	19	1979
CESR	Cornell	e^+e^-	8	1979

The luminosity (L) of powerful accelerators will reach $(1.5\text{–}3)\times 10^{31}$ $\mathrm{s}^{-1}\,\mathrm{cm}^{-2}$.

Plans of this sort for constructing powerful new accelerators with colliding e^+e^- beams predestine an extensive research program in QED through the year 2000.

At such high energies of the colliding beams, strong interactions will become more important. Until recently, all the experimental deviations from QED predictions at high energies could be explained satisfactorily by correcting for strong interactions, by using a dispersion-relation method, or by introducing model assumptions of one sort or another (e.g., assumptions based on the model of vector dominance). Quantum chromodynamics (QCD) is presently being offered as a candidate for the role of a theory of strong interactions. Although QCD has not yet been developed into a quantitative theory, it has already been successful in some significant ways. We will make systematic use of QCD in interpreting experimental data.

What do the words "test of QED at small distances" mean in the language of matrix elements?[2] We know that in QED one makes use of the concept of a point-like interaction of electrons with photons and that the physical processes which are studied experimentally are associated with the matrix elements of a perturbation theory. These matrix elements can in turn be represented in a graphic way as Feynman diagrams. The structural elements of these diagrams are the photon and electron propagators and vertices. Relativistic invariance allows the following generalizations of the propagator functions which are used in perturbation theory: for a photon propagator,

$$D(q^2) \sim F(q^2)/q^2\,; \tag{1.3}$$

and for an electron propagator,

$$S(p^2) = [G_1(p^2)\hat{p} + G_2(p^2)m]/(p^2 - m^2)\,. \tag{1.4}$$

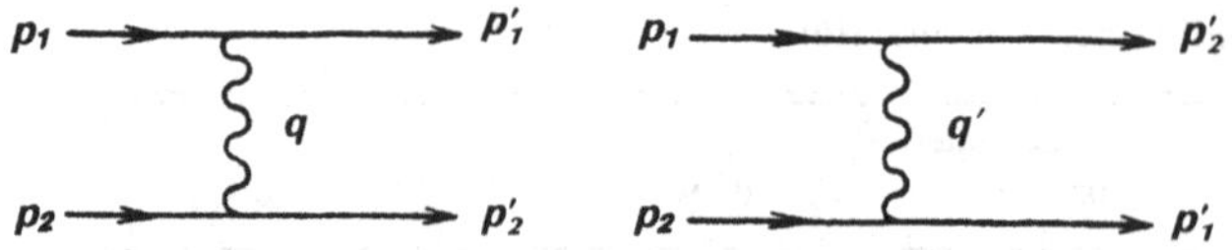

Figure 1.1. Feynman diagrams describing Møller scattering ($e^-e^-\to e^-e^-$).

Also allowed is a generalization of the vertex function for a spin-1/2 particle:

$$\Gamma_\mu(p,p',q)=e[\gamma_\mu\, F_1(q^2)+(i\mu/2m)\sigma_{\mu\nu}q_\nu\, F_2(q^2)]\,. \tag{1.5}$$

Here $F(q^2)$, $G_1(p^2)$, $G_2(p^2)$, $F_1(q^2)$, and $F_2(q^2)$ are arbitrary scalar functions; q is the momentum of a virtual photon; p is the momentum of a virtual particle with a spin; m is the mass of a particle with a spin, and μ is its anomalous magnetic moment; γ_μ are the Dirac matrices; and $\sigma_{\mu\nu}=\gamma_\mu\gamma_\nu-\gamma_\nu\gamma_\mu$. Photon and electron propagators or lines corresponding to free particles—electrons and photons—always come into a vertex. In a test of QED at small distances, the deviations from "point-like" QED are seen as deviations of the functions F, G_i, and F_i from unity. A test of a photon propagator differs substantially in its content from a test of an electron propagator.

As a specific example we consider the scattering of an electron by an electron (Møller scattering). In lowest-order perturbation theory, there are two Feynman diagrams which correspond to this scattering (Fig. 1.1).

In writing matrix elements corresponding to these diagrams, we will use a modified photon propagator (1.3), and in the vertex function (1.5) we will retain only the term with $F_1(q^2)$, since the term with $F_2(q^2)$ makes a negligibly small contribution to the differential cross section for the process (because of the small value of the ratio $\mu/2m$). In this case the factors $F_1(q^2)$ and $F_2(q^2)$ appear in the following combination in the matrix element for each of the diagrams:

$$\Phi(q^2)=F_1((p_1-p_1')^2F(q^2)F_1(p_2-p_2')^2)\,, \tag{1.6}$$

where $q^2=t=(p_1-p_1')^2=(p_2-p_2')^2$. We will also make frequent use of the notation $s=(p_1+p_2)^2$, $u=(p_1-p_2')^2$.

The matrix element for the scattering of an electron by an electron is written in the form[1]

$$\langle f|S^{(2)}|i\rangle=-\frac{ie^2(2\pi)^4}{4\sqrt{E_1E_2E_1'E_2'}}M\delta(p_1+p_2-p_1'-p_2')\,,$$

where E_i and E_i' are the energies of the ith electron in the initial and final states, and

$$M=\frac{\bar u(p_1')\gamma_\mu\, u(p_1)\bar u(p_2')\gamma_\mu\, u(p_2)}{t}\Phi(t)$$
$$+\frac{\bar u(p_2')\gamma_\mu\, u(p_1)\bar u(p_1')\gamma_\mu\, u(p_2)}{u}\Phi(u)\,.$$

The differential cross section is given by

$$\frac{d\sigma}{d\Omega}=\frac{e^4}{(2\pi)^2}\frac{1}{4E_2E_2'}\frac{1}{4}|M|^2 p_1'^2 dp_1' p_2'^2 dp_2' \frac{\delta(p_1+p_2-p_1'-p_2')}{\sqrt{(p_1p_2)^2-m^4}},$$

where $d\Omega=\sin\theta\, d\theta$, and θ is the electron scattering angle. After an integration over dp_1' and dp_2' [this integration is trivial because of the presence of a product of δ functions: $\delta(E_1+E_2-E_1'-E_2')\delta(\mathbf{p}_1+\mathbf{p}_2-\mathbf{p}_1'-\mathbf{p}_2')$] the expression for the cross section in the center-of-mass (c.m.) frame of the colliding electrons ($\mathbf{p}_1+\mathbf{p}_2=0=\mathbf{p}_2'+\mathbf{p}_1'$) becomes

$$\frac{d\sigma}{d\Omega}=\frac{e^4}{(2\pi)^2}\frac{1}{64E^2}|M|^2,$$

where $E=E_1=E_2=E_1'=E_2'$. Taking an average over the initial spin states and summing over the final spin states, we find the expression

$$\frac{d\sigma}{d\Omega}=\frac{\alpha^2}{2E^2}\left[\frac{(p_1p_2)^2+(p_1p_2')^2+2m^2p_1p_2-2m^2p_1p_2'}{t^2}\Phi^2(t)\right.$$
$$+\frac{(p_1p_2)^2+(p_1p_1')^2+2m^2p_1p_2-2m^2p_1p_1'}{u^2}\Phi^2(u)$$
$$\left.+2\frac{(p_1p_2)^2+2m^2p_1p_2}{ut}\Phi(u)\Phi(t)\right],$$

which can be written in terms of the variables s, u, and t ($s=2m^2+2p_1p_2$; $u=2m^2-2p_1p_2'$; $t=2m^2-2p_1p_1'$):

$$\frac{d\sigma}{d\Omega}=\frac{\alpha^2}{8E^2}\left[\frac{s^2+u^2-8m^4}{t^2}\Phi^2(t)\right.$$
$$\left.+\frac{s^2+t^2-8m^4}{u^2}\Phi^2(u)+2\frac{s^2-4m^4}{ut}\Phi(t)\Phi(u)\right]. \tag{1.7}$$

We assume that function (1.6) can be explained in a Taylor series around $q^2=0$. Retaining only the first two terms of this expansion, we write $\Phi(q^2)$ as

$$\Phi(q^2)=1+q^2/\Lambda_\gamma^2, \tag{1.8}$$

where q^2 is the 4-momentum transfer, and Λ_γ^2 is a positive and fairly large constant, which is to be determined from a comparison with experiment. If we ignore the mass of the electron in (1.7) (as we are almost always justified in doing at high energies), expression (1.7) simplifies. In this approximation the variables s, u, and t are related by the equation $s+u+t=0$. We then have

$$\Phi^2(t)=(1+q^2/\Lambda_\gamma^2)^2\approx 1+2t/\Lambda_\gamma^2;$$

$$\Phi^2(u)=\left(1+\frac{u}{\Lambda_\gamma^2}\right)^2\approx 1+\frac{2u}{\Lambda_\gamma^2}=1-\frac{2s}{\Lambda_\gamma^2}-\frac{2t}{\Lambda_\gamma^2};$$

$$\Phi(t)\Phi(u)=\left(1+\frac{t}{\Lambda_\gamma^2}\right)\left(1+\frac{u}{\Lambda_\gamma^2}\right)\approx 1+\frac{t+u}{\Lambda_\gamma^2}=1-\frac{s}{\Lambda_\gamma^2}.$$

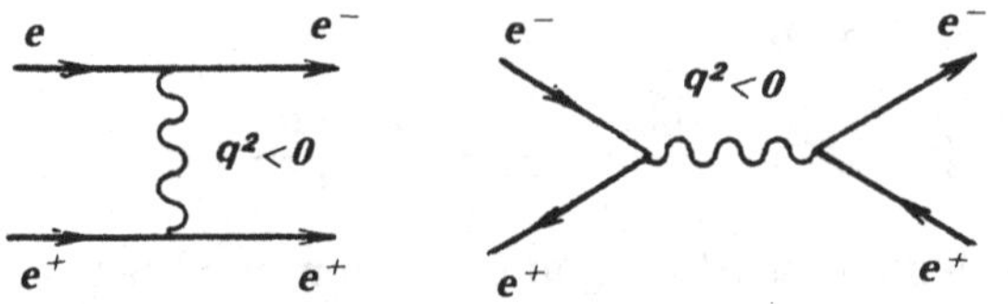

Figure 1.2. Diagrams describing Bhabha scattering ($e^-e^+ \rightarrow e^-e^+$).

We denote by $(d\sigma/d\Omega)_{\text{Møll}}$ the version of cross section (1.7) with $\Phi(t) = \Phi(u) = 1$. We construct the difference $d\sigma/d\Omega - (d\sigma/d\Omega)_{\text{Møll}}$, which is a measure of the model-independent deviation of the photon propagator from that adopted in the QED perturbation theory:

$$\frac{d\sigma}{d\Omega} - \left(\frac{d\sigma}{d\Omega}\right)_{\text{Møll}} = -\frac{\alpha^2}{4\Lambda_\gamma^2 E^2}\,\frac{u^3 + t^3 - 2s^3}{ut}.$$

Hence,

$$\frac{d\sigma/d\Omega - (d\sigma/d\Omega)_{\text{Møll}}}{\pm(\alpha^2/4E^2)(u^3 + t^3 - 2s^3)/ut} = \frac{1}{\Lambda_\gamma^2}. \tag{1.9}$$

The plus and minus signs in the denominator on the left-hand side of (1.9) have been introduced in order to keep the ratio on the left-hand side positive at all times, since Λ_γ^2 is by definition a positive quantity. Finally, we assign Λ_γ^2 plus- and minus-sign indices: $\Lambda^2_{\pm\gamma}$. We know that the differential cross section for e^-e^- scattering is measured within an experimental error, i.e., in a certain error corridor. The upper value of the experimental cross section corresponds to $\Lambda^2_{+\gamma}$, and the lower value to $\Lambda^2_{-\gamma}$. In each experiment, we thus obtain two different values of Λ_γ^2 for a given energy of the colliding e^-e^- beams. These two values determine limits on the applicability of QED. Using the uncertainty relation $\Delta x|\Delta \mathbf{p}| \gtrsim \hbar$, we can express the limiting values Λ_γ^2 or, alternatively, $(\Delta \mathbf{p})^2$ in centimeters and find the limits on the applicability of QED in customary units of length.

The main conclusion here is that the length Λ_γ is determined in a manner which does not depend on the nature of the function $\Phi(q^2)$, i.e., it is determined in a model-independent way.

The photon propagator can also be tested at positive values of q^2. For this purpose we should examine Bhabha scattering, i.e., the scattering of an electron by a positron, for which one of the diagrams contains a propagator with $q^2 > 0$ (Fig. 1.2).

To find an expression for the Bhabha scattering cross section analogous to (1.7), it is sufficient to make use of the fact that when we switch from the scattering process $e^-e^- \rightarrow e^-e^-$ to the process $e^+e^- \rightarrow e^+e^-$, the variable s is replaced by the variable t, and vice versa: $s \rightleftarrows t$. We thus find

$$\frac{d\sigma}{d\Omega}=\frac{\alpha^2}{8E^2}\left[\frac{t^2+u^2-8m^4}{s^2}\Phi^2(s)+\frac{s^2+t^2-8m^4}{u^2}\Phi^2(u)\right.$$
$$\left.+2\frac{t^2-4m^4}{us}\Phi(u)\Phi(s)\right]. \qquad (1.10)$$

At large values s, u, and $t \gg m^2$ we can ignore the masses of the electron and the positron. The analog of expression (1.9) for Bhabha scattering is

$$\frac{d\sigma/d\Omega-(d\sigma/d\Omega)_{\text{Bhabha}}}{\pm(\alpha^2/4E^2)(u^3+s^3-2t^3)/us}=\frac{1}{\Lambda^2_{\pm\gamma}}.$$

If we assume that the principles of crossing symmetry and analyticity hold for electromagnetic interactions (they do hold for strong interactions, as is proved by the agreement between experimental data and the predictions of the dispersion relations for πN, πK, and NN scattering and for several other processes), then for Møller scattering and Bhabha scattering these principles would be manifested in the following way: In a test of QED for the same values of $|q^2|$, the value of Λ_γ should be the same in both processes, because of crossing symmetry ($s \rightleftarrows t$). This conclusion must be tested experimentally.

The modification

$$\frac{1}{q^2}\to\frac{1}{q^2}\pm\frac{1}{q^2-\Lambda^2}\approx\frac{1}{q^2}\left(1\mp\frac{q^2}{\Lambda^2}\right) \qquad (1.11)$$

of the photon propagator, which is sometimes used, reduces in the first approximation to modification (1.8). It may be thought of as a manifestation of a coupling of a heavy neutral boson (or heavy photon) with an electron (or muon). In this case, for Bhabha scattering, for example, we find the cross section

$$\frac{d\sigma/d\Omega}{(d\sigma/d\Omega)_{\text{QED}}}=1\mp\frac{12E^2}{\Lambda^2_\mp}\frac{\sin^2\theta}{3+\cos^2\theta}. \qquad (1.12)$$

The minus and plus signs on the right-hand side correspond to the plus and minus signs in (1.11); $\Lambda^2_\mp$ characterizes the limits within which QED is valid; and θ is the angle to which the electron (or positron) is scattered in the c.m. frame.

The photon propagator can also be tested in the process $e^+e^-\to\mu^+\mu^-$ (only a single Feynman diagram corresponds to this process in lowest-order perturbation theory; see Fig. 1.2b). Modification (1.11) leads to the following ratio of cross sections for this process:

$$\frac{d\sigma/d\Omega}{(d\sigma/d\Omega)_{\text{QED}}}=1\mp\frac{8E^2}{\Lambda^2_\mp}.$$

For the process $e^+e^-\to\mu^+\mu^-$, however, it is vastly more important to test μ-e universality with respect to electromagnetic interactions than to test for a deviation from QED.

Testing an electron propagator is more complicated than testing a photon propagator. The expression for an electron propagator includes arbitrary

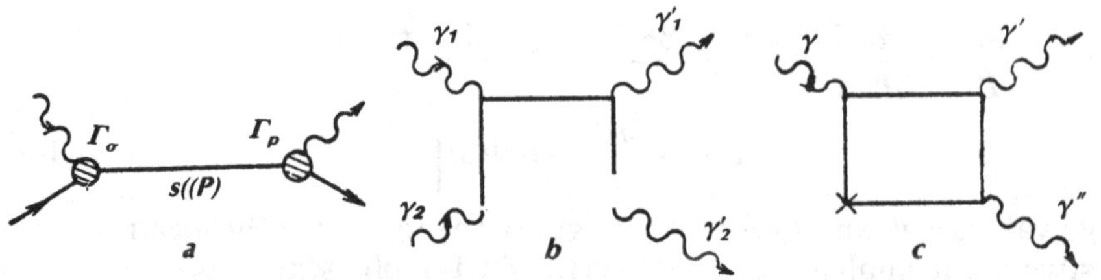

Figure 1.3. (a) Diagram describing Compton scattering by an electron; (b, c) examples of diagrams with closed electron loops.

scalar functions G_1 and G_2. The structure of an arbitrary matrix element is such that the propagator $S(p)$ must be multiplied by vertex functions Γ_σ and Γ_ρ (Fig. 1.3a). If all the lines at a vertex are virtual lines, the vertex will contain 12 arbitrary invariant functions. For a single virtual end we have four arbitrary invariant functions, which must be multiplied by two arbitrary functions which appear in the expression for the electron propagator. It can be shown that charge conservation and gradient invariance in the lowest-order Feynman diagrams, which contain unclosed electron lines (Fig. 1.3a), require that any complications of the electron propagator and any resulting change in the vertex functions due to the Ward identity must be eliminated completely. In diagrams of this sort it is not possible to modify the electron propagator since the matrix element for such a process is determined completely by the pole of the electron propagator in the limit $p^2 \to m^2$. Two-photon vertices ("seagull diagrams") or multiphoton vertices, however, can be modified in lowest-order perturbation theory and can make an observable contribution.

The electron propagator can be modified in lowest-order perturbation theory and can be tested for diagrams with closed electron loops. Some examples of processes of this type are $\gamma\gamma$ scattering (Fig. 1.3b) and the disintegration of a photon at a nucleus (Fig. 1.3c). Here it is necessary to integrate over the interior lines in the matrix element for the process under consideration. This can be done explicitly only by assuming some specific functional dependence of G_1 and G_2 on their argument; accordingly, in contrast with the case of the photon propagator, it is necessary in this case to introduce a model dependence of the invariant functions G_1 and G_2 on the momentum transfer. This is the primary distinction between a test of an electron propagator and a test of a photon propagator. For a test of an electron propagator in the simplest diagrams one introduces the following modification of the propagator, which is an analog of (1.11):

$$\frac{1}{\hat{p}-m} \to \frac{1}{\hat{p}-m} + \frac{1}{\hat{p}-m-\Lambda_e^*}. \tag{1.13}$$

This modification is interpreted as the incorporation into the problem of a heavy lepton with a mass Λ_e^*. Even in a model modification of this sort, however, a test of QED (a test of the electron propagator) has a completely definite

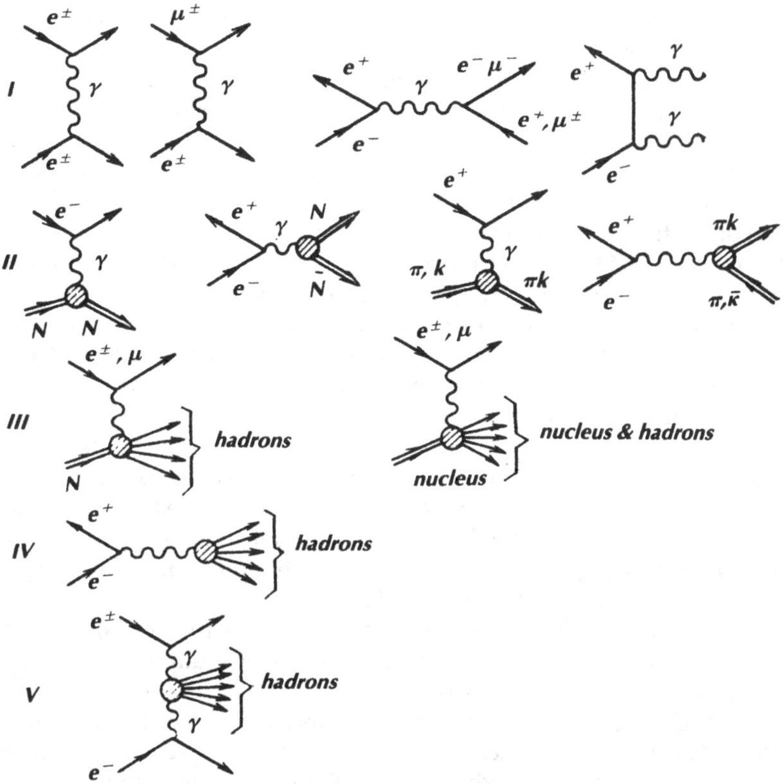

Figure 1.4. Various classes of diagrams illustrating the list of physical processes which are presently being studied actively at electron accelerators.

meaning, since it answers the question of the boundaries on the applicability of QED at small distances. Testing the electron propagator for Feynman diagrams of higher order in α with closed electron loops may be impossible, since the contribution of such diagrams to the cross sections for the processes under consideration is small. The experimental data would accordingly have to be highly accurate—beyond present capabilities—to make it possible to distinguish the contributions of these diagrams from the data and then determine Λ_e.

All the successes of QED at high energies over the past 10–15 years have been related to research on the physical processes described by the diagrams in Fig. 1.4. The list of specific physical processes which involve exclusively electrons, positrons, and photons—in which QED can presently be tested at high energies—is short (Fig. 1.4, class I). The processes described by diagrams of classes II–V (Fig. 1.4) are of considerable interest for testing QED and for determining the nature of the electromagnetic interaction of electrons, photons, and muons with hadrons (nucleons, pions, and K mesons). The list of such processes is again short.

The assignment of processes to separate classes corresponding to the diagrams of classes I–V was not done in a haphazard way. In the processes of class I the testing of QED is carried out under conditions of "clean" electromagnetic interactions, while in the processes of classes II–V there are hadrons, i.e., strongly interacting particles, present.[2] In the study of these processes information on the $e\gamma$ vertex and on the photon and electron propagators is taken from data in studies of the processes of class I, i.e., from data on tests of QED for clean electromagnetic interactions. In the processes of classes II–V, QED is therefore a foundation for studying the following topics: the electromagnetic structure of nucleons, pions, and K mesons (class II); hadron structure functions in deep inelastic scattering of leptons (class III); and the nature of the fragmentation of a photon into hadrons. As we will see below, QED is also the foundation for the erection of the framework of QCD and for testing its positions (class IV).

A diagram of class V is a diagram of the next higher order in the electromagnetic coupling constant α. This diagram, which describes the process $e^+e^- \to e^+e^- +$ hadrons, is interesting in that it presents the possibility of studying a $\gamma\gamma$ interaction. Two virtual photons can, under certain kinematic conditions, have approximately zero square 4-momenta; i.e., they can simulate the interaction of two real photons.

When we turn from the processes of class I to those of class V, we are turning from a study of simple processes to a study of more-complex processes. The contents of this book follow that order. A separate chapter is devoted to each class of diagrams. Chapters 3 and 4 could probably be interchanged according to taste. Historically (over 20 years), the research on these processes has evolved roughly in accordance with the order chosen here.

All of the QED processes which are accessible experimentally at existing accelerators and which have been studied in varying degrees of rigor and depth are thus reflected in this book.

Tests of QED at low energies are also being carried out extremely actively. Measurements in this range are fantastically precise. For example, the hyperfine splitting of levels of the hydrogen atom is being measured within an error in the 12th significant digit. This precision surpasses that of theoretical calculations incorporating strong interactions.

In one way or another, the overwhelming majority of the results at low energies have been obtained in experiments involving bound states of elementary particles: atoms, positronium, muonium, $\pi\mu$ atoms, etc. In delving into the bound-state problem we are switching from the nonrelativistic problem of determining the levels in atoms to the problem of the fine structure of the levels, to the problem of dealing with radiative corrections, and to the problem of the hyperfine splitting of levels. As the accuracy of the experimental data on the levels improves, one introduces corrections for the reduced mass, for the recoil of the nucleus, for the effect of the nuclear structure, for the nuclear polarizability, and for the electron screening of the nucleus. The structure of the correction terms is interesting and specific. This branch of physics is exceedingly deep in content and deserves a separate examination.[3,4]

Here it is important to note the profound interrelationship in the world of elementary particles. It might seem natural to expect that strong interactions would come into play in QED as we moved up to higher energies, where the production of hadrons in electromagnetic interactions becomes significant. It turns out, however, that even at low energies, beginning at a certain level of measurement precision, when radiative corrections are taken into account, it is necessary to introduce (for example) corrections for the hadronic polarization of vacuum. The physicist specializing in this area needs to know the quantitative limits within which strong interactions must be taken into consideration. With this goal in mind we turn to a brief review of results which have been found in tests of QED at low energies.

2. Tests of QED at low energies

2.1. Determination of the fine-structure constant

All the physical quantities which are calculated in QED contain the fine-structure constant $\alpha = e^2/\hbar c$, so we need to know the numerical value of this constant as accurately as possible. It is generally believed that the fine-structure constant can be found most accurately from the expression

$$\alpha^{-1} = \left[\frac{1}{4\mathrm{Ry}} \frac{1}{\gamma_{p'}} \frac{\mu_{p'}}{\mu_{\mathrm{B}}} \frac{2e}{h} \frac{c\Omega_{\mathrm{nbs}}}{\Omega_{\mathrm{NBS}}}\right]^{1/2}, \tag{1.14}$$

where Ry is the rydberg constant, $\gamma_{p'}$ is the gyromagnetic ratio of the proton (in water),[3] $\mu_{p'}/\mu_{\mathrm{B}}$ is the magnetic moment of the proton (in water), expressed in units of the Bohr magneton, $\Omega_{\mathrm{nbs}}/\Omega_{\mathrm{NBS}}$ is the ratio of the absolute ohm to the standard ohm (NBS means the U.S. National Bureau of Standards), and $2e/h$ is a quantity which is determined from the Josephson effect.[4] All the quantities in (1.14) are determined without the use of the assumptions embodied in QED.

The constant α can also be determined by comparing experimental values with values calculated from expressions containing α. One can then compare the various values found for α from (1.14) with those found by other methods. In principle, such a comparison cannot solve the problem of an accurate determination of α. However, a statistical least-squares analysis of data yields most-probable boundaries on the range of the fine-structure constant. A determination of the values of $\gamma_{p'}$ (Ref. 5) and $2e/h$ from the Josephson effect in 1975 led to the value $\alpha^{-1} = 137.035\,987\,(29)$. Yet another value was found for α just recently[1,5]:

$$\alpha^{-1} = 137.035963(15)\,. \tag{1.15}$$

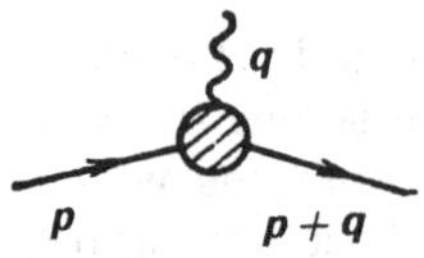

Figure 1.5. Diagram describing a vertex function.

2.2. Anomalous magnetic moment of the electron

A free electron has a spin angular momentum $s_e = \hbar/2$ and a magnetic moment $\mu_e = e\hbar/2m_e c$, where m_e is the mass of the electron, and the magnetic moment vector μ_e is directed opposite the spin vector. We thus have

$$|\mu_e/s_e| = e/m_e c\,.$$

The modulus of the ratio of the orbital magnetic moment of the electron to the orbital mechanical moment is $e/2m_e c$, or half the ratio $|\mu_e/s_e|$. It is called the "magnetic anomaly of the electron spin."

The ratio of the magnetic moment of the free electron to the Bohr magneton, μ_B, is the "g factor of the electron":

$$g = 2\mu_e/\mu_{\text{B}}\,.$$

It would appear that $g/2$ should be equal to unity, but actually we have

$$g/2 = 1 + a_e\,, \tag{1.16}$$

where a_e is an anomalous magnetic moment. The quantity

$$g_s = 2(1 + a_e) \tag{1.17}$$

is called the "total g factor of the electron."

The hypothesis that the electron has an anomalous magnetic moment was advanced in 1948 and confirmed experimentally soon thereafter.

The appearance of an anomalous magnetic moment can be related to a spatial distribution of the magnetic moment.

For any spin-1/2 charged particle the vertex function describing the interaction of the particle with an electromagnetic field (Fig. 1.5) can be written most generally as

$$e\left[\gamma_\mu F_1(q^2) + i\, a_e \frac{\sigma_{\mu\nu}}{2m} q_\nu F_2(q^2)\right]. \tag{1.18}$$

The functions $F_1(q^2)$ and $F_2(q^2)$ in (1.18) are called the "Dirac and Pauli form factors," respectively, and are normalized by $F_1(0) = 1$ and $F_2(0) = 1$, so that the total g factor can be written [see (1.17)]

$$g_s = 2[1 + a_e]\,. \tag{1.19}$$

[In (1.5) the normalization $F_2(0) = 1$, $a_e = \mu$, has been used.] When we go from the momentum representation $F_2(q^2)$ to the spatial representation $F_2(x)$ in the limit of small q^2, the function $F_2(x)$ can be interpreted as the spatial distribution of the magnetic moment (Ref. 6, for example). We will come back to this interpretation in Sec. 2 of Chap. 2.

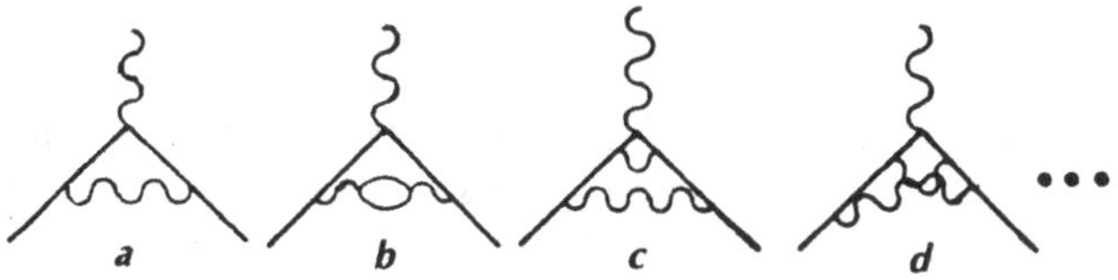

Figure 1.6. Example of a set of diagrams describing the contribution to the anomalous magnetic moment of the electron (or muon). In diagram b, the loop may consist of the pairs e^+e^-, $\mu^+\mu^-$, and $\pi^+\pi^-$ and pairs of other particles.

The form factors $F_i(q^2)$ can be calculated in QED in any order of perturbation theory (through the use of a renormalization of the divergent expressions). A theoretical value of the anomalous magnetic moment is found by distinguishing the parts proportional to $F_2(q^2)$ from the radiative corrections to the vertex function (Fig. 1.6, a–d). It is written as a series in powers of the fine-structure constant:

$$a_e = a_1(\alpha/\pi) + a_2(\alpha/\pi)^2 + a_3(\alpha/\pi)^3 + a_4(\alpha/\pi)^4 + \ldots, \tag{1.20}$$

where $a_1, a_2, a_3, a_4, \ldots$ are constants which can be calculated exactly in QED. In particular, Fig. 1.6b is the diagram corresponding to the vacuum-polarization component of the anomalous magnetic moment of a lepton, which may in principle stem from e^+e^-, $\mu^+\mu^-$, and $\pi^+\pi^-$ pairs and pairs of other, heavier particles (a fourth-order vertex diagram). The value which has been found at this point is[7]

$$a_e^{\text{theo}} = (1/2)(\alpha/\pi) - 0.328478445(\alpha/\pi)^2 + 1.184(7)(\alpha/\pi)^3 . \tag{1.21}$$

The coefficient a_3 in (1.20) was calculated on the basis of all 72 sixth-order diagrams.

Measurements of the anomalous magnetic moment of the electron have been carried out since 1948. In 1977, a value of surprising precision was found[8]:

$$a_e^{\text{expt}} = 1\,159\,652\,410(200)\times 10^{-12} . \tag{1.22}$$

Substituting the value of α from (1.14) into (1.21), we find

$$a_e^{\text{theo}} = 1\,159\,652\,578(155)\times 10^{-12} . \tag{1.23}$$

A new value has recently been found for the anomalous magnetic moment of the electron[9]:

$$a_e^{\text{expt}} = 1\,159\,652\,200(40)\times 10^{-12} . \tag{1.24}$$

The precision of this value surpasses that of the theoretical calculations, in which it now becomes necessary to consider terms proportional to $(\alpha/\pi)^4$. Contributing in this order of magnitude in α are 891 Feynman diagrams[10]; the expected value of the contribution is 30×10^{-12}. An error of no more than 20% in the calculation of a_4 would apparently make possible a comparison with experimental data at this point. In fourth order in α, the contributions of strong interactions to the polarization of vacuum become appreciable.

2.3. Anomalous magnetic moment of the muon

It is assumed that the Hamiltonians of the electromagnetic interaction of electrons and muons are of the same form ($e\mu\tau$ universality with respect to electromagnetic interactions is discussed in Sec. 3 of Chap. 1). Consequently, those Feynman diagrams which are taken into account in a calculation of the anomalous magnetic moment of the electron will contribute to the anomalous magnetic moment of the muon, a_μ. However, the correction for the polarization of vacuum by the e^+e^- pair contributes significantly to a_μ, since it contains additional factors of the type $\ln(m_\mu/m_e)$, while the analogous contribution of the $\mu^+\mu^-$ pair to the electron vertex is negligible. Furthermore, the contribution of the hadronic polarization of vacuum to the anomalous magnetic moment of the muon is significantly larger than the corresponding contribution to a_e because the mass of the muon is greater than that of the electron. As a result, in closed loops (Fig. 1.6b) momenta of virtual particles which are close in value to $m_\mu c$ also turn out to be close to the valence of $m_\pi c$. At the present experimental accuracy there is no need to consider the contribution of weak interactions to a_μ or a_e. Here are the experimental values of the anomalous magnetic moments of the μ^+ and μ^- mesons as of 1977 (Ref. 11):

$$a_{\mu^+}^{\text{expt}} = 1\,165\,910(12)\times 10^{-9}; \quad a_{\mu^-}^{\text{expt}} = 1\,165\,936(12)\times 10^{-9}. \tag{1.25}$$

For a_μ we have the theoretical prediction[12]

$$a_\mu^{\text{theo}} = 1\,165\,919(10)\times 10^{-9}, \tag{1.26}$$

which agrees surprisingly well with the experimental data. The hadronic contribution to (1.26),

$$a_\mu^{\text{had}} = 66.7(9.4)\times 10^{-9}, \tag{1.27}$$

is very important; if it is ignored, the values in (1.25) and (1.26) cannot be reconciled.

2.4. Lamb shift of levels in the hydrogen atom

From the nonrelativistic theory of the hydrogen atom we know that its energy levels E_{nlm} are classified on the basis of the principal quantum number $n \geqslant l+1$ and the orbital angular momentum $l = 0, 1, \ldots, n-1$. For each l there are $2l+1$ values of m, which is the projection of the orbital angular momentum l onto some particular axis, e.g., the z axis.

Relativistic quantum mechanics incorporates both the velocity dependence of the electron mass and the level splitting caused by the electron's spin. In the latter case one introduces the total angular momentum of the electron, $\mathbf{j} = \mathbf{l} + \mathbf{s}$, with the result that each level l splits in two: $l \pm 1/2$. The projection of the angular momentum $\mathbf{j}$ onto some particular axis, e.g., the z axis, now takes on $2j+1$ values: $m_j = m_l + m_s = m_l \pm 1/2$;

These relativistic corrections are summed to form a single correction to the energy levels in the nonrelativistic approximation:

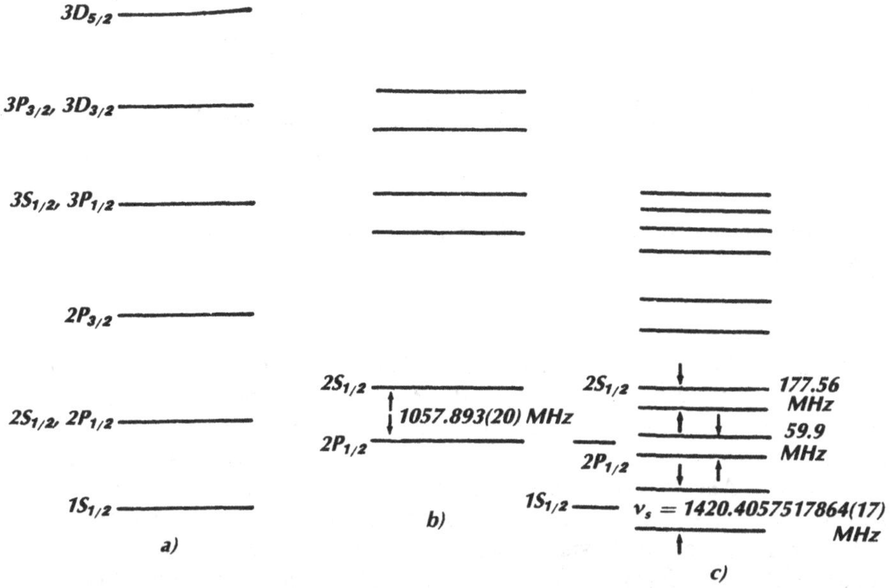

Figure 1.7. Level scheme of the hydrogen atom. (a) Fine structure of levels; (b) Lamb shift of levels (the fine splitting of levels); (c) hyperfine splitting of levels.

$$\Delta E_{nlj} = \alpha^2(3/4n - 1/j(j+1))(Z^4/n^3)\text{Ry}\,, \tag{1.28}$$

where Z is the atomic number of the atom. This is the so-called fine splitting of levels. Each nonrelativistic level nl now splits into two components: n, $j+1/2$, and $n, j-1/2$. The fine splitting of levels has an interesting aspect: For identical values of n, two adjacent values of the orbital angular momentum l give the same value of j and thus identical shifts (1.28), because of the spin. For $n=1$ there is only a single level, with $l=0$ and $j=1/2$. Beginning at $n=2$ we have

$$n=2\begin{cases} l=0, & j=1/2; \\ l=1, & j=1/2, \quad 3/2; \end{cases}$$

$$n=3\begin{cases} l=0, & j=1/2; \\ l=1, & j=1/2, \quad 3/2; \\ l=2, & j=3/2, \quad 5/2. \end{cases} \tag{1.29}$$

It can be seen from (1.29) that in the hydrogen atom the $2S_{1/2}$ and $2P_{1/2}$ levels should coincide; the $3S_{1/2}$ and $3P_{1/2}$, $3P_{3/2}$ and $3D_{3/2}$ levels should also coincide, since they have identical values of n and j (Fig. 1.7).

An experiment carried out by Lamb and Rutherford in 1947 to measure the fine structure in the hydrogen and deuterium atoms[13] yielded an unex-

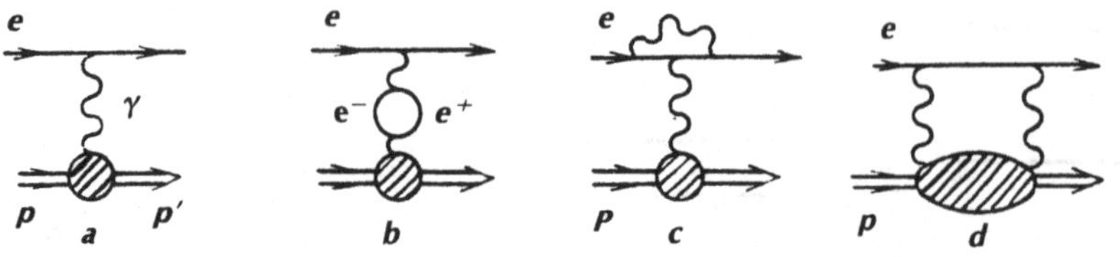

Figure 1.8. Diagrams whose contributions lift the degeneracy of the $2S_{1/2}$ and $2P_{1/2}$ levels in the hydrogen atom.

pected result which proved to be exceedingly important for the development of QED: It was found that the $2S_{1/2}$ and $2P_{1/2}$ levels do not coincide. Furthermore, the $2P_{1/2}$ level, which was expected on the basis of general considerations ($l = 1$) to lie above the $2S_{1/2}$ level ($l = 0$), turned out to lie below the $2S_{1/2}$ level (Fig. 1.7). Also in 1947, Bethe offered an explanation for the Lamb shift of levels.[14] He showed that the Lamb shift of the levels in the hydrogen atom could be explained on the basis of a change in the self-mass of a bound electron as a result of the emission and subsequent absorption of a virtual photon by the electron. In order to eliminate the divergence of the self-mass which arises in such calculations, he appealed to the mass renormalization principle which had been introduced in the same year by Kramers. The degree of the divergence reduced from linear to logarithmic. Those calculations were carried out in the nonrelativistic approximation. In the expansion of the vector potential of the electromagnetic field in photon creation and annihilation operators, the summation over the polarization indices of the photons was reduced to a summation over only two transverse directions. In order to obtain finite values for the self-mass of the electron instead of the logarithmic divergence, Bethe introduced a cutoff at $pc \sim mc^2$ in the integration over the momentum of the virtual photon. The shift of the self-mass found by this approach made it possible to explain the Lamb shift of levels. That study was important in that it proved the validity of using the Kramers principle for mass renormalization within the framework of QED.

The idea of a renormalization of mass and charge was subsequently utilized to remove all of the ultraviolet divergences in QED in a relativistically covariant formulation in studies by Dyson, Feynman, and Schwinger.

The degeneracy of the $2S_{1/2}$ and $2P_{1/2}$ levels in the hydrogen atom can be lifted by correcting the potential for the recoil of the proton, for the finite dimensions of the proton (Fig. 1.8a), for the polarization of vacuum (Fig. 1.8b), for the self-energy of a bound electron (Fig. 1.8c), for the appearance of resonance states of the proton in the exchange of two photons (Fig. 1.8d), etc. Any modification of the Coulomb potential, we might note, will have a greater effect on the shift of the $2S_{1/2}$ level, which lies close to the atomic nucleus (close to the proton in the case of hydrogen), than on the shift of the $2P_{1/2}$ level, so that these levels will exhibit an unusual shift.

Measurements of the Lamb shift of the hydrogen levels by two different methods (the subscripts are the first letters of the surnames of the researchers who measured this shift[15]) yielded approximately equal results:

$$\nu_{L-P} = 1057.893(20)\text{MHz}; \quad \nu_{A-N} = 1057.862(20)\text{MHz}. \qquad (1.30)$$

Unfortunately, the theoretical values found for the Lamb shift in Refs. 16 and 17 do not agree:

$$\nu_{[16]}^{\text{theo}} = 1057.916(10)\text{MHz}; \quad \nu_{[17]}^{\text{theo}} = 1057.864(14)\text{MHz}. \qquad (1.31)$$

In order to decide in favor of one of these values, it is necessary to reduce the experimental uncertainty by a factor of at least 3; this refinement will presumably become possible within a few years.

A further improvement in the precision of the theoretical value of the Lamb shift will depend in particular on the precision of the measurement of the mean-square value of the electromagnetic radius of the proton, $\langle r^2\rangle^{1/2}$. The dimensions of the proton stem from the strong interactions of the proton with the cloud of virtual particles (Chap. 2). The recoil of the proton (which is the nucleus in the hydrogen atom) caused by the motion of an electron along its orbit leads to a function $F_1(q^2)$ which is not equal to unity in the vertex function of the proton [see (1.5)]. This function is described approximately by

$$F_1(q^2) \approx 1 - (1/6)\langle r^2\rangle \mathbf{q}^2 \qquad (1.32)$$

[cf. (1.8)], where the parameter $\langle r^2\rangle$ is found through a comparison with experimental data on the elastic scattering of electrons by protons (this topic is discussed in more detail in Chap. 2), and $\mathbf{q}^2$ is the square of the recoil momentum of the proton. The coefficient of 1/6 arises formally from the definition of the mean-square value of the radius of the spatial distribution of the electric charge. The uncertainty regarding the value of $\langle r^2\rangle^{1/2}$ is already having a significant effect on theoretical estimates of the Lamb shift. For example, the value $\langle r^2\rangle^{1/2} = 0.87\ (2)$ causes a level shift $\Delta\nu = 0.148\ (16)$ MHz, while the value $\langle r^2\rangle^{1/2} = 0.80\ (2)$ causes a level shift $\Delta\nu = 0.125\ (6)$ MHz. The difference between these shifts is greater than the error of the theoretical calculations [see (1.31)].

2.5. Hyperfine splitting of levels in the hydrogen atom and muonium

The hyperfine splitting of levels in the hydrogen atom is caused by the interaction of the electron with the magnetic moment of the proton. Since the interaction in the S state is determined by short-range forces (the S orbit comes close to the proton), it is particularly sensitive to the spatial distribution of the magnetic moment. As in Sec. 1 of Chap. 1, it is clear that the physics of low energies and the physics of high energies meet in the phenomenon of the hyperfine splitting of the levels of the hydrogen atom, since the structure function corresponding to the distribution of the magnetic moment of the proton can be determined in experiments on the elastic and deep inelastic scattering of leptons by protons at high energies (Chaps. 2 and 3).

The validity of this interpretation of the effect of the structure of hadrons on the hyperfine splitting of levels is confirmed by, for example, comparative estimates of the components of this splitting which are caused by the polarizability of the proton and that of deuterium. The polarizability of the proton or of deuterium is determined in the following way: When a bound electron interacts with a proton, the diagram in Fig. 1.8d must be taken into account in addition to that in Fig. 1.8a as the experimental accuracy improves. The structure of the $\gamma pp'$ vertex in Fig. 1.8a is described by (1.5). In the approximation of small values of $\mathbf{q}^2$, the function ${}^sF_i(q^2)$ from (1.5) can be written in the form in (1.32). The quantities $\langle r^2\rangle_i^{1/2}$ are interpreted as the mean-square radii of the spatial distributions of the electric charge and of the magnetic moment of the proton.

The amplitude for virtual Compton scattering, $\gamma p \to \gamma' p'$ (Fig. 1.8d), is more complicated. Let us consider the actual Compton scattering of a photon. At small values of $\mathbf{q}^2$, the differential cross section for the process $\gamma p \to \gamma' p'$ can be written as an expansion in the photon energy ω (Ref. 18):

$$\frac{d\sigma}{d\Omega} = \frac{d\sigma_p}{d\Omega} - e^2 m \left(\frac{\omega}{m}\right)^2 [\alpha(1+\cos^2\theta) + 2\beta\cos\theta\,] \times \left[1 - 3\frac{\omega}{m}(1-\cos\theta)\right] + o(\omega^4)\,, \tag{1.33}$$

where $d\sigma_p/d\Omega$ is the cross section for the scattering of a photon by a point proton. The additional term in the first set of square brackets, which is proportional to ω^2, depends on two unknown structure constants, one of which (α) describes the correction to the cross section for scattering by a point proton for the electric dipole moment of the proton. This constant is called the "electric polarizability" of the proton. The other constant (β) describes the correction to the cross section for scattering by a point proton for the magnetic dipole moment of the proton and is called the "magnetic polarizability" of the nucleon. The same terms arise when we incorporate the amplitude for virtual Compton scattering of the photon by a nucleon and, therefore, for the e^-p interaction in the hydrogen atom.

As an electron revolves in an S orbit, it thus changes the instantaneous distribution of charge in the nucleon, leading to corrections to the hyperfine splitting of the S level. Since deuterium is presently believed to have a structure more "porous" than that of the proton, the correction for the effect of the polarizability in deuterium should be significantly greater than that in the proton. Calculations support this suggestion completely. Consequently, experiments carried out to determine the hyperfine splitting of levels in atoms should involve a study of the structure of the proton and of atomic nuclei. The hyperfine splitting of the S level in the hydrogen atom is measured fantastically precisely:

$$\nu_S = 1420.4057517864(17)\,\text{MHz}\,. \tag{1.34}$$

The polarizability component of ν_S has been calculated within a few mil-

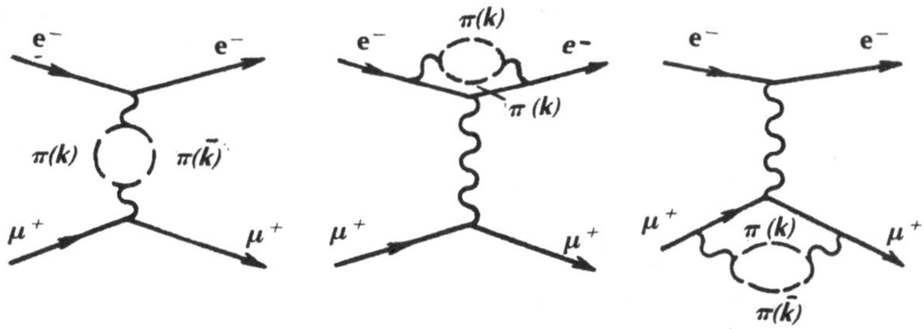

Figure 1.9. Polarization of vacuum by $\pi\pi$, $K\bar{K}$ hadron pairs (α^2 and α^3 corrections) in the $e^-\mu^+$ system.

lionths of a megahertz and is important in interpreting experimental data. Figure 1.7 shows the values of the splitting for other levels of the hydrogen atom. Assuming that the validity of QED is an accurately established fact, we can attempt to use the difference $\nu^{\text{expt}} - \nu^{\text{theo}}$ to reconstruct the structures of the proton and of nuclei (admittedly, for a limited interval of values of the square of the 3-momentum transfer).

We can escape the difficulties in dealing with the structure of elementary particles and nuclei in the hyperfine splitting of levels, and we can return to a test of the validity of QED, by turning to the bound state of such a "purely electromagnetic" system as muonium ($e^-\mu^+$). We have used quotation marks here because as the precision of experiments and calculations improves, it will unavoidably become necessary to deal with the polarization of vacuum caused by the hadronic states (Fig. 1.9). Calculations of the hyperfine splitting of the ground state of muonium yield the value[16]

$$\nu^{\text{theo}} = 4463.2979 \text{ MHz}\,, \tag{1.35}$$

which agrees satisfactorily with that obtained from experimental data,[19]

$$\nu^{\text{expt}} = 4463.30235(52)\text{MHz}\,. \tag{1.36}$$

In a subsequent experiment,[20] the error in the measurement of ν^{expt} was reduced to 3×10^{-8}. This reduced error means that we need new, more precise theoretical values of ν. Interestingly, the weak-interaction component of the hyperfine structure of muonium is expected to be on the order of 10^{-8}, which is not much smaller than the presumed error of future experiments.

2.6. Fine structure of the ground state of positronium[7,21,22]

Positronium should be recognized as one of the purest QED bound systems. It can be formed, for example, in the interaction of a beam of positrons with an electron shell of an atom. In the course of the interaction, a positron may capture one of the electrons of the shell and form either para-positronium (a system with oppositely directed spins) or ortho-positronium (one with identi-

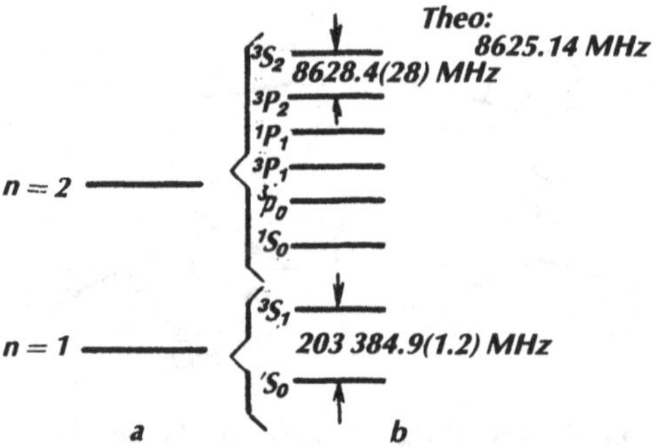

Figure 1.10. Level structure of positronium. (a) Ground levels; (b) fine splitting of levels.

cally directed spins). Para-positronium decays into two photons and has a lifetime of about 1.25×10^{-10} s; ortho-positronium decays into three photons and has a lifetime some three orders of magnitude longer (about 1.4×10^{-7} s).

In a static magnetic field the levels of positronium split (the Zeeman effect) and acquire the fine structure shown in Fig. 1.10b.

Although the measurements of the level splitting in positronium are less precise than those for the other systems discussed above, a comparison of the theoretical and experimental values in this case is particularly important for QED. The reason is that positronium can be used for an electrodynamic test of the Bethe-Salpeter equation, which gives a relativistically covariant formulation of the bound-state problem in QED. The kernel of the equation, which is an analog of the potential for the interaction of two particles (in our case, the electron and the positron), can be written with any prescribed accuracy by perturbation theory. The Bethe-Salpeter equation is thus regarded as giving an exact relativistic QED description of a bound system. The theoretical level difference

$$\nu^{\text{theo}} = {}^3S_1 - {}^1S_0 = 203400.3 \text{ MHz} \tag{1.37}$$

does not include all contributions proportional to α^2. Nevertheless, the value of ν^{theo} given by (1.37) already agrees satisfactorily with experimental data (Fig. 1.10, a and b). We should point out, however, the difficulties in the relativistic description which stem from the use of the concept of a four-dimensional wave function, which has no clear physical interpretation. In this regard, the quasipotential method of Logunov and Tavkhelidze[23] has some important advantages. That method combines the simplicity and transparency of the three-dimensional description of nonrelativistic quantum mechanics with the covariant apparatus of field theory. The quasipotential method has been used successfully to calculate the lowest-order Lamb shift $\alpha(Z\alpha)^4$ for an arbitrary mass of the particles and to calculate the fine and hyperfine structures of hydrogen-like atoms.[3]

2.7. Test of the *CPT* theorem

The precision with which the anomalous magnetic moments of the electron and muon are measured can be turned to advantage to test the *CPT* theorem in QED. As it pertains to this case, the *CPT* theorem asserts that the kinetic energy of an electron or μ^- meson in a magnetic field should be exactly equal to the kinetic energy of a positron or μ^+ meson, respectively, when the spin direction is changed ($\mathbf{s} \to -\mathbf{s}$). This energy is proportional to the g factor of the electron or muon. It follows that if the *CPT* theorem is not violated, then the g factors of the electron and the positron should be equal, as should the g factors of the μ^+ and μ^- mesons.

Using the experimental values in (1.25), we can find limits on the applicability of the *CPT* theorem:

$$(g_{\mu-} - g_{\mu+})/g = 2.6\times10^{-8}\,. \tag{1.38}$$

It can be shown that roughly the same values follow from measurements of the difference between g_{e-} and g_{e+}.

It is our purpose in this subsection to demonstrate the agreement between QED and experimental data at low energies and the need to correct for strong interactions beginning at a certain level of measurement precision. Quantum electrodynamics has also been tested in muonic atoms (μHe, μPb, etc.) and in such an exotic entity as the $\pi\mu$ atom. In atoms of this sort, the S orbit of the μ meson is very close to the nucleus (the π meson in the case of the $\pi\mu$ atom), and the structure of the nucleus or the π meson, the polarization of vacuum, and other effects of the dynamics of the nucleus or π meson begin to have a particularly strong effect. In several cases (e.g., in a study of the hyperfine structure of the ground state of muonium), the need for corrections for weak interactions has been seen within the framework of QED.

Up to this point we have not taken up a possible structure of leptons. Leptons have been regarded as point particles. The form factors $F_i(q^2)$, which should in principle be included in the lepton vertex functions, have been assumed equal to unity for arbitrary values of the squares of the 4-momentum transfers. The increase in the number of quarks and leptons and the rapid growth of their mass (e.g., the τ lepton is about 3500 times heavier than the electron) suggest that leptons may be composite particles. If they are, then electrons and muons are extremely unusual composite systems. Specifically, a test has shown that QED is valid (i.e., the interaction of electrons and muons with photons is a point interaction) down to distances 10^{-16} cm (Sec. 3 of Chap. 1).

The length scale 10^{-16} cm corresponds to masses $m^* \geqslant 100$ GeV (according to the relation $\lambda \geqslant \hbar/mc$). We are thus left with the assumption that if leptons are composite particles, then they are comparatively tightly bound states with a very small spatial size. Using the Gerasimov-Drell-Hearn sum rules, one can show that for a composite lepton the contribution to the anomalous magnetic moment is proportional to the mass ratio[24]:

$$\delta a_e \sim o(m_e/m_e^*)\,. \tag{1.39}$$

Table 1.2. Values found for the fine-structure constant from various measurements.[6]

Measurement	α^{-1}
Fine structure of hydrogen	137.03572(23)
Hyperfine structure of hydrogen	137.03597(22)
Fine structure of helium	137.03608(13)
Hyperfine structure of muonium	137.03592(12)
g_e-2	137.035981(29)
Josephson effect [see (1.14)]	137.035963(15)

Substituting (1.23) and (1.24) into (1.39), we find

$$|\delta a_e| = |a_e^{\text{QED}} - a_e^{\text{expt}}| = (378 \pm 161)\times 10^{-12} \text{ or} |\delta a_e| \lesssim 5\times 10^{-10}. \quad (1.40)$$

Let us assume that relation (1.39) holds exactly:

$$|\delta a_e| = m_e/m_e^*.$$

We then find $m^* \geqslant 10^6$ GeV and $\hbar/m^*c \lesssim 2\times 10^{-20}$ cm. We thus arrive at the striking conclusion that low-energy physics indicates surprisingly large masses for the constituents of leptons! Does this conclusion mean that relation (1.39) is wrong, or could it mean that it would be more reasonable to reject the model of composite leptons and go back to treating the electrons and muons as point particles? Or does nature have yet another surprise for us?

As we have already mentioned, at low energies QED is in excellent agreement with all the experimental data. As a rule, the additional incorporation of strong interactions leads to the agreement in precision required. Nevertheless, at low energies we run into some serious difficulties, and there is by no means any simple way to overcome them.

In the first place, the precision of experimental data in several measurements is at least an order of magnitude better than that of theoretical calculations. As we move to higher orders in α, the theoretical calculations become so complex that even resorting to computers is not always of help. There is an obvious need for simpler computational methods. It is totally unclear, however, just what these methods should be. All that is clear is that an infinite number of Feynman diagrams cannot be evaluated by ordinary methods.

Second, we have considered only two particles from the family of leptons: the electron and the muon. We now know that there is a third particle in this family: the τ lepton. Each of these particles has a corresponding type of neutrino. The breeding of leptons gives rise to some new problems.

Third, the value given above for α [see (1.14)] is not the same as the values found from other measurements (Table 1.2).

Interestingly, the possible discrepancy between the values found for α from the Josephson effect and from measurements of $g_e - 2$ may raise doubt that the values of the charges in the superconductivity phenomenon and in the physics of elementary particles are identical.

Fourth, we need a good quantitative theory for strong interactions, since ever more frequently it is becoming necessary to make corrections for strong interactions in comparisons with experimental data. So far, the only candidate for the status of a theory of strong interactions is QCD, constructed in complete analogy with QED. Unfortunately, QCD has many unresolved problems of its own, so that it may not always be useful for testing QED at this point.

3. Tests of QED at high energies; electroweak interactions; $e\mu\tau$ universality

3.1. Tests of QED and the structure of the nucleon

About 20 years ago accelerators with colliding e^+e^- beams offered beams of such low energy that corresponding tests of QED could not be of much interest, since the values of the parameter Λ achieved at those accelerators were small. Values of the momentum transfer even smaller than those at e^+e^- colliders could be achieved in the ordinary scattering of electrons (or positrons) by electrons of atoms at rest.

In this situation it seemed promising to test QED in experiments on the interactions of electrons, positrons, muons, and photons with nucleons.[25] In the processes

$$\gamma + N \to N + e^- + e^+ \tag{1.41}$$

(the production of wide-angle pairs),

$$e + N \to e + N + e^+ + e^- , \tag{1.42a}$$

$$\mu + N \to \mu + N + \mu^+ + \mu^- \tag{1.42b}$$

("trident" processes, with three leptons in the final state), and

$$e + N \to e + N + \gamma \tag{1.43}$$

(large-angle bremsstrahlung), for example, it would be possible under certain kinematic conditions to achieve large values of the 4-momentum of a virtual lepton or photon. It would thus become possible to substantially reduce the distances down to which QED could be tested. In experiments of this sort, however, the structure of the nucleon comes into play at a fairly large momentum transfer. Fortunately, in several cases which we will discuss below, the structure of the nucleon can be dealt with quite accurately by comparing the contribution of this structure to the cross section of the process of interest with the contribution made by the structure of the nucleon to the cross section for the elastic scattering of electrons by nucleons for the same values of the momentum transfer. With increasing energy of the incident lepton or photon, it becomes necessary to consider Feynman diagrams of the Compton type (Fig. 1.12, a–c) in addition to the Feynman diagrams which contain struc-

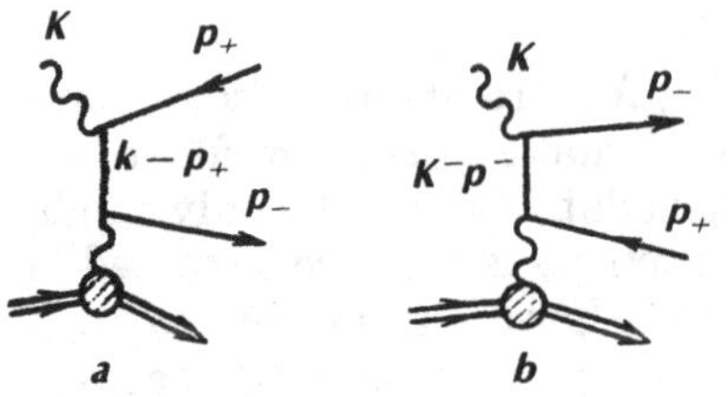

Figure 1.11. Diagrams which dominate the cross section for the production of wide-angle pairs at low energies of the impinging photons.

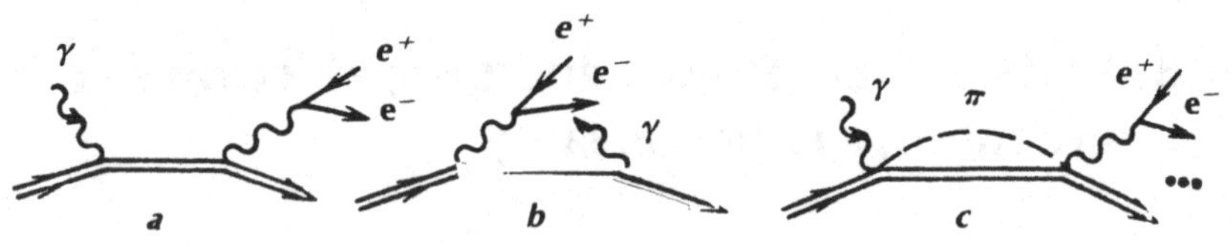

Figure 1.12. Diagrams of the Compton type which contribute to the production of wide-angle e^+e^- pairs.

ture functions of the nucleon (Fig. 1.11, a and b), information on which is extracted (as we just mentioned) from studies of the elastic scattering $e + N \to e + N$. Incorporating diagrams of the Compton type is difficult but necessary, since they may be extremely significant under certain kinematic conditions (see the discussion below of the bremsstrahlung of electrons). In this case a test of QED depends strongly on the accuracy with which strong interactions are taken into account.

In all of the processes listed above, the idea underlying the test of QED was not that physicists wished to be convinced of the validity of QED, although that consideration was also present. It was assumed at the outset that QED was valid and that deviations from it could be explained in terms of the existence of structure functions of the nucleons and of contributions from strong interactions. Rather, the idea was that new data on these processes could be obtained from the experiments, so that new limits could be established on the applicability of QED.

Nowhere in this chapter of the book do we go into the meaning of the concept of "structure of a nucleon"; the structure is determined by introducing form factors for the nucleon in the same manner as was done above for the electron [see (1.3)–(1.5)], since nucleons, like electrons, are spin-1/2 particles, and the electromagnetic interaction of electrons and nucleons is described by the same interaction Hamiltonian. In this chapter we have not introduced the concept of a quark-gluon structure of nucleons and K mesons; this will be done in the following chapters.

Let us take a slightly more detailed look at the idea of testing QED for all three of the processes mentioned above [(1.41)–(1.43)].

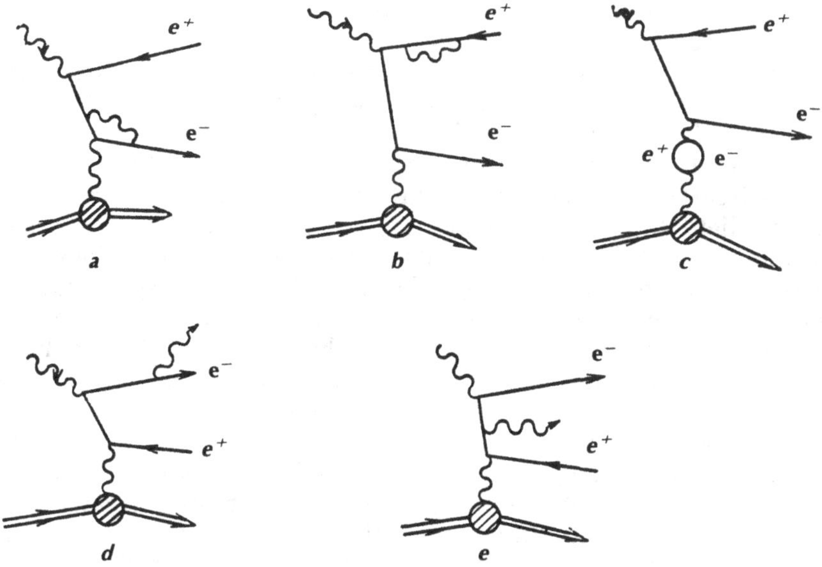

Figure 1.13. Diagrams describing radiative corrections for the emission of (a,b) virtual photons and (d,e) real photons and (c) the polarization of vacuum for the production of wide-angle pairs.

Production of wide-angle pairs.
It has been suggested that the cross section for the photoproduction of the e^+e^- pair at hydrogen will be dominated by the diagrams in Fig. 1.11, a and b (as we will show below).

For a virtual lepton line we can obtain a large 4-momentum transfer by choosing a positron scattering angle close to 90°. In this case, for the diagram in Fig. 1.11a, we have

$$(k-p_+)^2 = k^2 + p_+^2 \; -2kp_+ \approx -2kp_+$$
$$= -2kE_+ + 2|\mathbf{k}||\mathbf{p}_+|\cos\theta_+ = -2kE_+ \, ,$$

where p_+ and E_+ are the 4-momentum and energy of the positron, and $\mathbf{k}$ and k are the 3-momentum and energy of the photon.

For the other diagram (Fig. 1.11b) the positron is scattered through a large angle by the proton, and the lepton line in the intermediate state corresponds to a small square momentum:

$$(k-p_-)^2 \approx -2|\mathbf{k}_-|E-(1-\beta_- \cos\theta_-) \, ,$$

where $\beta = v/c$, and θ_- is the angle through which the electron is scattered.

In addition to the diagrams in Fig. 1.12, the calculations must include corrections for the recoil of the proton, for the form factors of the proton, and for radiation. This statement means that the calculations must incorporate the diagrams in Figs. 1.12, a–c, and 1.13, a–e (and also several other diagrams in this order in the constant α, similar to those shown in Fig. 1.13).

Let us examine experiments of two types: those in which one and two leptons are detected in the final state.

In the first case one observes a positron emitted at an angle of about 90°, and an integration is carried out over the entire range of the electron emission angle. It is convenient to choose a photon energy below 140 MeV, i.e., below the threshold for π-meson production, in order to eliminate the background from the decay of π^0 mesons and the contribution of a resonance in the intermediate state in diagrams of the type shown in Fig. 1.12c. The resonance contribution may become significant as the energy of the incident photons increases. The restriction to energies $E_\gamma \lesssim 140$ MeV of course results in a restriction on the distances down to which QED is tested. Since the square of the momentum transfer corresponding to the virtual lepton line in Fig. 1.11a is large, that diagram is sensitive to a modification of the electron propagator. The other diagram incorporates the virtual lepton line corresponding to a small momentum transfer and is relatively insensitive to such a modification. However, the diagram in Fig. 1.11b makes a contribution to the cross section of the process of interest here which is significantly greater than that of the diagram in Fig. 1.11a, since in it the propagator of the virtual lepton is small (it contains a factor of $1-\beta_- \cos\theta_-$), and an additional factor of $\ln(E_-/m) \approx 4$ arises when an integration is carried out over all electron emission angles. The diagram in Fig. 1.11b thus contributes about 80% of the cross section, while that in Fig. 1.11a contributes about 20%. This ratio of contributions mandates accurate calculations of the matrix elements corresponding to the diagrams in Figs. 1.11–1.13. A simple model for the modification of the electron propagator,

$$(p^2-m^2c^2)^{-1} \to (p^2-m^2c^2)^{-1}-(p^2-m^2c^2-\Lambda_e^2)^{-1} \tag{1.44}$$

[an analog of (1.12)], leads to the result that a 5% error in the calculation of the cross section for the process would make it possible to test QED down to distances of 0.7×10^{-13} cm.

This estimate of the error is a sum of several errors. The nucleon form factors contained in the nucleon vertex (Fig. 1.11, a and b) are completely the same as the form factors which are measured in the elastic scattering of electrons by nucleons at the same values of the 4-momentum transfer (see the diagram in Fig. 1.8a, where the electron should now be regarded as free, rather than bound). These form factors were measured in studies by Hofstadter.[26] The correction for the form factors is about 2%. The contribution to the cross section from the diagrams in Fig. 1.12, a–c, turns out to be negligible at impinging-photon energies below 140 MeV, because the nucleon propagator in the denominator contains the large mass of a nucleon. The contribution of the interference term from the interference of the diagrams in Figs. 1.11 and 1.12 is no greater than 1%. The radiative corrections (Fig. 1.13) also turn out to be small: about 1%. Correcting for the recoil of the proton consists of retaining terms of the type $(k/Mv)^n$ in the equations, where the quantity k has the dimensionality of a momentum, Mv is the momentum of the nucleon in the final state, M is the mass of a nucleon (which is finite here, not infinitely

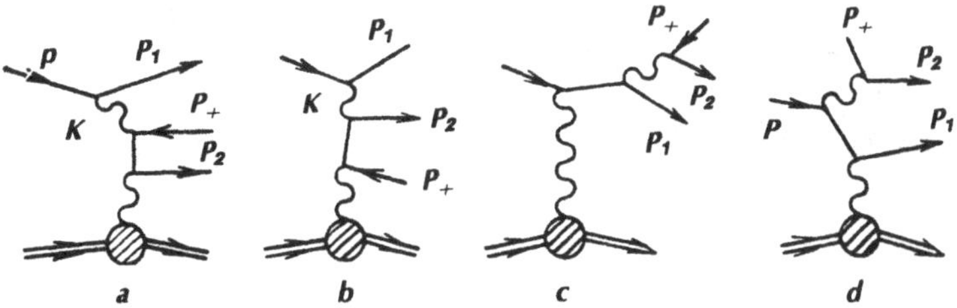

Figure 1.14. Diagrams describing processes with three leptons in the final state.

large; this is, in part, the point in incorporating the recoil of the nucleon), and n is an integer. The error in the calculation of this correction is less than 0.5%.

In experiments of the second type one detects an electron and a positron which are emitted at large angles; as a result, both diagrams (Fig. 1.11, a and b) become sensitive to a modification of the lepton propagator.

When the electron and positron have identical energies and are emitted at identical angles from the direction of the photon beam, the interference term vanishes by virtue of the requirement that the amplitude for the process be invariant under the operation of charge conjugation. Diagrams of the Compton type (Fig. 1.12, a and b) contribute negligibly little to the cross section, as they did for experiments of the first type. The structure of the nucleon is unimportant here. An experiment of the second type, in which both of the scattered leptons are detected, if carried out with an error on the order of 10%, makes it possible to test QED down to distances of about 3×10^{-14} cm.

An experiment of the first type was carried out with a photon beam with an average energy of 115 MeV (Ref. 27). The diagram in Fig. 1.11a contributed about 30% of the cross section for process (1.41). Bjorken and Drell[28] found a ratio

$$\sigma_{\text{expt}}/\sigma_{\text{theo}} = 0.96 \pm 0.14\,. \tag{1.45}$$

The modification of the electron propagator reduced to the introduction of a function

$$\Phi(q^2) = 1 - 2\mathbf{q}^2/\Lambda_e^2$$

in the expression for the cross section; this function differs from (1.8) by a factor of 2. It was concluded from a comparison with experimental data that the distances down to which QED could be tested in such an experiment would be about 0.9×10^{-13} cm.

Processes with three leptons in the final state ("tridents").[28,29]

Processes (1.42a) and (1.42b) are described by fourth-order diagrams (Fig. 1.14, a–d). We need to add to these diagrams four more, in which the momenta p_1 and p_2 should be interchanged. The form factors of the proton must be taken into account in these diagrams. In addition to the diagrams which we

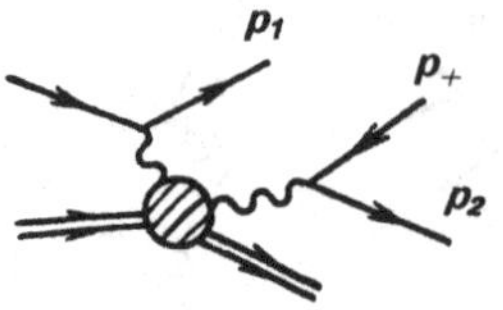

Figure 1.15. Diagram describing the Compton scattering of a virtual photon by a photon.

have listed, there is yet another in this order of perturbation theory; it is shown in Fig. 1.15. We should also take into account radiative corrections in each of these diagrams which include the emission of virtual and real photons. The interest in tridents stems from the possibility of testing the photon propagator. It has been suggested that the electron propagator is not modified. The square of the 4-momentum of the photon, k^2, shown in Fig. 1.14a,

$$k^2 = (p - p_1)^2 \approx -2pp_1 = 2EE_1(1 - \cos\theta)\,,$$

can be chosen quite large in a certain kinematics. In an experiment searching for tridents there are various possibilities for detecting the final state, as there are in experiments involving the production of wide-angle pairs:

(1) Two final electrons are detected in coincidence in a geometry which reduces the contributions from diagrams of the type which describe a Compton scattering of a virtual photon (the initial electron and the two final electrons lie in a common plane, and the final electrons are emitted at identical angles with respect to the direction of the incident electron; these angles must be less than 90° but much larger than m_e/E).

If that deviation from the QED predictions which is caused by the modification of the photon propagator is written as a factor $\Phi^2(k^2) = (1 + k^2/\Lambda^2)^2$, with which we are to multiply the cross section for the process, then a 10% deviation of the cross section at $\theta = 20°$ and at an energy $E = 500$ MeV of the impinging electron will correspond to distances of about 0.2×10^{-13} cm. For these kinematic conditions, with $k \approx 300$ MeV/c, a 10% discrepancy between the measured cross section σ_{expt} and the theoretical cross section σ_{theo} can be described by

$$\left[\sigma\,(1 + \frac{k^2}{\Lambda^2})^2\right]_{\text{theo}} / \sigma_{\text{expt}}(E,\theta) \approx 1.1\,,$$

from which we find $|k/\Lambda| \approx 0.3$ and $\Lambda \approx 1000$ MeV/c.

From the uncertainty relation $\Delta x \times \Lambda \gtrsim \hbar$, we find

$$x > \hbar/\Lambda \approx \frac{10^{-27}\times3\times10^{10}}{1000\times10^6\times1.6\times10^{-12}}\,\text{cm} \approx 0.2\times10^{-13}\,\text{cm}\,.$$

If the experimental error were zero, and the theoretical curve coincided precisely with the experimental curve, then the ratio

$$\sigma_{\text{theo}}(E,\theta)(1 + k^2/\Lambda^2)^2/\sigma_{\text{expt}}(E,\theta)$$

would be unity, the value of Λ would be infinitely large, and we would find $\Delta x \geqslant 0$ from the uncertainty relation. The meaning would be that the representation of a point interaction in QED was correct.

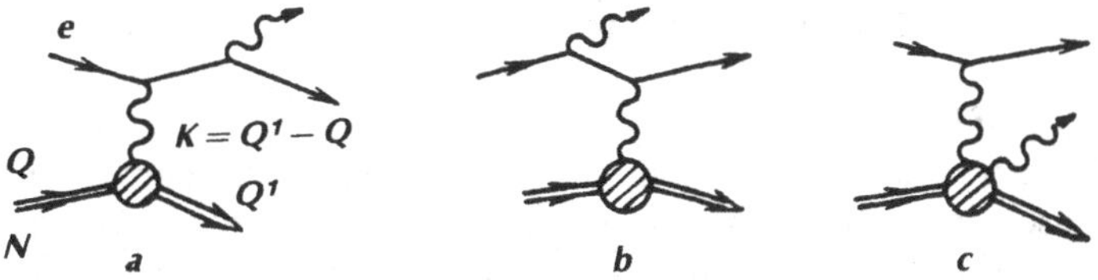

Figure 1.16. Set of diagrams considered in the calculation for process (1.43).

As the experimental error is reduced, and k^2 is increased, progressively larger values of Λ can be found from experimental data.

(2) A positron and one of the electrons are detected in coincidence. For example, if we detect an electron with a momentum p_1 and a positron with a momentum p_+, then we can accurately measure the square of the 4-momentum of the virtual photon (Fig. 1.14a). In this case we can use an experimental arrangement which is sensitive to a modification of the photon propagator. Writing that deviation which is caused in the cross section by the modification of the photon propagator (and the vertex) in the form in (1.8),

$$\Phi^2(k^2) = (1 + k^2/\Lambda^2)^2,$$

we can, with a 10–15% deviation of the experimental cross section, test QED at distances down to 0.5×10^{-13} cm in an experiment of this type. Calculations of the radiative corrections and of corrections for diagrams of the type which describe a Compton scattering of a virtual photon may roughly double this estimate of the error in Λ.

Corresponding arguments can be made in the case of muon tridents; in that case there is the additional problem of μe universality.

Bremsstrahlung.[30,31]

Let us examine bremsstrahlung for the case in which the impinging electron has an energy of about 500 MeV. Bethe and Heitler carried out calculations on the bremsstrahlung of electrons [(1.43)], and the production of e^+e^- pairs in the scattering of photons by protons [(1.41)]. They used lowest-order perturbation theory in the approximation of a point proton (they ignored the form factors) which was infinitely heavy (they ignored recoil). In the approximation of an infinitely heavy proton, the diagram in Fig. 1.16c contributes nothing to the cross section. As we mentioned earlier, as the energy of the impinging electrons and photons increases, it becomes necessary to deal with the recoil of the nucleon and its structure. In fact, Hofstadter's experiments[26] on the scattering of 500-MeV electrons by hydrogen showed that the form factors $F_i(q^2)$ are significantly different from unity. Consequently, the set of Feynman diagrams in Fig. 1.16, a–c, must be taken into consideration in a calculation of bremsstrahlung.

In the diagrams in Fig. 1.16, a and b, there is no difficulty of any sort in incorporating the effect of the structure of the proton, since this structure should correspond exactly to the structure of a nucleon which is measured in elastic *ep* scattering for identical values of the 4-momentum transfer. A diagram of the type in Fig. 1.16c may, under certain kinematic conditions, make a significant contribution, but it is an extremely complicated matter to take these conditions into account. In the production of wide-angle e^+e^- pairs the contribution from such diagrams is negligible because of the limitation on the energy of the incident photon.

The matrix element corresponding to the sum of diagrams in Fig. 1.16, a and b, is

$$M(a+b) = \frac{i\,e^3}{(2\pi)^{7/2}} \frac{Mm\delta(q+p_0-p-k-q')}{\sqrt{2k_0\varepsilon\varepsilon' E_0 E}} \frac{1}{(p_0-p)^2}$$

$$\times [\overline{w}^s(\mathbf{p})\Gamma^\mu w^{s_0}(\mathbf{p}_0)] \left\{\overline{u}^\sigma(\mathbf{q}')\left[\hat{l}\,\frac{1}{(\mathbf{q}+\mathbf{p}_0-\mathbf{p})-m}\,\gamma_\mu\right.\right.$$

$$\left.\left. +\,\gamma_\mu \frac{1}{\mathbf{q}-\mathbf{k}-m}\,\hat{l}\right] u^{\sigma_0}(\mathbf{q})\right\}.$$

Here M is the mass of the nucleon; m is the mass of the electron; $p_0(E_0, \mathbf{p}_0)$ and $p(E, \mathbf{p})$ are the 4-momenta of the initial and final states of the nucleon; $q(\epsilon, \mathbf{q})$ and $q'(\epsilon', \mathbf{q})$ are the 4-momenta of the initial and final states of the electron; $k(k_0, \mathbf{k})$ is the 4-momenta of the photon; the spinor $w^s(\mathbf{p})$ describes a nucleon with a momentum $\mathbf{p}$ and a spin s; $u^\sigma(\mathbf{q})$ is the corresponding spinor of the electron; $\hat{l} = \gamma^\alpha l_\alpha$ (l_α is the polarization 4-vector of the photon); and the Dirac matrices γ_μ satisfy the condition $\gamma^\mu\gamma^\nu + \gamma^\nu\gamma^\mu = 2g^{\mu\nu} I$, where I is the unit matrix

$$g^{\mu\nu} = \begin{pmatrix} 1 & 0 & 0 & 0 \\ 0 & -1 & 0 & 0 \\ 0 & 0 & -1 & 0 \\ 0 & 0 & 0 & -1 \end{pmatrix};$$

$$\Gamma^\mu = \gamma^\mu F_1(|p_0-p|^2) + (\mu/2M)\sigma^{\mu\nu}(p_0-p)_\nu F_2(|p_0-p|)^2,$$

[$F_i(|p_0-p|^2)$, $i=1,2$ is a form factor which depends on the recoil of the nucleon; $F_i(0)=1$; μ is the anomalous magnetic moment of the proton, expressed in nuclear magnetons; $\sigma^{\mu\nu} = (1/2)(\gamma^\mu\gamma^\nu - \gamma^\nu\gamma^\mu)$; and $pq = g^{\mu\nu}p_\mu p_\nu$]. We are using a system of units with $h = c = 1$ and $e^2/4\pi = 1/137$.

In the frame of reference in which the nucleon is initially at rest, the differential cross section for bremsstrahlung is[30]

$$d\sigma=\frac{e^6}{8(2\pi)^5}\frac{dk_0}{k_0ME}\frac{1}{|df(\varepsilon')/d\varepsilon'|}\frac{|\mathbf{q}'|}{|\mathbf{q}|}\frac{d\Omega_q d\Omega_k}{[(\varepsilon-\varepsilon'-k_0)^2-Q^2]}$$
$$\times\Bigg\{\Bigg[\bigg(\frac{\mathbf{q}'^2\sin^2\theta'}{(\varepsilon'-|\mathbf{q}'|\cos\theta')^2}+\frac{\mathbf{q}^2\sin^2\theta}{(\varepsilon-|\mathbf{q}|\cos\theta)^2}$$
$$-\frac{2|\mathbf{q}||\mathbf{q}'|\sin\theta\sin\theta'\cos\varphi}{(\varepsilon-|\mathbf{q}|\cos\theta)(\varepsilon'-|\mathbf{q}'|\cos\theta')}\bigg)(2m^2-Q^2+(\varepsilon'+k_0-\varepsilon)^2)$$
$$-2k_0^2\frac{\varepsilon'-|\mathbf{q}'|\cos\theta'}{\varepsilon-|\mathbf{q}|\cos\theta}-2k_0^2\frac{\varepsilon-|\mathbf{q}|\cos\theta}{\varepsilon'-|\mathbf{q}'|\cos\theta'}\Bigg][F_1(|p_0-p|^2)$$
$$+\mu F_2(|p_0-p|^2)]^2(M^2-ME)+M^2\Bigg[\frac{\mathbf{q}'^2\sin^2\theta'}{(\varepsilon'-|\mathbf{q}'|\cos\theta')^2}[(\varepsilon'+k_0+\varepsilon)^2-Q^2]$$
$$+\frac{\mathbf{q}^2\sin^2\theta}{(\varepsilon-|\mathbf{q}|\cos\theta)^2}[(\varepsilon+\varepsilon'-k_0)^2-Q^2]-\frac{2|\mathbf{q}||\mathbf{q}'|\sin\theta\sin\theta'\cos\varphi}{(\varepsilon'-|\mathbf{q}'|\cos\theta')(\varepsilon-|\mathbf{q}|\cos\theta)}$$
$$\times((\varepsilon+\varepsilon'+k_0)(\varepsilon+\varepsilon'-k_0)-Q^2+2k^2)+2k_0^2\frac{\mathbf{q}'^2\sin^2\theta'+\mathbf{q}^2\sin^2\theta}{(\varepsilon'-|\mathbf{q}'|\cos\theta')(\varepsilon-|\mathbf{q}|\cos\theta)}\Bigg]$$
$$\times\left[F_1^2(|p_0-p|^2)+2\left(\frac{\mu}{M}\right)^2F_2^2(|p_0-p|^2)(ME-M^2)\right]\Bigg\},\qquad(1.46)$$

where $f(\varepsilon')=E(\varepsilon')+\varepsilon'+k_0-\varepsilon-M$;

$$Q^2=\mathbf{q}'^2+\mathbf{q}^2+\mathbf{k}^2-2|\mathbf{q}||\mathbf{q}'|(\cos\theta\cos\theta'+\sin\theta\sin\theta'\cos\varphi)$$
$$+2|\mathbf{q}'||\mathbf{k}|\cos\theta'-2|\mathbf{q}|\mathbf{k}|\cos\theta;$$

ε' and $E(\varepsilon')$ are found from the conservation law $q+p_0=p+q'+k$; θ is the angle between the vectors $\mathbf{q}$ and $\mathbf{k}$; θ' is the angle between $\mathbf{q}'$ and $\mathbf{k}$; φ is the angle between the vectors $[\mathbf{k}\times\mathbf{q}]$ and $[\mathbf{k}\times\mathbf{q}']$; and $d\Omega_q\,d\Omega_k=\sin\theta'\,d\theta'\,d\varphi'\sin\theta\,d\theta\,d\varphi$.

In order to find the Bethe–Heitler formula from (1.46), we need to take the limit $M\to\infty$, assume that the nucleon is a point particle ($F_i=1$), and set $\mu=0$. The energy conservation law then takes the form

$$\varepsilon=\varepsilon'+k_0\,.$$

An expression for the differential cross section for the production of pairs at nucleons can be found from (1.46) through the replacements

$$\varepsilon'\to-\varepsilon_+;\quad\varepsilon\to\varepsilon_-;\quad|\mathbf{q}'|\to|\mathbf{q}_+|;\quad|\mathbf{q}|\to|\mathbf{q}_-|;\quad\theta'\to\pi-\theta_+;\quad\theta\to\theta_-;$$
$$\varphi_+\to\varphi-\pi$$

and

$$\frac{dk_0}{|df(\varepsilon')/d\varepsilon'|}\frac{|\mathbf{q}'|}{|\mathbf{q}|}d\Omega_k d\Omega_q\to-\frac{d\varepsilon_-}{k_0^2}\frac{|\mathbf{q}_+||\mathbf{q}_-|}{|d\psi(\varepsilon_-)/d\varepsilon_-|}d\Omega_{q+}\,d\Omega_{q-}\,,$$

where the plus sign is associated with the positron, the minus sign is associated with the electron, and

$$\psi(\varepsilon_-)=E(\varepsilon_-)+\varepsilon_-+\varepsilon_+-M-k_0\,;$$
$$d\Omega_{q_+}d\Omega_{q_-}=\sin\theta_+d\theta_+d\varphi_+\sin\theta_-d\theta_-d\varphi_-\,.$$

Expression (1.46) for the differential cross section for bremsstrahlung contains a complicated dependence on the recoil of the nucleon. To get a clear idea of the effects of the recoil, of the form factor, and of the anomalous magnetic moment of the proton (μ) on the bremsstrahlung cross section, we should compare (1.46) with the Bethe–Heitler formula.

It turns out that if we consider the recoil of the nucleon alone we find a redistribution of the Bethe–Heitler differential cross section. With $\varphi = 0$, $\theta' = 30°$, and $6° \leqslant \theta \leqslant 150°$, for example, the cross section incorporating recoil is smaller than the Bethe–Heitler cross section everywhere, while with $\varphi = 150°$, $\theta' = 90°$, and $60° \leqslant \theta \leqslant 120°$ the cross section incorporating recoil is everywhere greater than the Bethe–Heitler cross section. In contrast with the incorporation of recoil, the incorporation of the form factor always reduces the value of the cross section, since the nucleon form factors are always less than unity in the region of space-like momentum transfer. The differential cross section should be integrated over the angles φ and θ', and an integral bremsstrahlung cross section should be found as a function of E, k_0, and θ. We write this integral cross section as

$$d\sigma(E,k_0,\theta) = \sin\theta\, d\theta\, dk_0\, F(E,k_0,\theta)\,.$$

The corresponding expression for the integrated Bethe–Heitler formula takes the form

$$d\sigma_{BH}(E,k_0,\theta) = \sin\theta\, d\theta\, dk_0\, F_{BH}(E,k_0,\theta)\,.$$

Let us compare the values of the functions F and F_{BH} (Ref. 30) for $E = 550$ MeV and $k_0 = 250$ MeV:

θ (deg)	F/F_{BH}	θ (deg)	F/F_{BH}
10	1.39	90	0.40
20	1.51	120	0.48
30	1.25	150	0.95
60	0.61		

For $\theta < 10°$ we have $F(E,k_0,\theta) \to F_{BH}(E,k_0,\theta)$, so that even at $\theta = 5°$ the difference between the two functions does not exceed 3%; i.e., the effect of recoil and of the form factors becomes negligible.

The diagram in Fig. 1.16c must be evaluated by the method of dispersion relations. It can be seen from this diagram that the electron vertex can be factorized, written out exactly, and calculated, while the remainder of the diagram is a generalized Compton-effect diagram with a single virtual end. The existence of dispersion relations for the Compton scattering of a virtual photon (in the case at hand, for bremsstrahlung and for e^+e^--pair production) was proved in Ref. 32. The use of the dispersion-relation method to evaluate diagrams of this type has made it possible to extend the region in which beams of electrons and photons can be used to test QED up to energies of 500–600 MeV, since the method provides the accuracy required in the evaluation of the diagram in Fig. 1.16c and thus in the calculation of the differential cross section for the process.

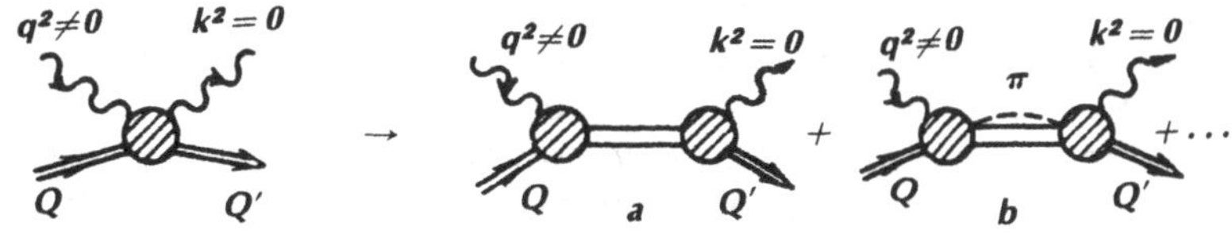

Figure 1.17. Set of diagrams considered in the description of the contribution of the Compton scattering of a virtual photon to bremsstrahlung.

We will not go into a lengthy description of the dispersion-relation method, which has been studied in detail in a monograph by Bogolyubov and Shirkov.[33] In Secs. 3 and 5 of Chap. 2 we will present some general principles for the use of this method in describing the electromagnetic form factors of hadrons. At this point we simply note that for the energy region in which we are interested here (not above 500 MeV) it is sufficient to incorporate in the calculation the two lowest-lying intermediate states: a one-nucleon state, with the nucleon form factors (Fig. 1.17a), and a one-pion state (Fig. 1.17b), which includes a nucleon, one π meson, and the form factors at the vertices of the diagram (they are shown in the figure as hatched circles and can be estimated through a comparison with the cross sections for the photoproduction of π mesons at nucleons).

Using the dispersion-relation method we can show that exact nucleon form factors should be introduced at the vertices of the diagram in Fig. 1.17a. This is one of the important advantages of the dispersion-relation method over the perturbation-theory method, which is ordinarily used. The dispersion relations substantially simplify the evaluation of the diagram in Fig. 1.17b.

Let us summarize the final results of calculations for the bremsstrahlung process. Diagrams of the Bethe–Heitler type (Fig. 1.16, a and c) have been taken into account exactly. In the diagrams corresponding to the one-nucleon approximation (Fig. 1.17a) the term proportional to the contribution of the electric charge has been calculated. In the interference term of the sum of the diagrams in Fig. 1.16, a and b, and Fig. 1.17a, the contributions proportional to the electric charge and to the first power of the anomalous magnetic moment have been retained. All of these approximations introduce an error of 5% for an impinging-electron energy near 500 MeV for angles $\theta', \theta \leqslant 90^\circ$. The diagram containing the one-pion intermediate state (Fig. 1.17b) has been taken into account in the approximation of a real Compton effect ($q^2 = 0$). For an impinging-electron energy of 500 MeV and for angles $\theta', \theta = 30^\circ$ (the energy of the virtual photon is about 250 MeV), the contribution of the one-pion state is less than that of the one-nucleon term. For angles $\theta, \theta' = 60^\circ$ ($E_\gamma \approx 270$ MeV), the contribution of the one-pion state is some five or six times larger than that of the one-nucleon term. For angles $\theta, \theta' = 90^\circ$ ($E_\gamma \approx 300$ MeV) the contribution of the one-pion state is about 15 times that of the one-nucleon term and becomes comparable to the contribution of the Bethe-Heitler cross section.

If we accept a possible 10% error in the determination of the experimental bremsstrahlung cross section, we find that QED can be tested at distances down to $\Lambda \geqslant 3 \times 10^{-14}$ cm for angles $\theta, \theta' \leqslant 30°$, for an impinging-electron energy of 550 MeV, and for a photon energy of 250 MeV.

An interesting phenomenon was first pointed out in Ref. 31: In the interval of angles for which the square of the 4-momentum of the virtual photon is approximately zero, there is a peak in the one-nucleon and interference terms in the bremsstrahlung cross section. The height of this peak is three or four orders of magnitude greater than the Bethe-Heitler cross section. Its angular width is only a few tens of arcseconds. With decreasing E_γ, the peak becomes progressively sharper. This peak could evidently be seen experimentally only if the step in the measured angle did not exceed a few tens of arcseconds. Measurements with a large step will average out the contribution and erase the effect.

3.2. Tests of QED in purely electromagnetic interactions

The startup of the SPEAR II and DORIS accelerators, and then the PETRA and PEP (Table 1.1), has made it possible to test QED at high energies in such purely electromagnetic processes as

$$\begin{array}{ll} e^-e^- \to e^-e^- & \text{(Møller scattering),} \\ e^+e^- \to e^+e^- & \text{(Bhabha scattering),} \\ e^+e^- \to \mu^+\mu^- & \text{(production of a muon pair),} \\ e^+e^- \to \gamma\gamma & \text{(annihilation of an } e^+e^- \text{ pair accompanied} \\ & \text{by the formation of two photons).} \end{array} \tag{1.47}$$

Radiative corrections become particularly important for these processes. As the energy of the colliding e^+e^- beams is increased, the effect of weak interactions and of the vacuum polarization caused by hadrons also becomes important.

Electromagnetic interactions cannot be separated from strong and weak interactions, and a test of QED at very high energies essentially reduces to (a) asserting that QED is valid and (b) explaining the observed deviations from it by means of corrections for strong and weak interactions.

Before we look at the results of experimental tests of QED, let us write differential cross sections for processes (1.47) and examine the characteristic features of the behavior of these cross sections. The reader can find a detailed derivation of an expression for the differential cross section for any of processes (1.47) in Ref. 33 and also in the monograph by Akhiezer and Berestetskiĭ which we mentioned above. In lowest-order perturbation theory, Møller scattering is described by two Feynman diagrams (Fig. 1.1). The differential cross section for the scattering of electrons by electrons, with allowance for the form factors [see (1.7)], is

$$\frac{d\sigma}{d\Omega}=\frac{\alpha^2}{2s}\left[\frac{q'^4+s^2}{q^4}\,\Phi^2(q^2)+\frac{2s^2}{q^2q'^2}\,\Phi(q^2)\Phi(q'^2)+\frac{q^4+s^2}{q'^4}\,\Phi^2(q'^2)\right], \quad (1.48)$$

where $t=q^2=-s\cos^2(\theta/2)$, $u=q'^2=-s\sin^2(\theta/2)$, and $s=4E^2$ is the square of the total energy of the colliding particles (θ is the angle through which the electron is scattered with respect to the direction of the beams of colliding electrons). Expression (1.48) agrees within terms of order m^4 with expression (1.7).

The photon propagator is modified in accordance with (1.11), and the form factors are chosen in the form

$$\Phi(q^2)=1\mp q^2/(q^2-\Lambda_{\pm}^2)\,.$$

The function $\Phi(q^2)$ is essentially the same as (1.8). If there are no deviations from a point-like interaction, the form factors can be assumed equal to unity, and (1.48) can be written

$$\frac{d\sigma}{d\Omega}=\frac{\alpha^2}{2s}\left(\frac{u^2+s^2}{t^2}+\frac{2s^2}{ut}+\frac{t^2+s^2}{u^2}\right). \quad (1.49)$$

It follows from (1.48) and (1.49) that the angular distribution has clearly expressed peaks at $\theta=0$ and 180°. For these peaks, however, the values of q^2 and q'^2 are small. The test of QED, in contrast, is carried out at large values of q^2 and q'^2 ($\theta\sim 90°$), where cross section (1.48) becomes small. An experimental test of QED thus runs into some additional difficulties associated with the need to build up a statistical base.

Incorporating radiative corrections is an important part of the problem of testing the applicability of QED at large momentum transfer. The corrections to processes (1.47) for the emission of real photons ($ee\to ee\gamma$, $ee\to\mu\mu\gamma$, $ee\to 3\gamma$) are usually made in two parts, associated with the emission of low-energy and high-energy photons. The corrections of the first type are factorized in the cross section and written as a factor $1+\delta_R$, which is used to multiply the differential cross section for the process of interest. The emission of a high-energy photon is often difficult to distinguish from the processes in which we are interested ($ee\to ee, \mu\mu, \gamma\gamma$), since the photon frequently cannot be detected. The correction for the emission of a high-energy photon is not factorized. Taking this correction into account is a rather complicated matter: It is necessary to allow for the angular distribution of three particles in the final state and to know the experimental geometry well. Corrections for the emission of high-energy photons have been calculated in α^3-order perturbation theory in several papers.[34]

With respect to Møller scattering, the correction for the emission of low-energy photons (Fig. 1.18) is taken into account (as we just mentioned) by multiplying cross section (1.48) or (1.49) by a factor of $1+\delta_R$. The diagrams of the next higher (fourth) order in α (Fig. 1.19, a–e) describe, in addition to the corrections to the process of interest here, $e^-e^-\to e^-e^-$, the new process $e^-e^-\to e^-e^-+$ hadrons, which is associated with the process $\gamma\gamma\to$ hadrons (Fig. 1.19e) (the subject of Chap. 5). In the kinematic region of energies and angles in which the cross section for the scattering $e^-e^-\to e^-e^-$ is large, the

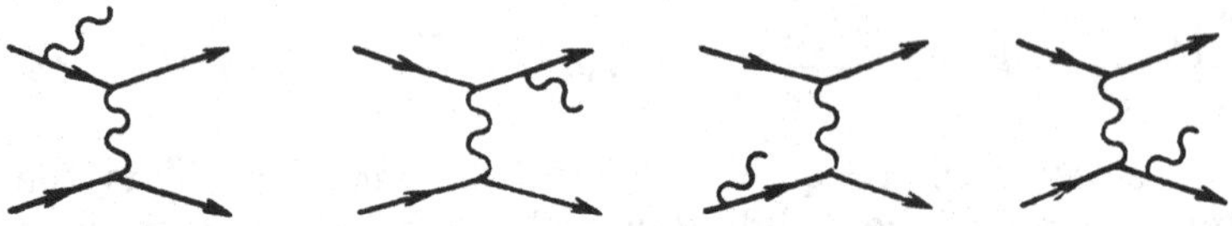

Figure 1.18. Radiative corrections of order α^3 to Møller scattering.

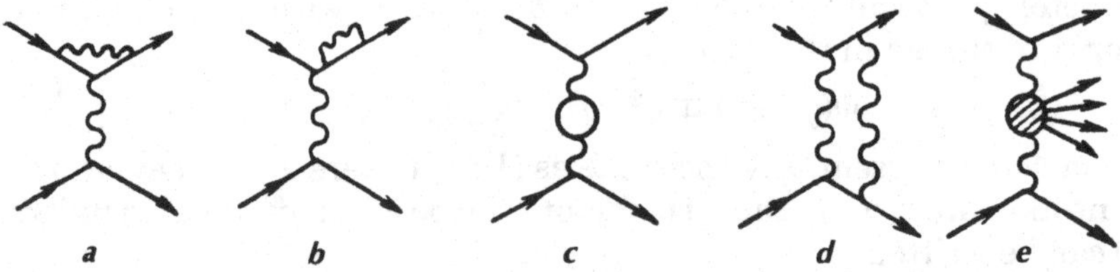

Figure 1.19. Set of diagrams of order α^4 describing corrections to Møller scattering.

contribution of the cross section for the process $e^-e^- \to e^-e^- +$ hadrons is negligible, and vice versa. The processes are thus well separated kinematically.

Radiation corrections of the type in Fig. 1.19, a and b, and also lepton loops ($ee, \mu\mu$; Fig. 1.19c) are calculated by standard perturbation-theory methods. Hadron loops (the hadronic polarization of vacuum), for which the intermediate state has $\pi\pi$, $K\overline{K}$, or $N\overline{N}$ loops instead of ee or $\mu\mu$ loops in Fig. 1.19c (there may be other hadron loops also), are taken into account by means of a dispersion integral in a method similar to that which we will describe below for Bhabha scattering.

In lowest-order perturbation theory, Bhabha scattering is described by the diagrams in Fig. 1.2. The differential cross section incorporating the form factors is written [see (1.10)]

$$\frac{d\sigma}{d\Omega} = \frac{\alpha^2}{8E^2}\left[\frac{t^2+u^2}{s^2}\Phi^2(s) + \frac{t^2+s^2}{u^2}\Phi^2(u) + \frac{2t^2}{us}\Phi(u)\Phi(s)\right] \qquad (1.50)$$

or

$$\frac{d\sigma}{d\Omega} = \frac{\alpha^2}{4s}\left[\frac{10+4x+2x^2}{(1-x)^2}\Phi^2(u) - \frac{2(1+x)^2}{1-x}\Phi(u)\Phi(s) + (1+x^2)\Phi^2(s)\right], \qquad (1.51)$$

where $x = \cos\theta$; $u = -s\sin^2(\theta/2)$; and $s = 4E^2$, $t = -s\cos^2(\theta/2)$.

If we set $\Phi(u) = \Phi(s) = 1$ in (1.50), we find

$$d\sigma/d\Omega = (\alpha^2/4s)(3+x^2)^2/(1-x)^2\,. \qquad (1.52)$$

If the leptons are not distinguished in the final state, expressions (1.50) and (1.51) acquire a more complicated form, which we will not reproduce here. In Bhabha scattering, in contrast with Møller scattering, there is only a single,

Table 1.3. Various corrections (%) to the $e^+e^- \to e^+e^-$, cross section.

	$s=1$		$s=10$		$s=30$	
θ,deg	δ_R	δ	δ_R	δ	δ_R	δ
20	0.1	0.1	2.8	1.3	3.8	2.5
90	− 8.1	0.6	− 8.0	3.0	− 8.5	4.6
160	− 8.1	− 0.3	− 6.1	3.4	− 5.4	5.1

sharply defined peak in the differential cross section, near $\theta = 0$, and there is a minimum near $\theta = 180°$.

As for Møller scattering, the radiative corrections for the emission of low-energy photons give rise to an additional factor of $1 + \delta_R$ in the expression for $d\sigma/d\Omega$. The set of diagrams responsible for these radiative corrections is equivalent to the set shown in Figs. 1.18 and 1.19. The corrections for the hadronic polarization of vacuum can be calculated by means of a dispersion integral. The photon propagator is written

$$g_{\mu\nu}/s \to (g_{\mu\nu}/s)[1 - \Pi_{\text{had}}(s)];$$

$$\text{Re}\Pi_{\text{had}}(s) = \frac{s}{4\pi^2\alpha}\int_{4m_\pi^2}^{\infty}\frac{\sigma(s')}{s'-s}\,ds',$$

where $\sigma(s')$ is the total cross section for the process $\gamma\gamma \to$ hadrons. The correction for the hadronic polarization of vacuum enters the cross section for the process $e^+e^- \to e^+e^-$ in the same manner as the correction for the emission of low-energy photons: as a factor $1 + \delta_{\text{had}}(s)$, where $\delta_{\text{had}} = -2\,\text{Re}\,\Pi_{\text{had}}(s)$. Table 1.3 lists comparative estimates of the contributions of the radiative corrections δ_R and of the vacuum-polarization corrections for various scattering angles of the e^+e^- pair in the final state with respect to the axis of the primary e^+e^- beam.

The production of a $\mu^+\mu^-$ pair is described most simply in lowest-order perturbation theory. In this case we have only a single Feynman diagram (Fig. 1.20).

The matrix element, written in accordance with Feynman's rules, is

$$M = \frac{i\,e^2}{(2\pi)^6}\left(\frac{m^2\mu^2}{EE'E_\mu E'_\mu}\right)^{1/2}$$

$$\times(2\pi)^4\delta(p' + p - q - q')\frac{\bar{u}(q')\gamma_\mu u(q)\bar{w}(p')\gamma_\mu w(p)}{(q+q')^2},$$

where m is the mass of the electron and μ is the mass of the μ meson. The differential cross section is given by

$$d\sigma = \frac{1}{(2\pi)^2}\frac{e^4m^2\mu^2}{2\sqrt{(qq')^2 - m^4}}\,\tfrac{1}{4}\left|\sum \bar{u}(q')\gamma_\mu u(q)\bar{w}(p')\gamma_\mu \bar{w}(p)\right|^2$$

$$\times\frac{1}{(q+q')^4}\,\delta(p' + p - q - q')\frac{d\mathbf{p}'}{\mathbf{E}'_\mu}\frac{d\mathbf{p}}{E_\mu}.$$

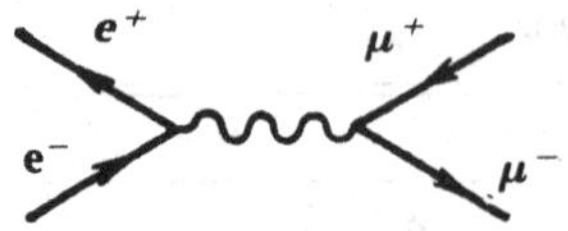

Figure 1.20. Feynman diagram describing $e^+e^- \to \mu^+\mu^-$ in the lowest order in α.

Here the summation sign means an average over the initial spin states and a summation over the final spin states of the e^+e^- and $\mu^+\mu^-$ pairs. These procedures are carried out with the help of projection operators for the electrons and positrons:

$$\Lambda(q) = (\hat{q} + m)/2m, \quad \Lambda(q') = (\hat{q}' - m)/2m$$

and corresponding projection operators for μ^+, μ^-.

We write

$$|\sum|^2 = \mathrm{Tr}\gamma_\mu \frac{\hat{q}+m}{2m}\gamma_\nu \frac{\hat{q}'-m}{2m}\mathrm{Tr}\frac{\hat{p}'-\mu}{2\mu}\gamma_\mu \frac{\hat{p}+\mu}{2\mu}\gamma_\nu\,;$$

$$\mathrm{Tr}\gamma_\mu \frac{\hat{q}+m}{2m}\gamma_\nu \frac{\hat{q}'-m}{2m} = \frac{1}{4m^2}\mathrm{Tr}(\gamma_\mu \hat{q}\gamma_\nu \hat{q}' - m^2\gamma_\mu\gamma_\nu)$$

$$= \frac{1}{m^2}[q_\mu q'_\nu + q'_\mu q_\nu - \delta_{\nu\mu}(qq' + m^2)]\,;$$

$$\mathrm{Tr}\gamma_\mu \frac{\hat{p}'-\mu}{2\mu}\gamma_\nu \frac{\hat{p}+\mu}{2\mu} = \frac{1}{\mu^2}[p_\mu p'_\nu + p'_\mu p_\nu - \delta_{\mu\nu}(pp' + \mu^2)]\,.$$

We can thus write

$$|\sum|^2 = \frac{1}{m^2u^2}[2(qp)(q'p') + 2(qp')(q'p) + 2\mu^2(qq') + 2m^2(pp') + 4m^2\mu^2]\,.$$

In the center-of-mass frame of the colliding e^+e^- beams, the following relations hold:

$$E = E' = E_\mu = E'_\mu; \quad \mathbf{q} = -\mathbf{q}'; \quad \mathbf{p} = -\mathbf{p}'\,;$$
$$qp = q'p'; \quad (qp)(q'p') = (qp)^2\,.$$

We use the definition $ab = a_0b_0 - \mathbf{ab} = a_0b_0 - ab\cos\theta$ of the product of two 4-vectors, where θ is the angle between the collision axis of the e^+e^- beams and the axis along the momenta of the outgoing $\mu^+\mu^-$ pair. We thus write (we are replacing $|\mathbf{q}|$ and $|\mathbf{p}|$ by simply q and p)

$$|\sum|^2 = \frac{2}{m^2\mu^2}[(E^2 - qp\cos\theta)^2 + (E^2 + qp\cos\theta)^2 + \mu^2(E^2 + q^2)$$

$$+ m^2(E^2 + p^2) + 2m^2\mu^2] = \frac{4E^4}{\mu^2m^2}\left(1 + \frac{p^2q^2}{E^4}\cos^2\theta + \frac{\mu^2 + m^2}{E^2}\right).$$

We ignore the mass of the electron, m, in the calculations; i.e., we assume $q^2 \approx E^2$; we conserve the ratio μ/E_μ; and we introduce $p/E_\mu = \beta_\mu$ and $s = 4E^2$.

We can then rewrite $|\sum|^2$ as

$$|\Sigma|^2 = \frac{s^2}{4\mu^2 m^2}\left[1 + \frac{p^2}{E^2}\cos^2\theta + \frac{\mu^2}{E^2}\right] = \frac{s^2}{4m^2\mu^2}(1 + \beta_\mu^2\cos^2\theta + 1 - \beta_\mu^2)$$
$$= \frac{s}{4\mu^2 m^2}[1 + \cos^2\theta + (1 - \beta_\mu^2)\sin^2\theta] .$$

We thus find

$$d\sigma = \frac{1}{(2\pi)^2}\frac{e^4}{2\sqrt{(qq')^2 - m^4}}\frac{1}{16}[1 + \cos^2\theta$$
$$+ (1 - \beta_\mu^2)\sin^2\theta]\frac{d\mathbf{p}'}{E'_\mu}\frac{d\mathbf{p}}{E_\mu}\delta(p' + p - q - q') ,$$

where

$$\delta(p' + p - q' - q) = \delta(E'_\mu + E_\mu - E' - E)\delta(\mathbf{p}' + \mathbf{p}' - \mathbf{q}' - \mathbf{q}) .$$

It is a trivial matter to carry out the integration over $d\mathbf{p}'$ by making use of the δ function $\delta(\mathbf{p}' + \mathbf{p} - \mathbf{q}' - \mathbf{q})$. We replace the integration over $d\mathbf{p}$ by an integration over dE_μ. From the equality $E_\mu^2 = \mathbf{p}^2 + \mu^2$ we have $2E_\mu\, dE_\mu = |\mathbf{p}|d|\mathbf{p}| = 2p\, dp$. We can now write $d\mathbf{p}/E_\mu$ as

$$\frac{d\mathbf{p}}{E_\mu} = \frac{p^2\, dp \sin\theta\, d\theta\, d\varphi}{E_\mu} = \frac{pE_\mu dE_\mu}{E_\mu} d\Omega; \quad d\Omega = \sin\theta\, d\theta\, d\varphi .$$

We thus have

$$\frac{d\sigma}{d\Omega} = \frac{1}{(2\pi)^2}\frac{e^4}{16s}\frac{p}{E'_\mu}[1 + \cos^2\theta + (1 - \beta_\mu^2)\sin^2\theta]\delta(E'_\mu + E_\mu - E - E')dE_\mu .$$

Here we have used the equality $\sqrt{(qq')^2 - m^4} = 2Eq \approx s/2$. We introduce the notation $\alpha = e^2/4\pi$. After a trivial integration over dE_μ we find

$$d\sigma/d\Omega = (\alpha^2/4s)\beta_\mu[1 + \cos^2\theta + (1 - \beta_\mu^2)\sin^2\theta] .$$

If we set $\beta_\mu = 1(p/E_\mu \approx 1)$, we find the usual expression for the differential cross section for the process $e^+e^- \to \mu^+\mu^-$:

$$d\sigma/d\Omega = (\alpha^2/4s)(1 + x^2) , \tag{1.53}$$

where $x = \cos\theta$. Integrating this equation over $d\Omega$, we find ($dx = \sin\theta\, d\theta$)

$$\sigma = \frac{\alpha^2}{4s} 2\int_0^1 dx(1 + x^2)\int_0^{2\pi} d\varphi = \frac{\alpha^2}{4s}4\pi\left(x + \frac{x^3}{3}\right)\Big|_0^1 ;$$

i.e.,

$$\sigma = (4\pi/3)\alpha^2/s . \tag{1.53a}$$

The calculations of the radiative corrections and of the corrections for the hadron polarization of vacuum is the same as the corresponding calculation for Bhabha scattering.

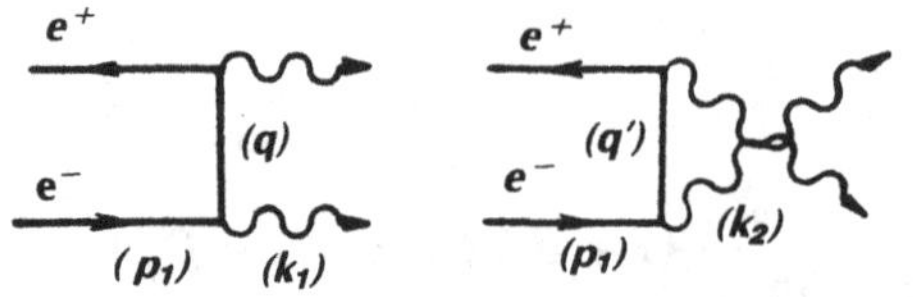

Figure 1.21. Test of the electron propagator in $e^+e^- \to 2\gamma$ in a first-order perturbation theory in α.

For $\theta = 90°$ and for various energies of the colliding e^+e^- beams ($\sqrt{s} = 2E$), the relative sizes of these contributions are roughly the same as for Bhabha scattering:

$\sqrt{s}$ (GeV)	1	10	30
δ_{had} (%)	-1.5	3.4	5.2
δ_R (%)	-4.6	-5.9	-6.4

The e^+e^- annihilation into two photons is described in lowest-order perturbation theory by two diagrams (Fig. 1.21). An electron propagator is tested in this process, in contrast with the first three processes in (1.47). The differential cross section corresponding to the diagrams shown in Fig. 1.21 can be written as follows, with radiative corrections:

$$\frac{d\sigma}{d\Omega} = \frac{\alpha^2}{2s}\left[\frac{q'^2}{q^2}|F(q^2)|^2 + \frac{q^2}{q'^2}|F(q'^2)|^2\right](1+\delta_R), \tag{1.54}$$

where $q^2 = (p_1 - k_1)^2 \approx -4E^2\sin^2(\theta/2)$ and $q'^2 = (p_1 - k_2)^2 \approx -4E^2\cos^2(\theta/2)$. If we choose the form factor $F(q^2)$ in the form $F(q^2) = 1 \pm q^4/\Lambda_\pm^4$, we can rewrite (1.54) as

$$\frac{d\sigma}{d\Omega} = \frac{\alpha^2}{2s}\left[\frac{q'^2}{q^2} + \frac{q^2}{q'^2} \pm \frac{4q^2q'^2}{\Lambda_\pm^4}\right](1+\delta_R). \tag{1.55}$$

If we assign the form factor the meaning of the exchange of a heavy electron, the expression for the differential cross section in (1.54) takes the slightly different form

$$\frac{d\sigma}{d\Omega} = \frac{\alpha^2}{2s}\left[\frac{q'^2}{q^2} + \frac{q^2}{q'^2} \pm \frac{2s^2 - 4q^2q'^2}{\Lambda_{e^*\pm}^4}\right](1+\delta_R). \tag{1.56}$$

Here is the standard procedure used for testing QED.

(1) The differential cross section of the process of interest is measured. Radiation corrections are incorporated in the expressions for the cross section. Corrections for the hadronic polarization of vacuum must be incorporated as the energy of the colliding beams increases ($2E \gtrsim 20$ GeV).

(2) The influence of weak interactions on the cross sections for electromagnetic processes is usually ignored. At beam energies $2E \gtrsim 30$ GeV, however, the terms which arise from the interference of the amplitudes of the electromagnetic and weak interactions may produce observable effects and should be taken into account. The expressions for the differential cross sections of these processes are correspondingly modified. In particular, an expression of this type will be written out explicitly for the process

Table 1.4. Limitations on the parameter $\Lambda_\pm$.[40]

Reaction	Λ_+, GeV	Λ_-, GeV
$e^+e^-\to e^+e^-$	16.1	22.6
$e^+e^-\to\gamma\gamma$	6.6	7.9
$e^+e^-\to\mu^+\mu^-$	11.9	22.8

$e^+e^-\to\mu^+\mu^-$. We will discuss the question of radiative corrections again, but now we will consider weak interactions.

(3) The photon and electron propagators are usually parametrized in the following way:

photon propagators

$$\Phi(q^2)=1\mp\frac{q^2}{q^2-\Lambda_\pm^2}\quad\text{for space-like } q^2<0,$$

$$\Phi(s)=1\mp\frac{s}{s-\Lambda_\pm^2}\quad\text{for time-like } s=q^2>0,$$

electron propagator $$F(q^2)=1\pm\frac{q^4}{\Lambda_\pm^4},\quad\text{or}\quad F(q^2)=1\pm\frac{q^4}{2\Lambda_{e^*\pm}^4}\sin^2\theta\,.$$

The second expression for the electron propagator is used in interpreting a deviation of QED from a point-like interaction in the case of the exchange of a heavy electron; $\Lambda_{e^*\pm}$ is the mass of the heavy electron. Limitations on the parameters $\Lambda_\pm$ and $\Lambda_{e^*\pm}$ are found from the error in the measurement of the cross sections (Sec. 1.1).

Until 1978 the highest energy attainable at a colliding-beam accelerator was no greater than 10 GeV in the c.m. frame (Table 1.1). By that time, several major experiments had been carried out to test QED in this energy range.[35–40] Beron *et al.*,[40] for example, measured the cross sections of the last three processes in (1.47) for $\theta\approx 90°$ at a resultant beam energy of 5.2 GeV. That study was valuable because QED was tested through simultaneous studies of all three reactions. This experiment was carried out at the SPEAR I accelerator. The cross sections for the reactions $e^+e^-\to e^+e^-$, $e^+e^-\to\gamma\gamma$ and $e^+e^-\to\mu^+\mu^-$ were measured within errors of $\pm 5\%$, $\pm 8\%$, and $\pm 15\%$, respectively. The deviations from QED were chosen in the form

$$\frac{1}{q^2}\to\frac{1}{q^2}\left(1\pm\frac{q^2}{q^2-\Lambda_\pm^2}\right)\quad\text{for the photon propagator},$$

$$\frac{1}{q^2-m^2}\to\frac{1}{q^2-m^2}\left(1\pm\frac{q^4}{\Lambda_\pm^4}\right)\quad\text{for the electron propagator}.$$

Table 1.4 shows the limitations found on the parameter $\Lambda_\pm$ in this experiment.

These results demonstrate that QED is valid down to distances of 4×10^{-15} cm. The cross sections for the reactions $e^+e^-\to e^+e^-$ and $e^+e^-\to\gamma\gamma$

were measured at the SPEAR II accelerator [Refs. 37–39] at an energy of 7.4 GeV and at $\theta \approx 90°$. The following limits were found on the values of $\Lambda_{\pm}$: for $e^+e^- \to e^+e^-$,[37]

$$\Lambda_+ = 38.0 \text{ GeV},$$
$$\Lambda_- = 33.8 \text{ GeV};$$

and for $e^+e^- \to \gamma\gamma$,[39]

$$\Lambda_+ = 10.7 \text{ GeV},$$
$$\Lambda_- = 9.0 \text{ GeV}.$$

The validity of QED in the region of a large momentum transfer (at small distances) was tested after the startup of the PETRA and PEP high-energy accelerators with colliding e^+e^- beams.

As was mentioned above, testing QED in this energy range is complicated by the circumstance that as the energy and the measurement accuracy increase, it is no longer legitimate to ignore weak interactions and the hadronic polarization of vacuum. In this case, QCD and the Weinberg-Salam theory of weak interactions must be incorporated in a theoretical interpretation of events. While the Weinberg-Salam theory agrees highly accurately with all the available experimental data over a broad energy range, $1 \leqslant E \leqslant 10^{10}$ eV, as has already been shown, QCD still requires a serious experimental test.

The past six or seven years have seen the discovery of the J/ψ particle, the family of Υ particles, the τ lepton, the F and D mesons, and other particles which are produced in electromagnetic interactions. It is necessary to test QED not only in the standard processes discussed above but also in other processes, involving new particles. For example, the τ meson was discovered[41] at the SPEAR II accelerator, and its existence was confirmed in experiments at the DORIS accelerator.[42] It has been established that the τ meson is a lepton with a spin of 1/2 and with properties similar to those of the μ meson. The family of leptons with a nonzero rest mass has now expanded to three particles (the electron, the μ meson, and the τ meson). It is therefore important (for example) to test the universality of the e, μ, and τ leptons with respect to electromagnetic interactions (more on this below).

Tests of QED at high energies frequently require allowance for the possible production of J/ψ, Υ, and other particles in an intermediate state.

It becomes a complicated matter to deal with the radiative corrections, especially when weak interactions are included. To get a clearer picture of the incorporation of weak interactions, let us examine the production of a muon pair, $e^+e^- \to \mu^+\mu^-$. The way in which the corrections are taken into account in this process is just the same as in a calculation of the corresponding corrections to the other processes in (1.47). The number of diagrams, however, is far smaller, so that the discussion of this question is simpler and clearer.

The basic diagram in lowest-order perturbation theory is shown in Fig. 1.20. The cross section calculated in this approximation is given by expression (1.53). The cross section is symmetric in the angle θ, specifically with respect to $\theta = 90°$. What types of corrections do we have to make to expression (1.53)?

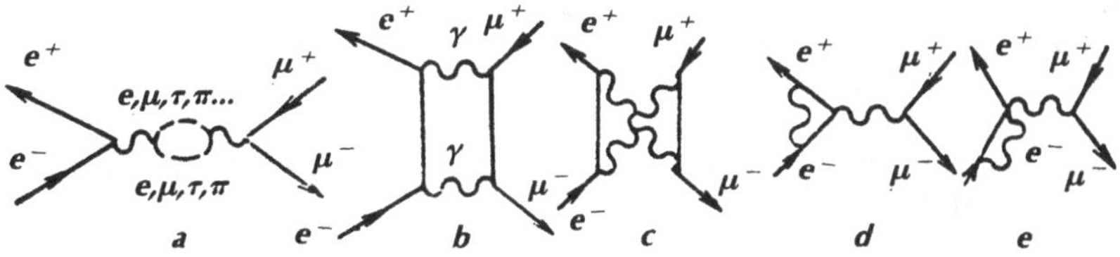

Figure 1.22. Examples of diagrams of order e^4.

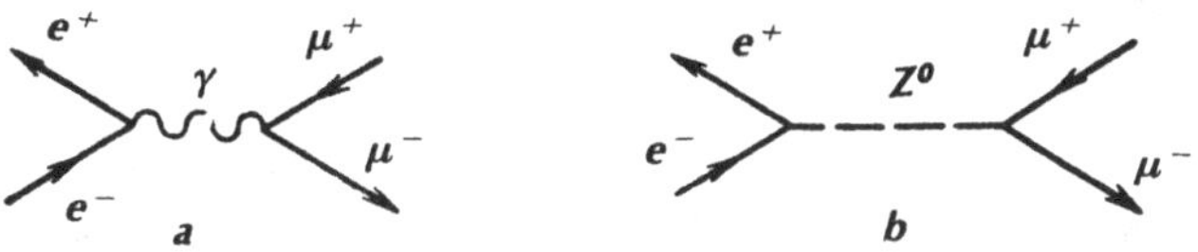

Figure 1.23. Diagrams used in describing $Z^0\gamma$ interference.

(1) Corrections for the bremsstrahlung of low-energy photons.
(2) Corrections for the bremsstrahlung of high-energy photons.
(3) Corrections for the interference of the amplitude corresponding to Fig. 1.20 with the amplitudes corresponding to all diagrams of order e^4 [ee, $\mu\mu$, and $\tau\tau$ loops; hadron loops (Fig. 1.22a); two-photon exchange (Fig. 1.22, b and c); vertex corrections (Fig. 1.22d); and the renormalization of external lines (Fig. 1.22e)].
(4) A correction for $Z\gamma$ interference. For the time being, incorporating weak interactions reduces to incorporating the interference of the amplitudes for the diagrams containing an exchange of a photon and a Z^0 boson (Fig. 1.23). The contribution of the diagram in Fig. 1.23b to the cross section for the process $e^+e^-\to\mu^+\mu^-$ at these energies ($\sqrt{s}=2E\lesssim 40$ GeV) is negligibly small because of the large mass of the Z^0 boson in the propagator of the Z^0 boson ($M_{Z^0}\approx 90$ GeV) and also because of the small value of the constant of the weak interaction. At energies $\sqrt{s}\gtrsim 30$ GeV, $Z^0\gamma$ interference introduces a forward-backward asymmetry in the angular distribution of μ mesons for the process $e^+e^-\to\mu^+\mu^-$. The asymmetry arises because the (electron and muon) lepton currents $j^{(l)}$ in the model of $SU(2)\times U(1)$ gauge theory of weak interactions contain terms with a γ_5 matrix[5]:

$$j^{(l)}\sim\bar{l}\gamma_\mu(g_v+g_a\gamma_5)l\,, \tag{1.57}$$

where τ and l are the spinors of electrons and muons, and g_v and g_a are the vector and axial constants of the weak interaction. The presence of terms with a γ_5 matrix in the Hamiltonian of the weak interaction disrupts the symmetry of the angular distribution with respect to the value $\theta=90°$.

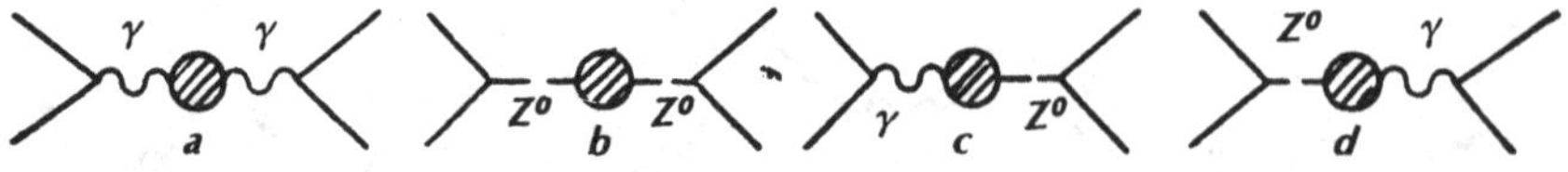

Figure 1.24. (a,b) Diagrams with single-loop corrections to the Z^0 and γ propagators; (c,d) diagrams with their mixing.

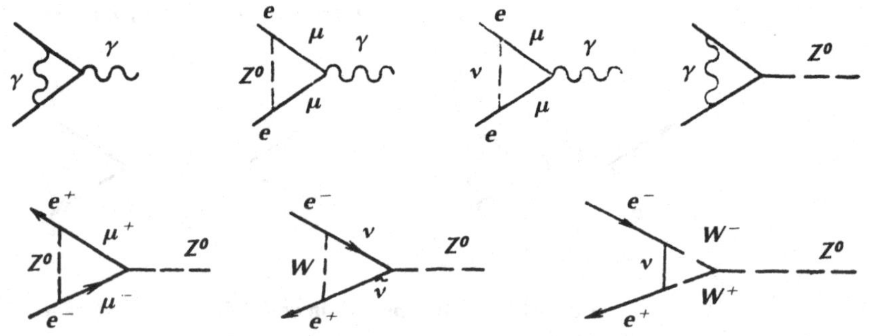

Figure 1.25. Corrections of the vertex type in the Weinberg-Salam model.

(5) Single-loop corrections[44] (Fig. 1.24). Infinite chains of single-loop diagrams of the same type are taken into account. The loops may be present in both the photon propagator and the Z^0 propagator. Quantum electrodynamics is used at the $\gamma \to e^+e^-$ vertex, while the Hamiltonian for the interaction of the Weinberg-Salam theory is used at the $e^+e^- \to Z^0$ vertex.
(6) Corrections of a vertex type (Fig. 1.25). An ordinary QED vertex is shown in the first diagram. The other diagrams are examples of vertices of various types which include electromagnetic and weak interactions.

The infrared divergences are eliminated by the usual QED methods: by incorporating a low-energy bremsstrahlung. This method is analyzed in detail in all of the monographs on QED. After the infrared divergences have been eliminated, one is left with a final expression which is invariant under changes in the mass of the low-energy photon.

The final expression for the differential cross section for the process $e^+e^- \to \mu^+\mu^-$ can be written

$$d\sigma = d\sigma^0 + d\sigma^{\text{brem}} + d\sigma^{\text{self}} + d\sigma^{\text{int}} + d\sigma^{\text{vert}} + d\sigma^{\text{ren}} + d\sigma^{\text{box}} + d\sigma^{\text{con}}, \qquad (1.58)$$

where $d\sigma^0$ is the cross section for (1.53); $d\sigma^{\text{brem}}$ is the correction for the emission by bremsstrahlung of low-energy photons, with allowance for weak interactions; $d\sigma^{\text{self}}$ are the self-energy corrections to the photon propagator and the Z^0 propagator; $d\sigma^{\text{int}}$ is the correction corresponding to $Z^0\gamma$ interference; $d\sigma^{\text{vert}}$ is the correction for vertices (see Fig. 1.25); $d\sigma^{\text{ren}}$ is the correction for the renormalization of external lines (Fig. 1.22e); $d\sigma^{\text{box}}$ is the contribution of box diagrams (Fig. 1.26); and $d\sigma^{\text{con}}$ is the contribution of the contraterms which arise upon renormalization—the elimination of the ultraviolet divergences.

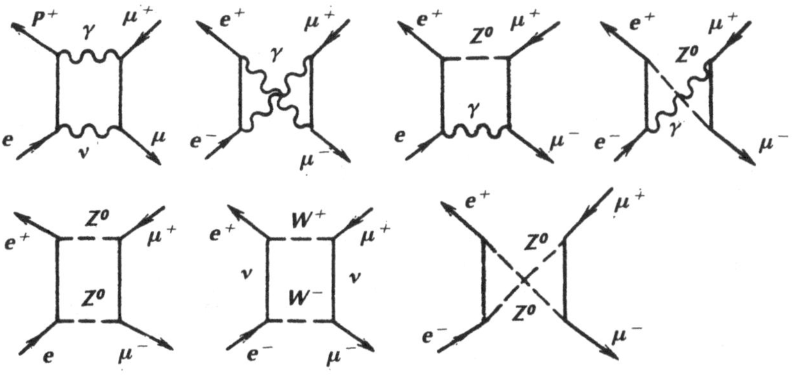

Figure 1.26. Examples of "box" diagrams and of crossing diagrams.

Corrections for the emission of high-energy photons, with allowance for weak interactions, have not yet been calculated.

We would like to offer a few comments of a general nature. We would expect weak interactions to have their greatest effect on the process $e^+e^- \to \mu^+\mu^-$ at colliding-beam energies close to the mass of the Z^0 boson (80–90 GeV). We do not yet have accelerators for such energies, but they will appear in the near future. At energies of the e^+e^- beams below the mass of the Z^0 boson, the radiative corrections for the single-loop diagrams are small, and where they are significant the corrections for the emission of high-energy photons become substantial. This relation between contributions has the consequence that if high-energy photons are ignored there is no basis for attempting to find an effect of weak interactions on the value of the differential cross section for the process $e^+e^- \to \mu^+\mu^-$ and to extract from electromagnetic interactions information on such parameters of the weak interaction as the Weinberg angle and the vector (g_v) and axial (g_a) coupling constants.

We will show below that the forward-backward angular asymmetry in the distribution of μ mesons would be completely observable.

Several specific features arise in calculations of radiative corrections near resonances.[46] For a nonresonant cross section the incorporation of the emission of low-energy photons in all orders of α leads to corrections of the type $(\Delta\omega/E)(4\alpha/\pi)\ln(2E/m)$, where $\Delta\omega$ is the energy resolution, and $2E$ is the energy of the e^+e^- beams in the c.m. frame. For narrow resonances such as the J/ψ particle, this factor takes analogous form: $(\Gamma/E)(4\alpha/\pi)\ln(2E/m)$. Near the position of the Z^0 boson, the corrections for the emission of low-energy photons are described by a complicated functional dependence on the variables E, M_R, $\Delta\omega$, and Γ, raised to a power of $(4\alpha/\pi)\ln(2E/m)$. They reach values of 50%.

The radiative corrections of all the types listed here should also be taken into account in analyses of the processes $e^\pm e^- \to e^\pm e^-$, $e^+e^- \to \gamma\gamma$, and $e^+e^- \to \tau^+\tau^-$. Single-loop radiative corrections with weak interactions were

Table 1.5. Limitations on the values of the parameters $\Lambda_{\pm}$.[45]

Parameter (GeV)	CELLO	JADE	MARK J	PLUTO	TASSO	MARK II
			$e^+e^- \to e^+e^-$			
Λ_+	83	112	128	80	140	—
Λ_-	155	106	161	234	296	—
			$e^+e^- \to \mu^+\mu^-$			
Λ_+	—	142	194	107	127	—
Λ_-	—	126	153	101	136	—
			$e^+e^- \to \tau^+\tau^-$			
Λ_+	139	111	126	79	104	—
Λ_-	120	93	116	63	189	—
			$e^+e^- \to \gamma\gamma$			
Λ_+	43	47	55	46	34	50
Λ_-	48	44	38	36	42	41

calculated for the process $e^+e^- \to e^+e^-$ in Ref. 47. For that process, the angular asymmetry in the distributions of electrons and positrons in the final state is zero. This is the picture of tests of QED at colliding-beam energies $2E \geqslant 30$ GeV.

The range of applicability of QED for the last three processes in (1.47) and for $e^+e^- \to \tau^+\tau^-$ has been studied at high energies at the PETRA accelerator ($2E \lesssim 37$ GeV) with the help of the CELLO, JADE, MARK J, PLUTO, and TASSO experimental installations and at the PEP accelerator ($2E \leqslant 30$ GeV) with the help of the MARK II installation. The limitations found on the parameters of $\Lambda_{\pm}$ and on the mass of a heavy electron (from the reaction $ee \to \gamma\gamma$) in these experiments are listed in Table 1.5. It can be concluded from this table that the leptons e, μ, and τ behave as point particles down to distances 2×10^{-16} cm.

Electroweak interactions.

As we mentioned earlier, the terms corresponding to an interference of the amplitudes of the electromagnetic and weak interactions can make observable contributions to the cross sections of electromagnetic processes at the exceedingly high energies which are being attained at the latest colliders. As the energy of the colliding beams is raised further, the weak interactions become dominant in e^+e^- annihilation, since the cross section for electromagnetic processes falls off in inverse proportion to E^2 as the beam energy is increased, while the cross section for weak interactions increases in proportion to E^2. We can write expressions for the differential cross sections for the processes $e^+e^- \to e^+e^-$ and $e^+e^- \to \mu^+\mu^-$, taking account of the interference of the amplitudes which are characterized by the diagrams in Fig. 1.27 and which are described by the weak-interaction Hamiltonian in the standard Weinberg-Salam [$SU(2) \times U(1)$] model with a single neutral boson Z^0.

The scattering amplitude corresponding to the diagrams in Fig. 1.27, a and b, is

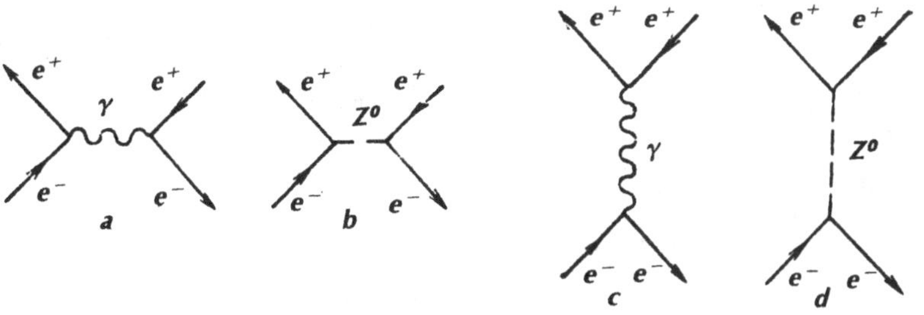

Figure 1.27. Diagrams used in calculating the interference of the amplitudes of the electromagnetic and weak interactions.

$$A \sim \frac{4\pi\alpha}{s}\bar{l}\gamma_\mu l\bar{e}\gamma_\mu\, e + \frac{1}{s - M_{Z^0}^2}\bar{l}\gamma_\mu(g_v + g_a\gamma_5)\bar{l}e\gamma_\mu(g_v + g_a\gamma_5)e\,. \quad (1.59)$$

Here $s = 4E^2$; $\bar{l}$ and l are the spinors of either the electron or the μ meson; $\bar{e}$ and e are the spinors of the electron; M_{Z^0} is the mass of the Z^0 boson; and g_v and g_a are the vector and axial constants of the weak interaction. The scattering amplitude corresponding to the diagrams in Fig. 1.27, c and d, differs from (1.59) only in the change of variables $s \to t = (q - q')^2$. The final expression for the cross section for Bhabha scattering[48] is

$$\frac{d\sigma}{d\Omega} = \frac{2\alpha^2}{s}\left\{\frac{1}{8}\frac{(3+x^2)^2}{(1-x)^2} + \frac{(3+x^2)}{(1-x)^2}[(3+x)Q - x(1-x)R]g_v^2\right.$$

$$-\frac{1}{1-x}[(7+4x+x^2)Q + (1+3x^2)R]g_a^2 + \left[\frac{16}{(1-x)^2}Q^2 + (1-x^2)^2R^2(g_v^2 - g_a^2)^2\right.$$

$$\left. + (1+x)^2\left[\frac{2}{(1-x)}Q - R\right]^2(g_v^4 + 6g_v^2g_a^2 + g_a^4)\right\}, \quad (1.60)$$

where $x = \cos\theta$ (θ is the polar angle with respect to the axis of the colliding beams); $g_v = -(1/2)(1 - 4\sin^2\theta)$ ($\sin^2\theta_W$ is the square of the sine of the Weinberg angle); $g_a = -1/2$; and $Q = \rho M_Z^2q^2/(q^2 - M_Z^2)$ and $R = \rho M_Z^2 s/(s - M_Z^2)$ [$\rho = G_F/8\sqrt{2}\pi\alpha = 4.49\times10^{-5}$ GeV^{-2}, $q^2 = -(s/2)(1 - \cos\theta)$, and M_Z is the mass of the Z^0 boson].

The differential cross section for the process $e^+e^- \to \mu^+\mu^-(\tau^+\tau^-)$ found in lowest-order perturbation theory and corresponding to the incorporation of the diagrams in Fig. 1.27, a and b, is given by

$$d\sigma/d\Omega = (2\alpha^2/s)\left\{(1+x^2)\left[\frac{1}{8} + g_v^2R + 2(g_v^2 + g_a^2)^2R^2\right]\right.$$

$$\left. + 2x(g_a^2R + 8g_v^2g_a^2R^2)\right\}. \quad (1.61)$$

If we set $R = Q = 0$ in (1.60) and (1.61), these expressions become (1.52) and (1.53), respectively. The terms proportional to Q and R describe an interference of the electromagnetic and weak interactions, while the terms propor-

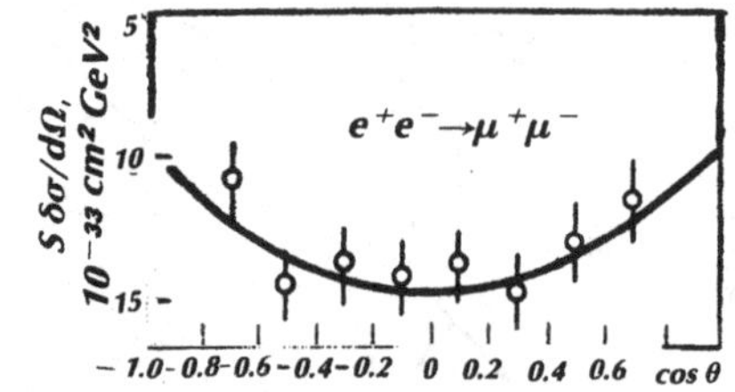

Figure 1.28. Combined experimental data obtained at the JADE, MARK J, PLUTO, and TASSO installations, which lead to the mean asymmetry $\langle A_{\mu\mu}\rangle = -(0.09 \pm 4.9)\%$.

tional to Q^2, QR, and R^2 describe contributions from weak interactions alone. A distinctive feature of cross section (1.61) is the presence of a term proportional to x, so that there is a significant forward-backward asymmetry in the distribution of the production of $\mu^+\mu^-$ and $\tau^+\tau^-$ pairs.

To illustrate the point, we write the following expression, which follows from (1.61):

$$A_{\mu\mu} = \frac{(d\sigma/d\Omega)_{x=1} - (d\sigma/d\Omega)_{x=-1}}{(d\sigma/d\Omega)_{x=1} + (d\sigma/d\Omega)_{x=-1}} = \frac{8(g_a^2 R + 8 g_v^2 g_a^2 R^2)}{1 + 8 g_v^2 R + 16(g_v^2 + g_a^2)^2 R^2}. \tag{1.62}$$

For $M_Z = 90$ GeV, $s = 1000$ GeV^2, $\sqrt{s} \approx 32$ GeV, $g_v = -0.04$, and $g_a = -0.5$, we have $A_{\mu\mu} \approx -0.1\,(-10\%)$. Experimentally the forward-backward asymmetry is expressed by

$$A_{\mu\mu} = \frac{F-B}{F+B} \approx \frac{3}{4}\frac{G_F}{\sqrt{2}\pi\alpha}\frac{sM_Z^2}{s-M_Z^2} g_a^2 , \tag{1.63}$$

where F and B are the numbers of μ and τ mesons with charge of a definite sign which are emitted into the forward (F) or backward (B) hemisphere. Since experimental installations usually cut out only some part of the forward or backward hemisphere, the theoretical values calculated from (1.63) will differ for different installations. The asymmetry $A_{\mu\mu}$ in (1.63) was measured at the PETRA accelerator. Here are the experimental data and corresponding theoretical predictions (for $\sqrt{s} \leqslant 32$ GeV):

DEVICE	JADE	MARK J	PLUTO	TASSO
$A_{\mu\mu}^{\text{expt}}$,(%)	-8 ± 9	0 ± 9	7 ± 10	-1 ± 12
$A_{\mu\mu}^{\text{theo}}$,(%)	-6	-5	-5.8	-6

If we combine the experimental data obtained by all the groups at PETRA (Fig. 1.28), we find a mean asymmetry (at a 95% confidence level)

$$\langle A_{\mu\mu}\rangle = (-0.9 \pm 4.9)\% ;$$

i.e., we can assert that essentially no asymmetry is observed. However, recent measurements of the coefficient $A_{\mu\mu}$ at PETRA at an energy of 36 GeV revealed that the asymmetry is nonzero at a level of three standard deviations[49]: $A_{\mu\mu} = (-7.7 \pm 2.4)\%$. This result must be verified. The relative change in $d\sigma/d\Omega$ which results from incorporating weak interactions,

$$\Delta = \frac{d\sigma/\delta\Omega - (d\sigma/d\Omega)_{\text{QED}}}{(d\sigma/d\Omega)_{\text{QED}}},$$

is so small (about 0.002) at $s = 1000\ \text{GeV}^2$ that it is essentially beyond observation.

In all of the experiments which have been designed to observe an effect of weak interactions in electrodynamic processes, an attempt has been made to obtain, if not accurate values of the weak-interaction parameters (the constants g_v, g_a, or $\sin^2\theta_W$), at least limitations on them. The limitations which are found provide upper limits on the values of parameters and are in agreement with the values generally accepted in the standard Weinberg-Salam model. As we mentioned earlier, however, it would hardly be possible to determine the weak-interaction constants without incorporating radiative corrections for the emission of high-energy photons.

Some models of electroweak interactions which have been proposed only recently allow an expansion of the sector of neutral vector mesons in the theory of gauge fields. In other words, these models expand the $SU(2)\times U(1)$ group to the $SU(2)\times U(1)\times U'(1)$ group[50] or the $SU(2)\times U(1)\times SU'(2)$ group.[51] In both of these models, the effective interaction Hamiltonian is written in the form

$$H = (4G_F/\sqrt{2})[(j^{(3)} - \sin^2\theta_w\, j_{\text{em}})^2 + Cj^2_{\text{em}}],$$

where $j^{(3)}$ is the third component of the weak isospin current, j_{em} is the electromagnetic current, and the parameter C describes the deviation from the standard Weinberg–Salam model. In the limit of small values of $|q^2|$, the models convert into the Weinberg–Salam model.

In comparisons of the predictions of the new models with experimental data, it is assumed that QED is valid and that the value of $\sin^2\theta_W$ is 0.23 (the generally expected value). In this case a comparison with experimental data can yield a limitation on the value of C. In models with two neutral vector mesons, C is expressed in terms of the masses of the new mesons, M_{Z_1} and M_{Z_2}, as follows:

$$C = \gamma(M^2_{Z_2} - M^2_{Z^0})(M^2_{Z^0} - M^2_{Z_1})/(M_{Z_1}M_{Z_2})^2, \tag{1.64}$$

where $\gamma = \cos^4\theta_W$ in the model of Ref. 50 or $\gamma = \sin^4\theta_W$ in the model of Ref. 51, and $M_{Z_2} > M_{Z^0} > M_{Z_1}$. Expression (1.64) with fixed values of C leads to a dependence of M_{Z_1} on M_{Z_2}. Figure 1.29 shows experimental limitations on the masses of the neutral vector mesons.

Increases in the energy of the colliding e^+e^- beams will undoubtedly result in some surprising new possibilities for studying the effects of weak and strong interactions on electromagnetic processes. As an illustration we might cite searches for a sixth quark (the t quark). The discovery of the family of Υ particles points to the existence of a fifth quark: the b quark (in addition to the known u, d, s, and c, quarks). Experimental data indicate that the mass of the b quark evidently lies in the interval $5.18 < m_b < 5.28$ GeV (Chap. 4). If the b quark does exist, then we should introduce a new doublet of quarks, $\binom{t}{b}$, in the Weinberg-Salam model; the sixth quark would be the t quark with a

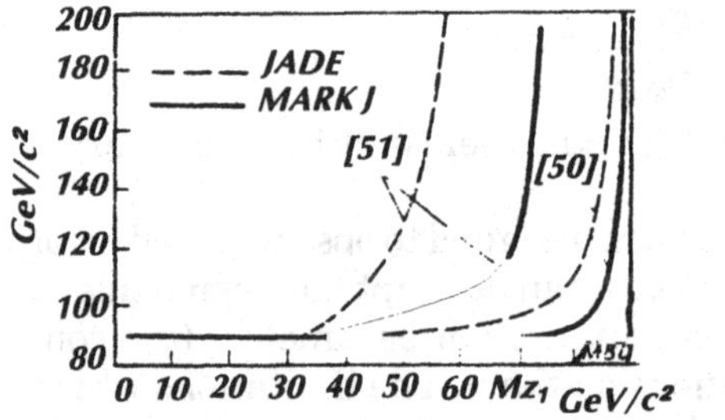

Figure 1.29. Experimental limitations on the masses (M_{Z_1} and M_{Z_2}) of neutral vector mesons in the models of Refs. 50 and 51 and the Weinberg-Salam (WS) model.

charge of $+2/3$. Data which have now been obtained at the PETRA accelerator indicate that if the t quark does exist, its mass would have to be greater than 18 GeV. One might suggest the existence of a $t\bar{t}$ bound state with quantum numbers $J^{PC}=1^{--}$, since such a state could be produced with the help of either a photon or a Z^0 boson in an e^+e^- annihilation process. In lowest-order perturbation theory for the process $e^+e^-\to\mu^+\mu^-$, for example, one can draw six diagrams in which a $t\bar{t}$ bound state could be present as an intermediate state. This set of diagrams is the same as the set shown in Figs. 1.23 and 1.24, when the loop in Fig. 1.24 is replaced by a $t\bar{t}$ bound state. This statement means that at high energies one could take up the problem of detecting a $t\bar{t}$ bound state in studies of the asymmetry ($A_{\mu\mu}$) of the distribution of $\mu^+\mu^-$ pairs.

3.3. $e\mu\tau$ universality; search for heavy leptons[52,53]

A test of the universality of electrons, muons, and τ leptons reduces to a test of QED at small distances. If the observed experimental cross sections for the processes $e^+e^-\to e^+e^-$, $e^+e^-\to\mu^+\mu^-$, and $e^+e^-\to\tau^+\tau^-$ are the same as those calculated in QED, then it is assumed that the electrons, muons, and τ leptons are universal with respect to electromagnetic interactions. It has been nearly a century since the discovery of the electron and half a century since the discovery of the muon; the τ lepton was not discovered[41] until 1975–1976. It turns out that the τ lepton—a spin-1/2 particle—decays weakly, and its properties are similar to those of a muon. A test of the universality of τ leptons would of course be of major interest. The reaction $e^+e^-\to\tau^+\tau^-$ has been detected on the basis of the final products of the decay of the τ lepton:

$$\tau\to\mu+\nu+\bar{\nu}\,(16\%);$$

$\tau\to$ lepton + one or several hadrons + several neutrinos (84%).

Figure 1.30 compares the experimental data with theoretical curves. The experimental data are seen to agree well with the QED predictions over a broad energy range for both the $\mu^+\mu^-$ pairs and the $\tau^+\tau^-$ pairs. In a gauge

Figure 1.30. Comparison of experimental data and theoretical curves on the scattering cross sections for the reactions (a) $e^+e^- \to \mu^+\mu^-$ and (b) $e^+e^- \to \tau^+\tau^-$.

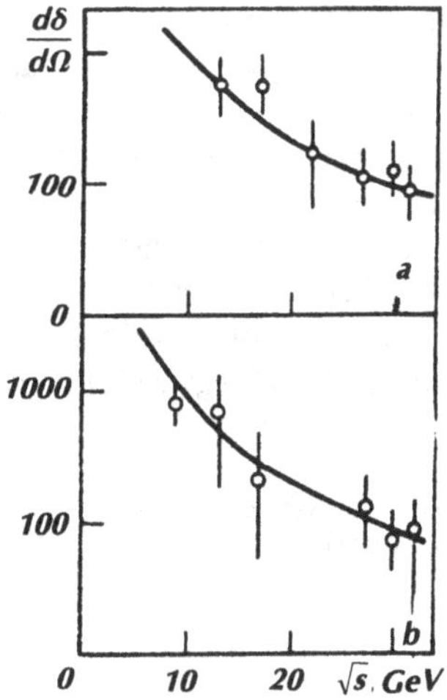

theory of electroweak interactions the presence of new heavy leptons is linked with the appearance of heavy quarks. A search for new heavy leptons is accordingly an important problem. An experiment carried out by the PLUTO group[53] was designed to search for spin-1/2 leptons with a point-like interaction. It was suggested that the new leptons might be produced by L^+L^- pairs and would decay by one of the two schemes

$$L \to l\nu_e\nu_L (l \text{ is a known lepton}) \text{ or}$$

$$L \to \nu_L + \text{hadrons}$$

(a corresponding problem has been solved by other experimental groups at the PETRA accelerator). None of the experiments has revealed heavy leptons within the limitation $m_L \lesssim 17$ GeV.

Here are the limitations on the masses of new heavy leptons:

Device	JADE	MARK J	PLUTO	TASSO
m_L(GeV)	17	16.0	14.5	15.5

* * *

Quantum electrodynamics survived the test in excellent shape: The theory of a point-like interaction of charged leptons with an electromagnetic field turned out to be correct down to distances on the order of 10^{-16} cm. All the leptons are universal with respect to electromagnetic interactions.

As colliding-beam accelerators with higher energies are started up, this entire research program to test QED at small distances will be played out again. As the energy increases, the radiative corrections of higher orders in α, the hadronic vacuum corrections, and the interference of electromagnetic and weak interactions become progressively more significant, and new opportunities open up for testing weak-interaction models in electroweak interactions. However, the perturbation-theory method as it exists today in QED (and QCD) presents serious computational difficulties which cast doubt on the possibility of direct calculations at higher orders in α (and α_s) with an arbitrary prespecified accuracy. Doubt is therefore cast on the possibility of testing QED at very high energies and very small distances, since the theory as it stands today has no other, more powerful computational methods.

Further progress in research on quantum electrodynamics will thus lean heavily on the development of computational methods in the theory of strong and electromagnetic interactions.

*The manuscript of this book was presented to the editorial office in 1982; plans and times of the construction of colliders have been changed since then.

Chapter 2 Electromagnetic form factors of elementary particles

1. Introductory comments

It was shown in Chap. 1 that leptons in electromagnetic interactions appear to be point-like down to distances of 10^{-16} cm. Nucleons, in contrast, do not behave as point particles in electromagnetic interactions, since their form-factor functions differ from unity even if the 4-momentum transfer q^2 is not very large.

Studying the behavior of the form-factor functions (we will say simply "form factors") of nucleons and other hadrons (e.g., π, K, and η mesons) as functions of q^2 in electromagnetic processes in which electrons and μ mesons are scattered elastically by these particles should be recognized as one of the most important problems in elementary-particle physics today.

The idea that elementary particles should have a certain structure is not new. It has been under discussion since the discovery of the electron. Today, however, the concept of the structure of elementary particles is more profound than the concept of the structure of an electron was at the beginning of this century (see the papers by Abraham, Lorentz, Poincaré *et al.*).

The parton model of the structure of elementary particles (Chap. 3), which presently holds sway, appeared in 1969. Before then it was assumed that each particle was constructed from virtual particles of all types, including some similar to itself. The bootstrap hypothesis arose: It is possible to write a system of equations of such a nature that in it the properties of each particle (mass, spin, parity, decay width, etc.) will be expressed in terms of the properties of all other particles. The solution of this system of equations should yield a theoretical description of the properties of all elementary particles. The bootstrap hypothesis was based on the mathematical method of dispersion relations, which proved very convenient, thanks to the use of the unitarity condition, both for writing bootstrap equations and for subsequent practical applications.

The concept of the structure of elementary particles is intimately related to the concept of their form factors, as we will be discussing in detail below. In this chapter we consider only the electromagnetic form factors which are determined from elastic scattering and annihilation processes. We now know quite a bit about the form factors of the π and K mesons and the nucleons. New information on the form factors of η and η' mesons has been extracted from the decays $\eta \to \mu^+\mu^-\gamma$ and $\eta' \to \mu^+\mu^-\gamma$. The experimental data on the scattering of electrons by pions, kaons, and nucleons will be interpreted in terms of dispersion relations. The existence of dispersion relations was first proved theoretically by Bogolyubov[54] in 1956, for the scattering of π mesons by nucleons. The existence of dispersion relations has been proved for several other interactions of elementary particles.

Despite quantum chromodynamics, the dispersion-relation method remains one of the most reliable computational methods in field theory, and it continues to generate successful explanations of experimental data. There can be no doubt that it will soon be used in QCD also.

Everywhere in this chapter we assume that it is valid to use dispersion relations to describe the form factors of elementary particles.

Unfortunately, QCD, presently a candidate for the role of a theory of strong interactions, still lags behind the dispersion-relation method in terms of the accuracy with which form factors can be described. The QCD approach to the description of form factors predicts only the asymptotic method of the form factors of hadrons:

$$F_h(Q^2) \to [\alpha_s(Q^2)/Q^2]^{n_h - 1}, \tag{2.1}$$

where n_h is the number of valence constituent quarks in the hadron ($n_h = 3$ for baryons; $n_h = 2$ for π and K mesons), and $\alpha_s(Q^2)$ is the coupling constant in QCD, given by

$$\alpha_s(Q^2) = \frac{12\pi}{(33 - 2f)\ln(Q^2/\Lambda^2)}.$$

Here f is the number of species (flavors) of quarks (e.g., if four types of quarks—u, d, s, and c—are taken into consideration, then $f = 4$), and Λ is the QCD constant [see Sec. 7 of Chap. 3, Eqs. (3.126) and (3.127)]. We have three comments regarding (2.1).

First, it reproduces the behavior of the form factors as functions of $q^2 = -Q^2$, which is predicted by the quark counting rules[55] to within powers of $\alpha_s(Q^2)$. We recall that the quark counting rules predict that the form factors will fall off with increasing Q^2 in accordance with

$$F(Q^2) \sim (1/Q^2)^{n-1},$$

where n is the number of constituent quarks ($n = 3$ for nucleons, $n = 2$ for π and K mesons, and $n = 1$ for leptons).

Second, (2.1) is approximate. It gives only the first term in the series from the exact expression for the meson form factor,

$$\frac{\alpha_s(Q^2)}{Q^2}\left(1 + \sum_{n=2,4,\ldots} C_n \left(\ln\frac{Q^2}{\Lambda^2}\right)^{-\gamma_n}\right),$$

and the first term in the series from the exact expression for the magnetic form factor of a nucleon, G_M,

$$\left(\frac{\alpha_s(Q^2)}{Q^2}\right)^2 \sum_{n,m} b_{n,m}\left(\ln\frac{Q^2}{\Lambda^2}\right)^{-\gamma_n-\gamma_m}.$$

Here $\gamma_{n,m}$ are anomalous dimensionalities (Sec. 7 of Chap. 3), and $b_{n,m}$ and C_N are coefficients. The dependence on $\ln(Q^2/\Lambda^2)$ is thus more complicated than in (2.1) [through the constant $\alpha_s(Q^2)$].

Third, within the framework of QCD it is not possible to find that the nucleon charge form factor G_E falls off with Q^2, as it must according to the quark counting rules.

The form factors G_E and G_M are linear combinations of the form factors $F_1(q^2)$ and $F_2(q^2)$ introduced earlier [see (1.5)] and will be discussed below.

We refer the reader interested in the details of the emergence of the concept of the "form factor of an elementary particle or nucleus" and the first studies of the form factors of particles and nuclei to the reviews in Refs. 56–58.

2. Form factors of elementary particles

There is no rigorous definition of a form factor. In most cases form factors are identified with generalized vertex functions.

The interaction of an electromagnetic field with nucleons (π mesons) is described by the vertex in Fig. 2.1a, while weak interactions are described by vertices of the type in Fig. 2.1b. The interaction of an electromagnetic field with a ρ meson is associated with the vertex in Fig. 2.1c. The strength of an interaction is characterized by coupling constants. Let us consider a three-particle interaction (Fig. 2.1a). We will consider only photons, π mesons, and nucleons here. Experimental data indicate that the coupling constant for the coupling of the π-meson field with the nucleon field is $G^2/4\pi \approx 15$, while that representing the coupling of the electromagnetic field with nucleons is $e^2/\hbar c = 1/137$. In describing the interaction of an electromagnetic field with a nucleon field (or a π-meson field), we need to recall the strong coupling of nucleons with π mesons. For this reason, the vertex in Fig. 2.1a, which describes the interaction of a photon with a nucleon, must be supplemented with the infinite set of diagrams shown in Fig. 2.2, in which π mesons are represented by dashed lines. This set of diagrams, along with the diagram in Fig. 2.1a, can be represented symbolically by the single diagram in Fig. 2.3, where the hatched circle means the contribution of all possible diagrams allowed by the corresponding Lagrangian. If one end of the diagram (Fig. 2.3) is a virtual photon ($q^2 \neq 0$), while the two others describe a free nucleon, then the vertex function F shown in Fig. 2.3 will depend on the single variable q^2: $F = F(\rho^2 = M^2, \rho'^2 = M^2, q^2 \neq 0)$. Such a function $F(M^2, M^2, q^2)$ is called the "electromagnetic form factor" of the nucleon. Actually, as we already saw in Chap. 1, the vertex which is responsible for the interaction of a proton or neutron with a photon contains two form factors, not the single form factor drawn (for simplicity) in Fig. 2.3.

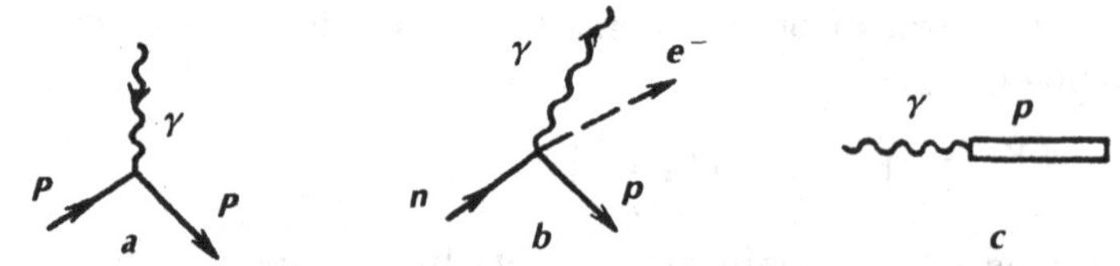

Figure 2.1. Simplest vertex functions.

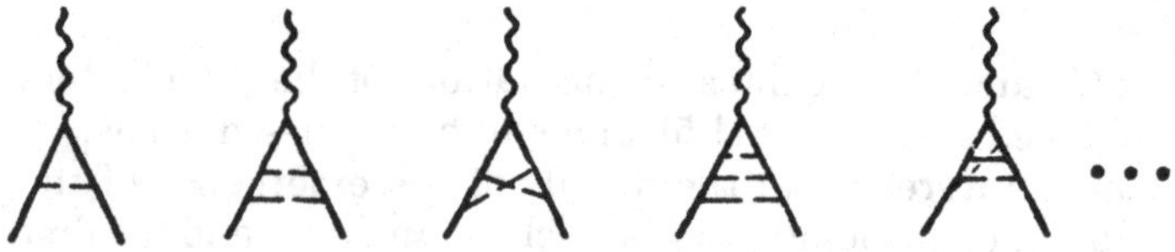

Figure 2.2. Various diagrams contributing to a generalized vertex function (see Figs. 2.3 and 2.6).

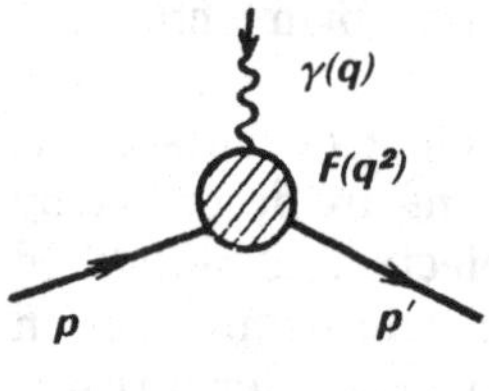

Figure 2.3. Generalized vertex function of a nucleon.

We have no methods for summing an arbitrary infinite number of the diagrams in Fig. 2.3, and the form factors are calculated approximately.

Form factors also appear in the case of vertices describing weak interactions (Fig. 2.1b). For example, in the decay $n \to p + e + \nu$, (Fig. 2.4), the hatched circle means the incorporation of all corrections for the strong interactions of the nuetron and the proton, and the functions $F(M_n^2, M_p^2, q^2)$ describing these interactions, where the variable q^2 is the difference between the 4-momenta of the proton and the neutron, are also called "form factors," although in this case all four of the particles (n,p,e,ν) lie on the mass shell. The term "form factor" (electromagnetic form factor of the π meson or axial form factor of the nucleon) is related to the virtual particle whose 4-momentum is used to study the dependence of the form factor and that real particle with which the given virtual particle forms a vertex. The term "axial form factor" which we used above is related to that part of the Hamiltonian of the weak interaction which contains the γ_5 matrix.

In QCD, the picture of the appearance of form factors of hadrons in electromagnetic interactions with leptons is different. For example, the form factors of the proton in elastic $ep \to ep$ scattering should appear as a result of a

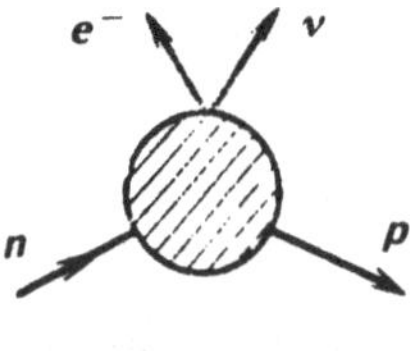

Figure 2.4. Generalized vertex function corresponding to the decay $n \to p + e + \nu$.

summation of all the diagrams describing the interactions which arise when an electron exchanges a photon with all the quarks of the nucleon. Here it is necessary to take into account the interaction among the quarks and gluons making up the nucleon. Since the coupling constant $\alpha_s(Q^2)$ depends logarithmically on Q^2 in QCD, becoming large at small values of Q^2, we cannot use perturbation theory in QCD in the region of Q^2 in which we are interested. For this reason, QCD presently gives us only a qualitative idea of the form factors and certain predictions regarding the behavior of the form factors as $Q^2 \to \infty$.

Form factors must not be interpreted as the real spatial distributions of the electric charge of the nucleon or of an anomalous magnetic moment, as we stressed more than once in Chap. 1. This statement follows from the circumstance that $F(x)$ is a four-dimensional Fourier transform of the function $F(q^2)$:

$$F(x) \sim \int d^4q \exp(i\,qx) F(q^2)\,.$$

The function $F(x)$ has a temporal distribution as well as a spatial distribution, and it is only at small values of q^2, under the assumption that $F(q^2)$ can be expanded in a Taylor series around $q^2 = 0$, and with a corresponding definition of the distribution of the electric charge or anomalous magnetic moment, that we can find a relationship between the form factors $F_{1,2}(q^2)$ and the mean-square radius of the nucleon, $\langle r^2 \rangle$, or the mean-square radius of the distribution of the magnetic moment.

We define the mean-square radius of the electric-charge distribution as the second moment of the fourth (charge) component of the nucleon current:

$$\langle r^2 \rangle = \int d^3r\, r^2 j_4(x)\,.$$

The matrix element

$$\langle p's' | j_\mu(0) | ps \rangle = \frac{ie}{(2\pi)^3} \frac{M}{\sqrt{EE'}} \bar{u}_{s'}(\mathbf{p}')[F_1(q^2)\gamma_\mu - F_2(q^2)\sigma_{\mu\nu} q_\nu]\, u_s(p)$$

is related to the matrix element of the nucleon current, $j_\mu(x)$, by

$$\langle p's' | j_\mu(x) | ps \rangle = \exp[-i(p'-p)x] \langle p's' | j_\mu(0) | ps \rangle\,.$$

Substituting the fourth component of the current from this relation into the relation for $\langle r^2 \rangle$, and going through some lengthy calculations, which we will not reproduce here, we find

$$\langle r^2\rangle = -6[\partial F_1/\partial q^2|_{q^2=0} - F_2(0)/2M] + \text{inconsequential terms.}$$

If we now determine the charge form factor of the nucleon in the form (we will return to this point in Sec. 6 of Chap. 2

$$G_E = F_1(q^2) - (q^2/2M)F_2(q^2)\,,$$

we easily see that

$$\langle r^2\rangle \approx -6\,\partial G_E/\partial q^2|_{q^2=0}\,.$$

It follows that an expansion of the nucleon charge form factor G_E in a series in q^2 at small values of q^2 can be written

$$G_E(q^2) = G_E(0) + q^2\left|\frac{\partial G_E}{\partial q^2}\right|_{q^2=0} + \ldots = 1 - \tfrac{1}{6}\langle r^2\rangle q^2 + \ldots,$$

since we have $F_1(0) = 1$ [see the discussion of (1.5)]. Here the space-like vector q is determined by the inequality $q^2 > 0$. Everywhere in this book, a space-like vector is determined by the inequality $q^2 < 0$, while a time-like vector is determined by the inequality $q^2 > 0$. The expansion for $G_E(q^2)$ should thus be rewritten with the opposite sign in front of $\langle r^2\rangle$:

$$G_E(q^2) = 1 + (1/6)\langle r^2\rangle q^2 + \ldots. \tag{2.2}$$

At small values of q^2 we can write the representation

$$G_E(q^2) = \int f(r)\exp(i\,\mathbf{q}\cdot\mathbf{r})d\mathbf{r},$$

where $f(r)$ is a Fourier 3-transform of the function $G_E(q^2)$, which describes the spatial distribution of the charge. Introducing the definition

$$\langle r^2\rangle = \int r^2 f(r)d^3r\,,$$

we find a graphic physical interpretation of the charge form factor of the nucleon: as the three-dimensional spatial distribution of charge within the nucleon. Again in the general case of large values of q^2, it is common to see the use of this interpretation of the form factor G_E, but that usage is not legitimate, as we now know.

Form factors can be interpreted most accurately as the effect of a virtual cloud of particles on some amplitude of a process in which we are interested (or on some matrix element in which we are interested). For example, the form factor shown in Fig. 2.3 is interpreted as the effect of the virtual π-meson cloud of a nucleon on the electromagnetic interaction of the nucleon. In QCD terms it would be equivalent to assert that form factors arise from the resultant interaction of all the quarks of the nucleon with an electron. In QCD we could also talk about a virtual cloud of quarks and gluons around a nucleon.

In this chapter we use the concept of a form factor exclusively for electromagnetic vertices (Fig. 2.1a) one of whose ends corresponds to a virtual photon, while the two other ends necessarily lie on the mass shell; i.e., we will use the concept of a form factor exclusively for the electromagnetic interactions of elementary particles.

Functions of the type $F_i(q_1^2, q_2^2)$, which characterize (for example) the amplitude for the Compton scattering of a virtual photon (Fig. 2.5), will be called "relativistic structure functions" or simply "structure functions."

In field theory, form factors arise in an analysis of the invariant properties of vertex functions. For example, the electromagnetic form factor of the π meson appears in an analysis of a generalized π-meson vertex (Fig. 2.6). It can be shown that the only nonvanishing vector is of the form

$$\langle q_2 | j_\lambda | q_1 \rangle = e \frac{(q_1 + q_2)_\lambda}{\sqrt{4\omega_1\omega_2}} F_\pi(t) T_3 ,$$

where $q^2 = t = (q_2 - q_1)^2; \omega^2{}_i = q^2{}_i + m^2{}_\pi$ (m_π is the mass of the π meson; $q^2{}_i$ is the square of the 3-momentum of the π meson), ω_1 and ω_2 are the energies of the π meson in the initial and final states, respectively, $F_\pi(t)$ is a scalar function of the invariant variable t, and T_3 is the third component of the isospin of the π meson. The function $F_\pi(t)$ is normalized to unity at the point $t = 0$. The expression on the right-hand side in the equation above is the most general expression which satisfies the requirements of relativistic invariance and which allows the presence of an arbitrary scalar function $F_\pi(t)$, which is shown by experiment to differ from unity. A nucleon vertex has a more complicated shape (Fig. 2.3), because of the spin and isospin of nucleons. An isospin was not introduced in (1.5) since we used that relation exclusively for leptons. Taking into account the invariance under Lorentz transformations, the gradient invariance, the requirement that free nucleons satisfy the Dirac equation, and the isotopic structure of the nucleons, we find that the most general expression for a matrix element of the nucleon current j_λ^N is

$$\langle p' | j_\lambda^N | p \rangle = \frac{e}{\sqrt{4E_f E_i}} \langle \bar{w}(p') | (F_1^s(t)\mathbf{1} + F_1^v(t)\tau_3)\gamma_\lambda + i(F_2^s(t)\mathbf{1}$$

$$+ F_1^v(t)\tau_3)\sigma_{\lambda\nu} k_\nu | \omega(p) \rangle , \qquad (2.3)$$

where E_f and E_i are the energies of the nucleon in the final and initial states; $\bar{w}(p')$ and $w(p)$ are the spinors of the nucleon in the final and initial states; the superscripts s and v specify the isotopically scalar and isotopically vector parts of the nucleon form factors: $t = k^2{}_\nu$; $\mathbf{1}$ is the unit matrix; $\tau_3 = \begin{pmatrix} 1 & 0 \\ 0 & -1 \end{pmatrix}$; and $k_\nu = (p' - p)\nu$.

In expression (2.3) the term

$$\langle \bar{w}(p') | (F_1^s(t)\mathbf{1} + F_1^v(t)\tau_3)\gamma_\lambda | w(p) \rangle = \langle \bar{w}(p') | F_1(t)\gamma_\lambda | w(p) \rangle$$

in the absence of a meson cloud describes the interaction of a nucleon with a charge e and a magnetic moment equal to the Bohr magneton with the electromagnetic field. In the limit $t \to 0$, this term describes the motion of a free, noninteracting proton; it follows that for a proton we have

$$F_{1p}(0) = F_1^s(0) + F_1^v(0) = 1 ,$$

while for a neutron we have

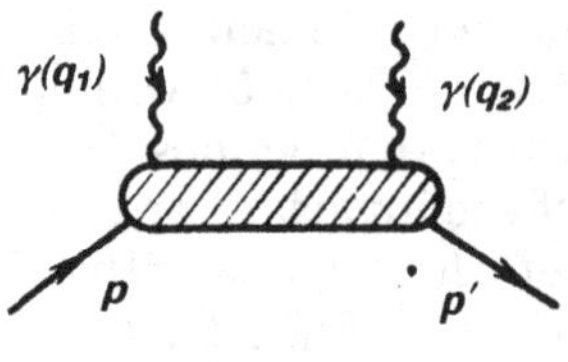

Figure 2.5. Generalized vertex diagram corresponding to the virtual Compton effect.

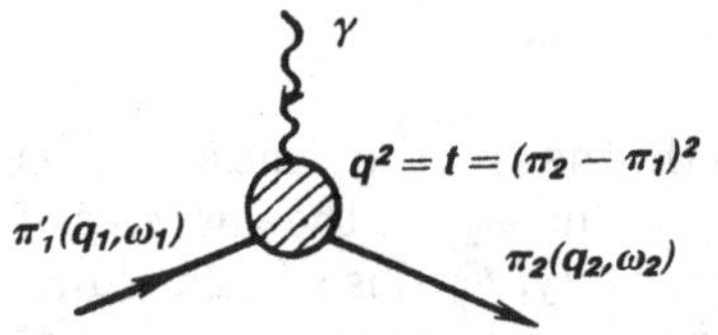

Figure 2.6. Generalized π-meson vertex function.

$$F_{1n}(0) = F_1^s(0) - F_1^v(0) = 0$$

or

$$F_1^s(0) = F_1^v(0) = 1/2\,.$$

The term

$$i\langle w(p')|(F_2^s(t)\mathbf{1} + F_2^v(t)\tau_3)\sigma_{\lambda\nu}k_\nu|w(p)\rangle = i\langle \overline{w}(p')|F_2(t)\sigma_{\lambda\nu}k_\nu|w(p)\rangle$$

in Eq. (2.3) is related to the anomalous magnetic moment of the nucleon. This term is occasionally called the "Pauli" term, since Pauli[59] showed that in a description of spin-1/2 particles one can add to the ordinary Dirac equation

$$\left(\gamma_\lambda \frac{\partial}{\partial x_\lambda} + M - ie\gamma_\lambda A_\lambda\right)\Psi = 0 \quad (\hbar = c = 1)\,,$$

which incorporates the interaction with the electromagnetic field (the term $-ie\gamma_\lambda A_\lambda$), a term $(ie\varkappa/2M)\gamma_\lambda\gamma_\nu F_{\lambda\nu}$, where $\varkappa$ is an arbitrary constant. This arbitrary constant may be interpreted as an additional magnetic moment of the nucleon. The quantity $F_{\lambda\nu}$ is the electromagnetic field tensor. In the limit $t\to 0$ the second term is thus normalized to the value of the anomalous magnetic moment of a free nucleon:

$$\begin{aligned} &i\langle \overline{w}(p')|(F_2^s(0)\mathbf{1} + F_2^v(0)\tau_3)\sigma_{\lambda\nu}k_\nu|w(p)\rangle \\ &= i\frac{1}{2M}\langle \overline{w}(p')|(\varkappa_s + \varkappa_v\tau_3)\sigma_{\lambda\nu}k_\lambda|w(p)\rangle\,. \end{aligned}$$

It follows that for the proton we have

$$F_{2p}(0) = F_2^s(0) + F_2^v(0) = (1/2M)(\varkappa_s + \varkappa_v) = (1/2M)\varkappa_p\,,$$

and for the neutron we have

$$F_{2n}(0) = F_2^s(0) - F_2^v(0) = (1/2M)(\varkappa_s - \varkappa_v) = (1/2M)\mu_n\,.$$

Since the anomalous magnetic moment of the proton is $\varkappa_p = 1.79$, while that of the neutron (which is also the total magnetic moment of the neutron, μ_n) is $\varkappa_n = \mu_n = -1.91$, we can write

$$F_2^s = (1/2M)(\varkappa_p + \varkappa_n)/2 = -0.06/2M;$$
$$F_2^v = (1/2M)(\varkappa_p - \varkappa_n)/2 = 1.85/2M.$$

We see that the isotopic-vector part dominates the proton and neutron form factors $F_2(q^2)$.

The π-meson and nucleon vertices presented above are parts of the diagrams which describe the scattering processes

$$e + \pi \to e + \pi, \tag{2.4a}$$
$$e + N \to e + N. \tag{2.4b}$$

In general, the complete set of structure functions for the amplitudes for the scattering in Eqs. (2.4a) and (2.4b) is larger than the number of form factors discussed above. The reason is that the total amplitude for scattering (2.4a) and (2.4b) is described not only by the diagrams in Figs. 2.3 and 2.6 but also by an infinitely large set of diagrams, which includes, for example, the exchange of two, three, etc., photons. As a result, in general, there will be an increase in the number of invariant relativistic structure functions. Their total number is determined by the number of free ends in the Feynman diagram and by the presence of spins and isospins of the particles described by the free ends.

The isotopic structure of the amplitudes for these processes is sought with the help of rules for combining the isospins of particles and with the help of the isospin conservation laws. For elastic scattering process $a + b \to a + b$, invariance under time reversal reduces the number of structures. In particular, for the process $e + N \to e + N$ the number of relativistic structure coefficients decreases from eight to six. When isotopic invariance is taken into account, the number of structure coefficients in these reactions increases to 12.

The vertex parts, shown in Figs. 2.3 and 2.6, may also be thought of as parts of Feynman diagrams describing the annihilation processes

$$e^- + e^+ \to \pi^- + \pi^+, \tag{2.5a}$$
$$e^- + e^+ \to N + \overline{N}. \tag{2.5b}$$

In this case the expressions for the vertex parts are

$$\langle 0|\, j_\lambda\,|\pi\bar{\pi}\rangle = e\,\frac{(q^+ - q^-)_\lambda}{\sqrt{4\omega_+\omega_-}}\,F_\pi(t), \quad t = (q^+ + q^-)^2 \geqslant 4m_\pi^2\,;$$
$$\begin{aligned}\langle 0|\, j_\lambda^N |N\overline{N}\rangle = e\,\frac{1}{\sqrt{4E_N E_{\overline{N}}}}\,&\langle \overline{w}_{\overline{N}}(p_{\overline{N}})|(F_1^s(t)|\mathbf{1} + F_1^v(t)\tau_3)\gamma_\lambda \\ &+ i(F_2^s(t)\mathbf{1} + F_2^v(t)\tau_3)\sigma_{\lambda\nu}k_\nu\,|w_N(p_N)\rangle, \\ &t = (p_N + p_{\overline{N}})^2 \geqslant 4M^2.\end{aligned} \tag{2.6}$$

Here the form factors of the π mesons and nucleons differ from those written in Eq. (2.3) in having a different range of the variable t.

In field theory it is asserted that the form factor $F_\pi(t)$ must be a common analytic function for scattering process (2.4a) and annihilation process (2.5a). A corresponding assertion has been made for the form factors of nucleons. The dispersion relations provide the necessary coupling of the form factors written in different t regions.

3. Electromagnetic form factor of the π meson

Studies[54,60] of the analytic properties of the vertex function of the π meson lead to the assertion that $F_\pi(t)$ is an analytic function in the t plane with a cut $4m_\pi^2 \leqslant t \leqslant \infty$. In other words, for the annihilation process $e^+ + e^- \to \pi^+ + \pi^-$ the form factor of the π meson is a complex quantity, while for the scattering $e + \pi \to e + \pi$ $[t<0, \operatorname{Im} F_\pi(t) = 0]$ it is real. Consequently, according to the Cauchy theorem from the theory of analytic functions, the form factor of the π meson can be written as the integral

$$F_\pi(t) = \frac{1}{\pi} \int_C \frac{\operatorname{Im} F_\pi(t')dt'}{t' - (t + i\varepsilon)}. \tag{2.7}$$

The integration contour is shown in Fig. 2.7. The imaginary increment $i\varepsilon$ in t means that the form factor $F_\pi(t)$ takes on observable values as it approaches the real axis from above. Everywhere below we will omit the imaginary increment $i\varepsilon$. Taking the limit in the denominator in the integrand in (2.7), we find[6)]

$$\operatorname{Re} F_\pi(t) = \frac{\mathscr{P}}{\pi} \int_C \frac{\operatorname{Im} F_\pi(t')}{t' - t} dt'. \tag{2.8}$$

The behavior of $\operatorname{Im} F_\pi(t)$ as $t \to \infty$ is not known. However, if we assume $\lim_{t\to\infty} \operatorname{Im} F_\pi(t) = \text{const}$, then we need to carry out one subtraction to make integral (2.7) converge. We carry out this subtraction at the point $t = 0$, at which we have $F_\pi(0) = 1$. The dispersion integral then becomes

$$F_\pi(t) - F_\pi(0) = \frac{t}{\pi} \int_{4m_\pi^2}^{\infty} \frac{\operatorname{Im} F_\pi(t')dt'}{t'(t'-t)}. \tag{2.9}$$

Relation (2.9), an identity, cannot be used for applications. To convert it into an equation, we need to use some additional information. The further calculation of the function $F_\pi(t)$ depends on the assumptions made regarding the behavior of $\operatorname{Im} F_\pi(t)$ at $t \geqslant 4m_\pi^2$. For this purpose we make use of the unitarity condition [see (2.6)]

$$\operatorname{Im}\langle 0| j_\nu |q^+ q^-\rangle = \sum_n \langle 0| j_\nu |n\rangle \langle n| T^+ |q^+ q^-\rangle, \tag{2.10}$$

where n runs over all of the intermediate states which are allowed by the conservation laws (states with a total angular momentum $J = 1$, since the photon is a spin-1 particle), and T^+ is the Hermitian-adjoint amplitude. The lowest state is one with two π mesons; a 3π state is forbidden, since we have

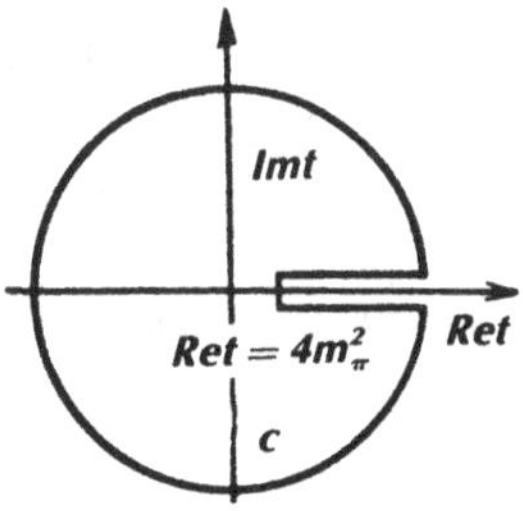

Figure 2.7. Integration contour in the plane of the complex variable t.

$\langle n|T^+|\pi^+\pi^-\rangle = 0$ according to a theorem analogous to the Furry theorem in electrodynamics. The next allowed state has four π mesons; etc. Among the states n there may be pairs $K\bar{K}$, $N\bar{N}$, $N\bar{N}\pi\pi$, etc. The threshold for their production in terms of t lies far from the beginning of the cut; this circumstance is frequently utilized in specific calculations.

Figure 2.8 is a graphical representation of relation (2.10). We recall that dispersion integrals (2.7) and (2.8), along with unitarity condition (2.10), make it possible in principle to carry out a summation of an infinite number of the diagrams in Figs. 2.3 and 2.6. For this purpose, we should write dispersion integrals for all of the amplitudes which enter unitarity condition (2.10) and solve the infinite system of nonlinear singular integral equations. This task is of course essentially impossible to carry out, and the calculations are restricted to the simplest cases.

Let us assume that we are interested in small values of t. If in the unitarity condition at $t' > t$ there are no states which make anomalously large contributions, the region of high energies makes a small contribution, which behaves as essentially a constant, since the denominator of the integrand in (2.9) can be written

$$t'(t'-t) = t'^2(1-t/t') \approx t'^2 .$$

Under these assumptions we can therefore restrict the analysis to the lowest lying of the n states. Specifically, we restrict the present analysis to a single state: the 2π state. Condition (2.10) then becomes

$$\mathrm{Im}\langle 0|\, j_\lambda\, |qq\rangle \approx \langle 0|\, j_\lambda\, |qq\rangle\ \langle qq|T^+|qq\rangle .$$

The vertex $\langle 0|\, j_\lambda\, |q^+q^-\rangle$ is described by a p wave alone, since, as was mentioned earlier, the total angular momentum in the intermediate state is $J=1$. The amplitude $\langle \pi\pi|T^+|q^+q^-\rangle$ should therefore also be described by a p wave alone. Working from the unitarity condition for the matrix S [see (2.10)], we find the expression

$$\mathrm{Im}\,F_\pi(t) = F_\pi(t)\exp[-i\delta_1(t)]\sin\delta_1(t) \qquad (2.11)$$

for the imaginary part of the form factor, since we have $\langle 0|\, j_\lambda\, |qq\rangle \sim F_\pi(t)$, and the p partial wave of the amplitude $\langle qq|T^+|qq\rangle$ is equal to $\exp(-i\delta_1)\sin\delta_1$.

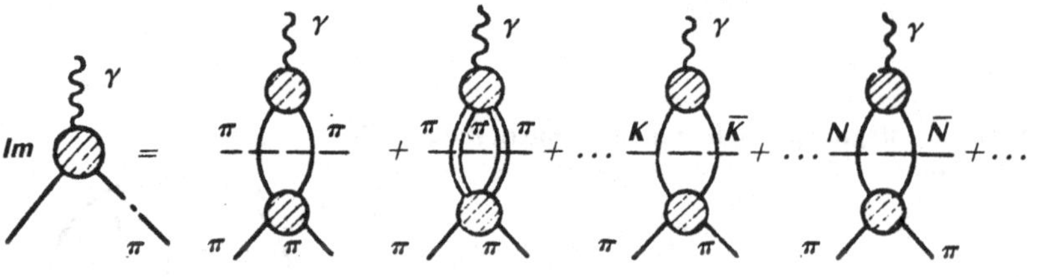

Figure 2.8. Graphical representation of the unitarity condition.

The minus sign on the argument of the exponential function arises because of the Hermitian nature of the operator T.

Expression (2.11) can be rewritten as

$$\operatorname{Im} F_\pi(t) = \operatorname{Re} F_\pi(t) \tan \delta_1(t)\,, \tag{2.12}$$

where the phase shift $\delta_1(t)$ in the $\pi\pi$ interaction corresponds to quantum numbers $I = 1, J = 1$. Substituting (2.12) into (2.9), we find a representation for the π form factor which is called a "dispersion relation":

$$\operatorname{Re} F_\pi(t) = 1 + \frac{t}{\pi}\,\mathscr{P}\int_{4m_\pi^2}^{\infty} \frac{\operatorname{Re} F_\pi(t') \tan \delta_1(t')}{t'(t'-t)}\,dt'\,. \tag{2.13}$$

The form factor thus satisfies a singular linear integral equation which can be solved by the Muskhelishvili-Omnes method.[61] The general solution of (2.13) is complicated, so for simplicity we write the solution of the integral equation without a subtraction:

$$F_\pi(t) = \exp\left\{\frac{t}{\pi}\,\mathscr{P}\int_{4m_\pi^2}^{\infty} \frac{\delta_1(t')dt'}{t'(t'-t)}\right\}. \tag{2.14}$$

By making various assumptions regarding the behavior of the phase shift δ_1 in the $\pi\pi$ scattering, we can find various expressions for $F_\pi(t)$. The behavior of the phase shift $\delta_1(t)$ in $\pi\pi$ scattering has been studied experimentally up to values $t \lesssim 2\ \mathrm{GeV}^2$. These results can be used in practical calculations, e.g., in calculating the form factors of π mesons and nucleons. The integration in (2.14) can be carried out explicitly with the help of the theory of residues if the phase shift δ_1 is expressed as the ratio of polynomials $P(k)$ and $Q(k)$ (Refs. 62 and 63):

$$k^3 \cot \delta_1(t) = P(k)/Q(k)\,,$$

where k is the 3-momentum of the π meson in the frame of reference in which the pair of π mesons is produced. The invariant variable in this frame of reference is equal to the square of the total energy of the two π mesons: $t = 4(k^2+1)\ (m_\pi = 1)$.

For the scattering of an electron by a π meson in the c.m. frame, the invariant variable t takes the following form in the case of backscattering:

$$t = -2\nu(1-\cos 180°) = -4\nu = -4(\omega^2-1), \quad \omega = ik\,.$$

After integrating relation (2.14), we find

$$F_\pi(\nu) = \prod_j \frac{i\omega - k_j}{i\omega + k_i} \frac{i + k_i}{i - k_j}, \quad \text{Im}\, k_i > 0, \quad \text{Im}\, k_j < 0\,,$$

where $k_{i,j}$ are the roots of the equation $1 + i \tan \delta_1 = 0$, $\nu = \omega^2 - 1$. In the case $\nu = 0$ the energy is $\omega = 1$, and we have $F_\pi(0) = 1$. In particular, if we choose the phase shift δ_1 to be of the form $ak^3 \cot \delta_1(k) = k^2{}_r - k^2$, which corresponds to the Breit–Wigner resonance formula

$$\exp(i\delta_1) \sin \delta_1 = ak^3/(k_r^2 - k^2 - i\, ak^3)\,, \tag{2.15}$$

and if we set $\varepsilon = ak_r < 1$, we find

$$F_\pi(\nu) = \frac{k_r + 1/(\varepsilon + k_r)}{k_r + \omega^2/(\varepsilon\omega + k_r)}\,. \tag{2.16}$$

If we expand the expression $1/(\varepsilon\omega + k_r)$ under the condition $\varepsilon\omega/k_r < 1$ in (2.16), we can write form factor (2.16) as

$$F_\pi(t) = \frac{t_r - 4a}{t_r - t - 4i\, ak^3}\,. \tag{2.17}$$

At the point $t = 0$ we have $k = i$ and $F_\pi(0) = 1$. Setting $a = 0$ in (2.17), we find [see (2.18) below] the usual expression for the π meson form factor in the pole approximation (ρ pole):

$$F_\pi(t) = (1 - t/t_\rho)^{-1}\,.$$

Form factor (2.16) enters the dispersion relations for the scattering of π mesons by nucleons. The resonant interaction of π mesons described by (2.15) is interpreted as the exchange of a ρ meson ($\pi + \pi \to \rho \to \pi + \pi$), with a mass $m_\rho = 750$ MeV. With $\varepsilon = 0.2$, we find a good description of the experimental data on the scattering of π mesons by nucleons: this agreement justifies the approximation used in the derivation of (2.16).

The form factor of the π meson can be measured experimentally in all reactions which are described by diagrams containing a vertex $\gamma \to 2\pi$. However, it is an exceedingly laborious matter to extract experimental data on the form factor of the π meson. The processes which are studied, e.g., the electroproduction of π mesons, $e + p \to e + p + \pi$ (more on this below), usually receive contributions from not only the form factor of the π meson but also several other unknown quantities, which must be determined from the same experiment. It is therefore difficult to unambiguously extract a form factor. Only in the scattering of electrons or μ mesons by π mesons (Fig. 2.9a) or in the annihilation processes $e^+ + e^- + \overset{+}{\pi^0} + \overset{-}{\pi^0}$, $\mu^+ + \mu^- \to \overset{+}{\pi^0} + \overset{-}{\pi^0}$ (Fig. 2.9b) is it possible to directly measure the π form factor.

The process $\pi + e \to \pi + e$ has been studied in the scattering of π mesons by the electrons of various nuclei (see Ref. 64, for example). At low π energies, however, the recoil of the electrons is small; stiff requirements are accordingly imposed on the experimental precision, since the form factor $F_\pi(t)$ differs only slightly from unity at a small space-like momentum transfer.

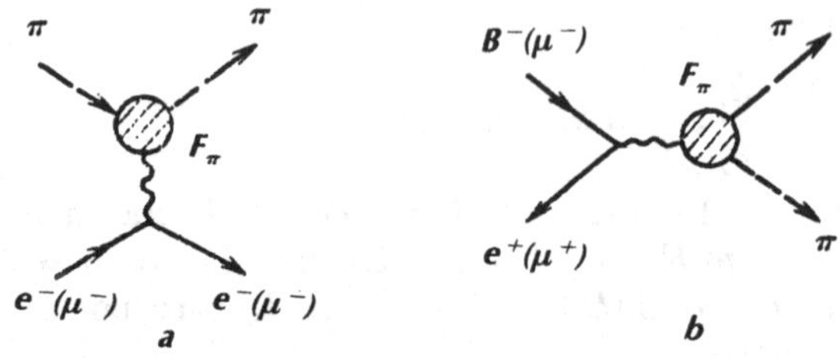

Figure 2.9. (a) Scattering of electrons (or μ mesons) by π mesons; (b) annihilation of electrons and positrons ($\mu^+\mu^-$ mesons) with the production of a pair of π mesons. The electromagnetic form factor of the π meson is measured in these reactions.

The π form factor in the region of space-like momentum transfer has been measured[65] in the electroproduction of the π meson (Fig. 2.10):

$$e^- + p \rightarrow e^- + N \rightarrow \pi .$$

The experimental points are afflicted with a rather large error, which stems from not only the experimental error but also the error in theoretical calculations. Figure 2.11 compares the experimental data with two theoretical curves. One curve is

$$F_\pi(t) = (1 - t/m_\rho^2)^{-1} \tag{2.18}$$

(this is the ρ-dominance model; see the dashed line in Fig. 2.11 and also Fig. 2.12). The other curve is

$$F_\pi(t) \equiv G_E^v \tag{2.19}$$

(the solid line in Fig. 2.11), where $G^v{}_E$ is the isotropic-vector part of the electric nucleon form factor (Sec. 6 of Chap. 2).

Expression (2.18) can be derived in a trivial way from (2.8) if we replace the imaginary part Im $F_\pi(t)$ by the expression $\pi g_{\rho\gamma} g_{\pi\pi\rho} \delta(t - m^2{}_\rho)$, i.e., if we choose Im $F_\pi(t)$ in the pole approximation with a zero decay width.[7)] The quantities $g_{\rho\gamma}$ and $g_{\pi\pi\rho}$ are the coupling constants describing the coupling of the ρ meson with the photon and that of the ρ meson with π mesons, respectively (Fig. 2.12). If, in the ρ-dominance model, we choose m_ρ to be 600 ± 80 MeV, we find a mean value of 0.80 ± 0.10 fm for the electromagnetic radius of the ρ meson, r_π [defined by (2.2)].

New measurements of the form factor of the π meson and of its electromagnetic radius in the reaction $e + p \rightarrow e + \pi^+ + n$ have been carried out at Stanford, Hamburg, and Saclay.[66] It turns out that determining the electromagnetic radius of the π meson is not a simple matter. Experimental data are acquired at relatively large values of q^2, while the radius is determined by the value of the derivative of the π form factor at the point $q^2 = 0$:

$$\langle r_\pi^2 \rangle = 6|\partial F_\pi(q^2)/\partial q^2|_{q^2=0} .$$

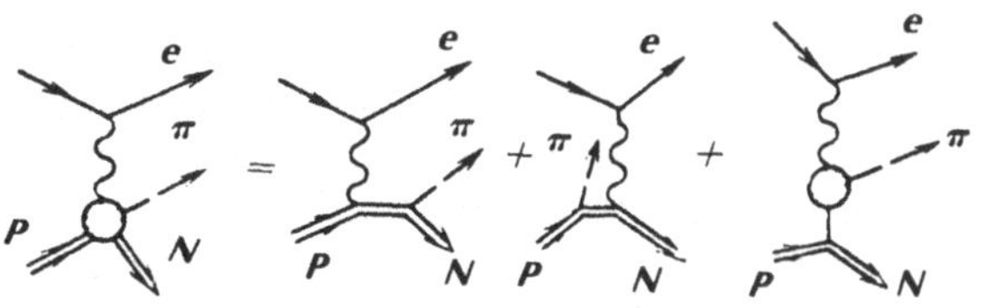

Figure 2.10. Contributions to the amplitude for the electroproduction of the π meson.

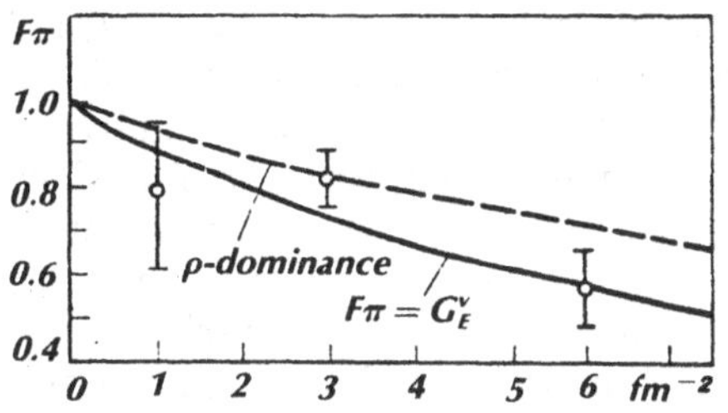

Figure 2.11. Comparison of experimental data and theoretical curves on the form factor of the π meson.

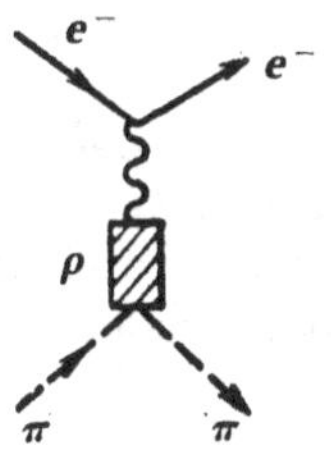

Figure 2.12. Diagram describing $e^- + \pi \to e^- + \pi$ in the ρ-dominance approximation.

Near $q^2 \approx 0$, the form factor $F_\pi(q^2)$ is determined by a function which approximates the experimental data in the region $q^2 \neq 0$. In this connection it would be natural to ask whether the shape of the form factor is determined in a unique way by the experimental data. In other words, can it be assumed that the experimental data unambiguously point to a simple pole or a dipole behavior of the form factor, etc? Does the value of $\langle r_\pi^2 \rangle$ change substantially in the various theoretical fits?

In an effort to answer these questions, the results of experimental observations have been subjected to a statistical analysis for the pole model,

$$F_\pi(q^2) = (1 - q^2/m_\rho^2)^{-1}$$

[see (1.8)]; for the dipole model,

$$F_\pi^D(q^2) = (1 - q^2/m_D^2)^{-2}$$

[see (2.39)]; and for the pole model with a correction for the finite width δ,

$$F_\pi^\delta = \frac{1}{1 - q^2/m^2} \frac{1 + \delta}{1 + (1 - q^2/m^2)\delta}.$$

The results of this analysis are shown in Table 2.1.

It follows from Table 2.1 that none of the models for describing the form factor of the π meson can be singled out as preferable. This conclusion means that a more accurate determination of $\langle r_\pi \rangle$ will require measuring the π form factor as close as possible to the value $q^2 = 0$.

Interestingly, the pole formula is very nearly the same as the formula describing the isovector part of the nucleon form factor.

When the powerful Soviet 76-GeV proton accelerator came on line (Institute of High-Energy Physics, Serpukhov), it became possible to carry out direct measurements of the pion form factor in the scattering of π mesons by the electrons of the nuclei of liquid hydrogen at significantly larger values of the space-like momentum transfer. Measurements were carried out[67] in a beam of π mesons with a momentum of 50 GeV/c over the interval $0.014 \leqslant |q^2| \leqslant 0.035$ (GeV/c)^2. The emission angles of the final particles were measured, and the coplanarity of the momenta of the scattered particles was tested in order to eliminate background processes. A correct estimate of the radiative corrections was important. To give an idea of the complexity of this experiment, we note that a determination of the π form factor from the expression for the scattering cross section,

$$d\sigma/dq^2 = (d\sigma/dq^2)_p |F_\pi(q^2)|^2 \tag{2.20}$$

[$q^2 \approx -4EE' \sin^2(\theta/2)$; where E and E' are the energies of the electron before and after the scattering, respectively], required 14 corrections which depended on q^2 and 12 which did not. The data were analyzed in accordance with the pole model:

$$|F_\pi(q^2)|^2 = N/(1 - Aq^2)^2 .$$

The constant N is normalized (to unity), and the radius of the π meson is determined from the expression

$$F_\pi(q^2) = 1 + (1/6)q^2\langle r_\pi^2 \rangle$$

(we again note that the space-like values of q^2 lie in the region $q^2 < 0$). Analysis of the experimental data thus yielded the value

$$\langle r_\pi^2 \rangle = (0.61 \pm 0.15)\text{fm}^2 .$$

This experiment was continued: a subsequent measurement of the form factor of the π meson was carried out in a beam of π^- mesons at a momentum of 100 GeV/c at the Fermilab accelerator (in the U.S.A; Ref. 68). The interval of the momentum transfer was doubled in this experiment: $0.03 \leqslant |q^2| \leqslant 0.07$ (GeV/c)^2. After the corrections were made, the corrected values of the form factors (Fig. 2.13) were used to determine the electromagnetic radius of the π meson. The q^2 dependence of the form factor $F_\pi(q^2)$ was chosen in the form

$$|F_\pi(q^2)|^2 = [1 - (1/6)\langle r_\pi^2 \rangle\, q^2]^{-2} .$$

These results yielded the value $\langle r_\pi^2 \rangle = (0.31 \pm 0.04)$ fm^2 or $\langle r_\pi^2 \rangle^{1/2} = (0.56 \pm 0.04)$ fm.

Table 2.1. Results of an analysis of experimental data on the q^2 dependence of the form factor of the π meson.

Model	m_ρ, m_D m, MeV	δ	$\langle r_\pi^2 \rangle^{1/2}$, fm	χ^2/DF
Pole	686	0	0.705 ± 0.013	53/51
Dipole	1134	0	0.603 ± 0.01	48/51
Pole corrected for	781	0.13*	0.654 ± 0.0011	38/51
finite width	787	0.14 ± 0.10	0.651 ± 0.076	38/50

*Analysis was carried out with the fixed value $\delta = 0.13$.

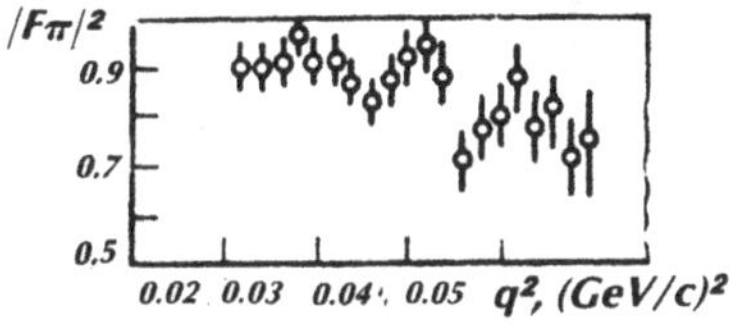

Figure 2.13. Pion form factor.

A fit of the experimental data by the dipole formula

$$|F_\pi(q^2)|^2 = [1 - (1/12)q^2 \langle r_\pi^2 \rangle]^{-4}$$

yields the same value of the radius $\langle r_\pi \rangle$.

The π form factor has also been studied at time-like momenta, $t = q^2 > 0$, i.e., in the process[69] $e^+ + e^- \to \pi^+ + \pi^-$. It is easy to see (Fig. 2.9b) that the cross section for this process is proportional to the square of the modulus of form factor F_π. Values of $|F_\pi|^2$ were obtained over the interval $300 \leqslant E \leqslant 500$ MeV, where E is the energy of the electron (or positron) in the c.m. frame for the reaction $e^+ + e^- \to \pi^+ + \pi^-$.

It is assumed that F_π is dominated by the ρ-exchange diagram (Fig. 2.14a), where the decay width of the ρ meson, Γ_ρ, is taken into account. A possible contribution of a $\rho\omega$ interference term to the diagram in Fig. 2.14a is taken into account by adding a diagram (Fig. 2.14b) with a factor $\xi \exp(i\alpha)$ to characterize the $\rho\omega$ mixing. The complete expression for the form factor F_π is written in the form[70]

$$F_\pi(s) = \frac{m_\rho^2 [1 + (\Gamma_\rho / m_\rho) d]}{m_\rho^2 - 4E^2 + i m_\rho \Gamma_\rho (p/p_0)^3 m_\rho / 2E} + \xi \exp(i\alpha) \frac{m_\omega^2}{m_\omega^2 - 4E^2 + i m_\omega \Gamma_\omega}, \tag{2.21}$$

where d is chosen in such a way that the ρ-meson part of the form factor is normalized (to unity) at a zero momentum transfer ($d = 0.48$); $p = \sqrt{E^2 - m_\pi^2}$; $p_0 = \sqrt{m_\rho^2/4 - m_\pi^2}$; $s = 4E^2$; ξ and α are adjustable parameters which describe the $\rho\omega$ interference. Figure 2.15 shows the results of a comparison of theoretical calculations with the experimental data. We see that a value $\xi \neq 0$ is preferable, although the possibility $\xi = 0$ cannot be ruled out entirely.

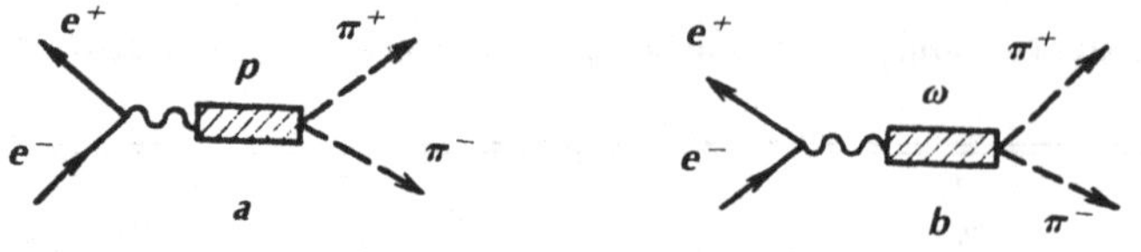

Figure 2.14. Interference of contributions to the form factor F_π. (a) ρ-exchange diagram, with allowance for the width of the ρ meson, Γ_ρ; (b) ω-exchange diagram.

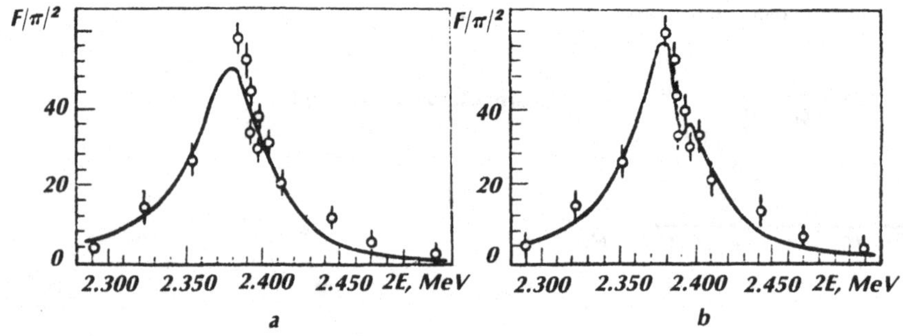

Figure 2.15. Comparison of theoretical curves and experimental data for (a) $\xi = 0$ and (b) $\xi = \xi(E,\alpha) \neq 0$ (Ref. 70).

New data on the π form factor in the region of time-like q^2 have been obtained at colliding-beam accelerators at Orsay, Frascati, Novosibirsk, and Hamburg (Fig. 2.16). In this region of q^2 values, the differential cross section is written in the form

$$d\sigma/d\Omega = (\alpha^2/8s)(p_\pi/E_\pi)^3 \sin^2\theta |F_\pi(s)|^2 ,$$

where p_π and E_π are the momentum and energy of the π meson, θ is the angle at which the π mesons are emitted, and s is the square of the total energy of the colliding beams. An expression which includes only the ρ_0 meson and allows for its decay width Γ,

$$F_\pi(s) = m_\rho^2/(m_\rho^2 - s - im_\rho \Gamma) ,$$

gives a good description over most of the energy range, except at $s \approx 1.2$ GeV. The Novosibirsk group suggests introducing a ρ' meson with a mass of about 1.2 GeV in order to explain the slight "shoulder" on the function $|F_\pi(s)|^2$ in this region.

Figure 2.17 shows all the values of $|F_\pi|^2$ for positive and negative q^2. It appears that the dot-dashed theoretical curve, which incorporates the contribution of the ρ meson, the $\rho\omega$ interference, the inelastic $\omega\pi$ channel, and a ρ' meson with a mass $M_{\rho'} \approx 1250$ MeV, gives the best description of the entire set of experimental data. The solid curve ignores the ρ' meson.

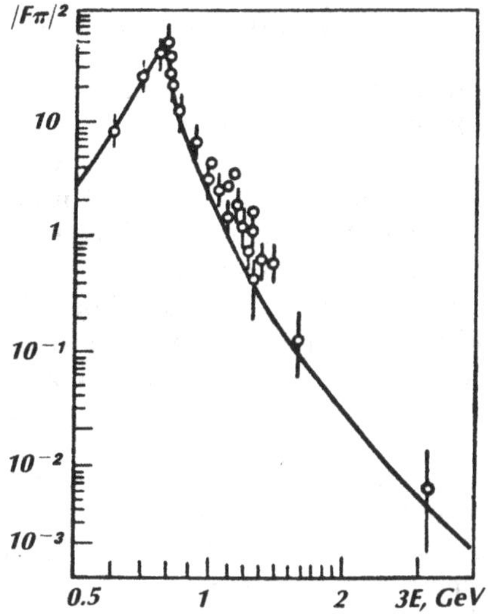

Figure 2.16. Square of the pion form factor vs the total energy in the c.m. frame (data obtained at Orsay, Frascati, Novosibirsk, and Hamburg).

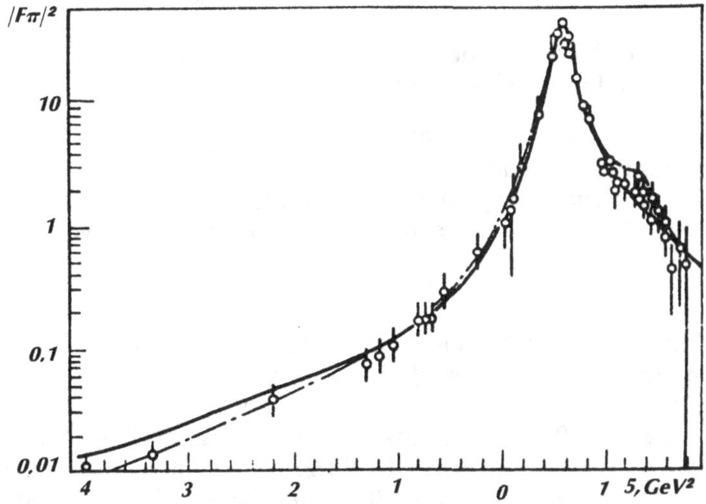

Figure 2.17. Square of the pion form factor for space-like and time-like s^2.

An analytic expression which incorporates all of the fundamental properties of the pion form factor and which describes the experimental data in the space-like and time-like regions of q^2 can be constructed by a conformal-mapping method.[71]

4. Electromagnetic form factor of the K meson

The first direct measurements of the form factor of the K meson in the space-like region of momentum transfer were carried out in 1979 at the Fermilab accelerator by a group of Soviet and American scientists.[72] These measurements became possible after the energies of the K-meson beams were raised to the point that the recoil which the K meson underwent after the scattering by the electron was large, and it became possible to detect a deviation of the form factor of the K meson from unity. The cross section for the reaction $K + e \to K + e$ is written in a form analogous to the expression for the cross section for the scattering of a π meson by an electron [see (2.20)]:

$$d\sigma/dq^2 = (d\sigma/dq^2)_p |F_K(q^2)|^2 . \tag{2.22}$$

The form factor of the K^- meson is found from the cross section for the scattering of negatively charged K^- mesons with a momentum of 250 GeV/c by electrons of nuclei of liquid hydrogen. The maximum permissible recoil of the electrons reached $|q^2| = 0.131$ (GeV/c)^2. This experiment was carried out over the interval $0.037 \leqslant |q^2| \leqslant 0.119$ (GeV/c)^2. The measured values of $|F_K(q^2)|^2$ are shown in Fig. 2.18. The pole model was selected for the form factor:

$$|F_K(q^2)|^2 = [1 - (1/6)q^2\langle r_K^2\rangle]^{-2} .$$

This model led to the value $\langle r_K^2\rangle = (0.28 \pm 0.05)$ fm^2. The use of the dipole formula for the K form factor

$$|F_K(q^2)|^2 = [1 - (1/12)q^2\langle r_K^2\rangle]^{-4} ,$$

yields approximately the same value of $\langle r_K^2\rangle$.

The charge radius measured for the neutral K^0 meson by the method of regeneration of K_s mesons[73] turned out to be negative:

$$\langle r_{K^0}^2\rangle = (-0.054 \pm 0.026)\ \text{fm}^2 .$$

This result can be defended theoretically in the $SU(3)$ quark model if it is assumed that the mass of the strange quark is greater than the masses of the u and d quarks.[74]

Figure 2.19 shows data on measurements of the K form factor as time-like q^2. At energies near 1 GeV there is a significant peak, apparently due to the production of the $\varphi(1020)$ meson. We see that the theoretical curve does not give a convincing description of the experimental data in the interval $1.2 \leqslant W \leqslant 1.8$ GeV.

The following general formula is valid for relating the electromagnetic correction to the mass of a particle to the amplitude for the Compton scattering of a virtual photon, $T_{\mu\nu}$ (Fig. 2.5):

$$\Delta m = \frac{1}{8\pi^2} \int \frac{T_{\mu\nu}(q^2,\nu) g^{\mu\nu}}{q^2 + i\varepsilon} d^4q , \tag{2.23}$$

where $\nu = pq/m$, q is the momentum of the photon, and p is the momentum of the particle.

Figure 2.18. Form factor of the *K* meson.

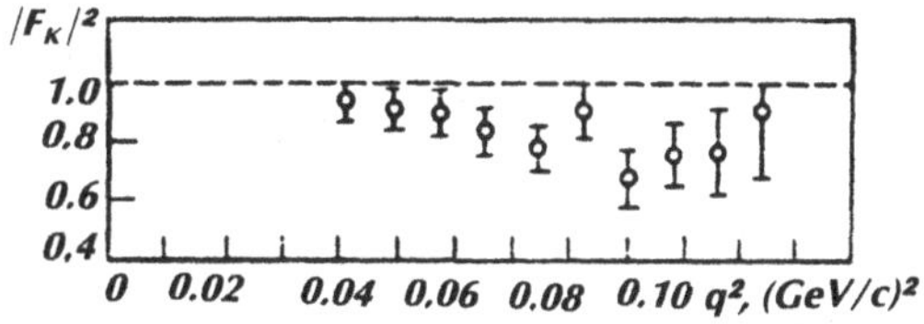

Figure 2.19. Square of the charge form factor of the *K* meson. The contributions from ω and φ mesons were taken into account in the plotting of the theoretical curve.

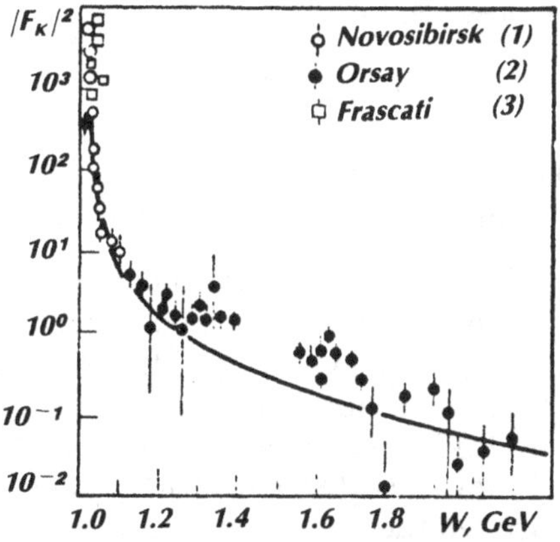

If we restrict the analysis to the single-meson contribution to the imaginary part of the amplitude for the scattering of photons by mesons, we can derive an approximate expression for the mass correction corresponding to the diagram in Fig. 2.20:

$$\Delta m^2 = \frac{i\alpha}{4\pi^3} \int \frac{d^4 q}{q^2 + i\varepsilon} F_\pi^2(q^2) \left(2 + \frac{q^2 - 4m^2}{q^2 - 2\rho q}\right). \tag{2.24}$$

If we use a pole approximation of the type in (2.18) for the form factor of the π meson, we find from (2.24) the correct mass difference for π mesons. However, the pole approximation for the form factors of the K mesons and expression (2.24) lead to the wrong sign for the mass difference for K mesons (instead of $\delta m_K = m_{K^+} - m_{K^0} = -4$ MeV, the theory predicts a positive value of δm_K).

There are two ways to improve the calculations: (1) discard (2.24) and switch to a more accurate approximation of the amplitude $T_{\mu\nu}$ in (2.23); (2) alter the form factor $F_K(q^2)$ under the assumption that (2.24) is sufficiently accurate.

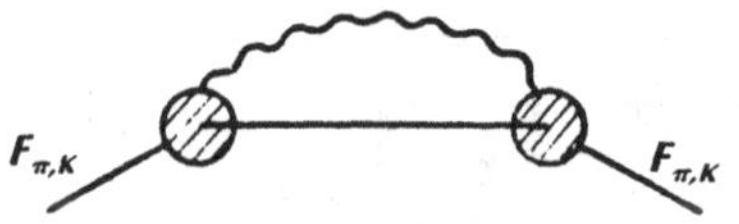

Figure 2.20. Electromagnetic correction to the mass of a meson in first-order perturbation theory.

Pursuing the first of these possibilities, one could incorporate in the amplitude for the Compton scattering of a virtual photon by K mesons contributions from K^* and other K-resonance states. The terms which arise in the expression for the mass difference here contain divergent integrals. The cutoff constant must be chosen at high energies, but it is possible to obtain the necessary mass difference.

An attempt might be made to find form factors $F_K^{v,s}$ with a more natural behavior at low energies (here the superscripts v and s specify the isovector and isoscalar parts of the form factor):

(a) F_K^v should have a pole corresponding to a ρ meson.

(b) F_K^s should have poles corresponding to ω and φ mesons.

(c) F_K^v should be related to the amplitude for the process $\pi\pi \rightarrow K\overline{K}$ by a two-particle unitarity condition.

(d) The form factors must be normalized correctly.

(e) The integral contribution of the high-energy region can be described through the use of the parametrization as^{-b}, whose form is reminiscent of a Regge amplitude.

The basic conclusion found under assumptions (a)–(e) is the following: The difference $\delta m_K = m_{K^+} - m_{K^0} = -4$ MeV can be found with an altered form factor $F^v{}_K$; specifically,

$$F_K^v = \frac{1}{1 - t/m_\rho^2} - c\int_{s_0}^{\infty} \frac{ds'}{s' - s}\left(\frac{s'}{m_\rho^2}\right)^{-b},$$

where c, s_0, and b are arbitrary parameters.

The form factor F_K^v is described well at low energies by the pole approximation, while at high energies there is a slowly decaying tail. The value of b must be small ($10^{-3} \leqslant b \leqslant 10^{-2}$), and c lies in the interval $10^{-5} \leqslant c \leqslant 10^{-2}$. Beginning at s_0, the contribution of the high-energy region is taken into account. The dependence of F_K^v on s_0 is weak.

It thus appears that by saturating the imaginary parts of the form factors of the π and K mesons with resonant states [see (2.9) and (2.10)] and by specifying the correct asymptotic behavior of the dispersion integral in (2.9), we can obtain a satisfactory description of the entire set of experimental data on the π and K form factors.

5. Description of nucleon form factors by means of dispersion relations

Dispersion relations for the nucleon form factors are written in complete analogy with the dispersion relation for the π form factor. We stipulate at the outset that we are writing dispersion relations for isotropic form factors (a scalar form factor F^s and a vector form factor F^v) with a single subtraction:

$$F^v_{1,2} = F^v_{1,2}(0) + \frac{t}{\pi}\int_{4m_\pi^2}^{\infty} \frac{\operatorname{Im} F^v_{1,2}(t')dt'}{t'(t'-t)}, \tag{2.25}$$

$$F^s_{1,2}(t) = F^s_{1,2}(0) + \frac{t}{\pi}\int_{9m_\pi^2}^{\infty} \frac{\operatorname{Im} F^s_{1,2}(t')dt'}{t'(t'-t)}. \tag{2.26}$$

The imaginary $\operatorname{Im} F^{s,v}_{1,2}$ for nucleon current (2.6) are found from the unitarity condition for the S matrix ($S = 1 + iT$):

$$\operatorname{Im}\langle 0|\, j_\lambda^{s,v}|N\bar{N}\rangle = \sum_n \langle 0|\, j_\lambda^{s,v}|n\rangle\, \langle n|T^+|N\bar{N}\rangle\,. \tag{2.27}$$

The physical threshold for this process sets in at an energy $t \geqslant 4M^2$. Intermediate states n may contain two π mesons, three π mesons, and several other low-lying states (Fig. 2.21). For the low-lying intermediate states, the values of $\operatorname{Im} F^{s,v}_{1,2}$ are nonzero in the nonphysical region $t < 4M^2$. For example, the isovector functions $\operatorname{Im} F^v_{1,2}$ are nonzero at $t \geqslant 4m_\pi^2$, and states n which are even in terms of the number of π mesons contribute to these form factors. The isoscalar functions $\operatorname{Im} F^s_{1,2}$ are nonzero at $t \geqslant 9m_\pi^2$, and they receive contributions from only states n which are odd in terms of the number of π mesons. This circumstance explains the different lower limits in (2.25) and (2.26). The use of double Mandelstam dispersion representations makes it possible to analytically continue the unitarity condition for the physical region $t \geqslant 4M^2$ into the region of nonphysical values, $4m_\pi^2 \leqslant t \leqslant 4M^2$, so that no fundamental difficulties arise in theoretical calculations.

From this point on we proceed in much the same way as we did in describing the form factor of the π meson. At low values of t we need retain in unitarity condition (2.27) only the single lowest-lying intermediate state for each of the form factors, under the assumption that the other intermediate states make a negligible contribution. This statement means that for the isovector form factors we are left with only a 2π intermediate state, and the expressions for the functions $\operatorname{Im} F^v_{1,2}$ take the form

$$\operatorname{Im} F^v_{1,2} = F_\pi(t)\langle \pi\pi|\, T^+_{1,2}\,|N\bar{N}\rangle\,. \tag{2.28}$$

where $F_\pi(t)$ is the form factor of the π meson, and $\langle \pi\pi|\, T^+_{1,2}\,|N\bar{N}\rangle$ is the amplitude for the process $\pi\pi \to N\bar{N}$, which contributes to the Dirac and Pauli form factors. For the imaginary parts of the isoscalar form factors we are left with only a three-meson intermediate state; the expression for them is written as a product of two amplitudes:

$$\operatorname{Im} F^s_{1,2}(t) = \langle \gamma|\, T|\pi\pi\pi\rangle\, \langle \pi\pi\pi|\, T^+_{1,2}\,|N\bar{N}\rangle\,, \tag{2.29}$$

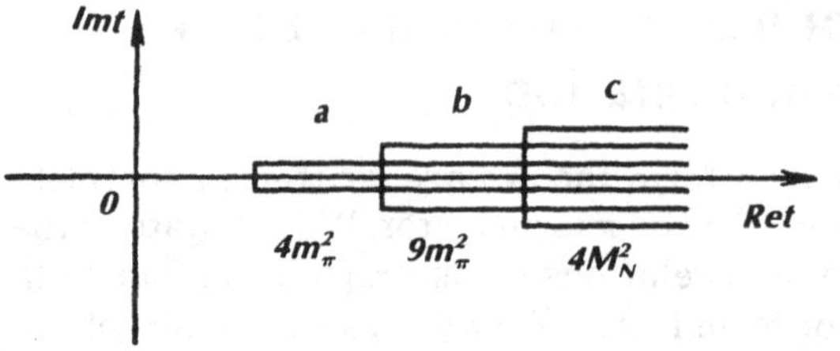

Figure 2.21. Beginning of the cuts in the unitarity condition for states n containing (a) two π mesons, (b) three π mesons, and (c) an $N\bar{N}$ pair (cf. Fig. 2.8).

where $\langle\gamma|T|\pi\pi\pi\rangle$ represents the $\gamma\to 3\pi$ vertex, and $\langle\pi\pi\pi|T^+_{1,2}|N\bar{N}\rangle$ is the amplitude for the process $\pi\pi\pi\to N\bar{N}$. By analogy with (2.28), this amplitude contributes to the Dirac and Pauli form factors.

Let us consider the case of the isovector form factor. We assume that the amplitudes $\langle\pi\pi|T^+_{1,2}|N\bar{N}\rangle$ are described well by a single resonance wave of the type in (2.15). In the limit of a very narrow resonance, and under the assumption that we have $\mathrm{Re}\langle\pi\pi|T^+_{1,2}N\bar{N}\rangle = 0$ at the point of the resonance, we find, near the resonance,[8)]

$$\langle\pi\pi|T^+_L N\bar{N}\rangle \approx \mathrm{Im}\langle\pi\pi|T^+_i|N\bar{N}\rangle \approx \frac{A_i(ak^3)^2}{(t_v-t)^2+(ak^3)^2} \approx \pi A_i(ak^3)\delta(t_v-t)\,, \tag{2.30}$$

where A_i is a normalization constant, t_v is the position of the resonance, and a is its width. The integration in (2.25), which is now simplified considerably, leads to the result

$$F^v_{1,2}(t) = F^v_{1,2}(0) + \mathrm{const}_{1,2}/(t_v/t-1)\,. \tag{2.31}$$

Expression (2.31) can be rewritten as the well-known Clementel-Villi formula[75]:

$$F^v_{1,2}(t) = F^v_{1,2}(0)\left(1 - a^v_{1,2} + \frac{t_v}{t_v - t}a^v_{1,2}\right). \tag{2.32}$$

If we allow a strong resonant interaction in a 3π intermediate state [of the type in (2.30)], then in complete analogy with the derivation of (2.32) we can derive a Clementel-Villi formula for isoscalar form factors also:

$$F^s_{1,2}(t) = F^s_{1,2}(0) + t\,\mathrm{const}_{1,2}/(t_s - t)\,. \tag{2.33}$$

or

$$F^s_{1,2}(t) = F^s_{1,2}(0)(1 - b^s_{1,2} + t_s b^s_{1,2}/(t_s - t))\,. \tag{2.34}$$

Using (2.31) and (2.33), we can easily write an expression for the charge form factor of the neutron:

$$F_{1n}(t) = F^s_1(t) - F^v_1(t) = \frac{t}{2}\left(\frac{b^s_1}{t_s - t} - \frac{a^v_1}{t_v - t}\right).$$

Experimental data indicate that at all values of t we have $F_{1n}(t)\approx 0$, or

$$b^s_1/(t_s - t) - a^v_1/(t_v - t) \approx 0\,.$$

Since this relation holds everywhere, it must hold near $t = 0$; hence

$$b_1^s/t_s \approx a_1^v/t_v = A .$$

To describe the form factors of the nucleons, we thus need not six parameters $(a^v{}_1, a^v{}_2, b^s{}_1, b^s{}_2, t_v, t_s)$ [see (2.32) and (2.34)] but only five $(A, a^v{}_2, t_v, b^s{}_2, t_s)$. Furthermore, the positions of the resonances, t_v and t_s, can be found from experimental data, so that the number of independent parameters in the Clementel-Villi model can be reduced to 3.

The main feature of considered form factors (2.32) and (2.34) is that all tend toward constant limits with increasing t. Until 1962 this circumstance was interpreted as evidence that the nucleons have a core. At that time, experimental data on the scattering of electrons by protons and neutrons were available for $|q^2| \leqslant 25\ \text{fm}^{-2}$ and were interpreted with the help of a single vector ρ meson and a single scalar meson.[76] As the energy of the impinging electrons and q^2 were increased, however, it was found that the form factors continued to fall off as $1/q^2$, and the new data did not agree with the existence of a core for the nucleons. It thus became necessary to reexamine the theoretical interpretation of the experimental data.

6. Review of experimental data on nucleon form factors; theoretical interpretation

All the conclusions regarding the behavior of form factors as functions of the momentum transfer t are based on the Rosenbluth formula[77] for the differential cross section for the scattering of electrons by protons. This formula was derived in the approximation of single-photon exchange:

$$d\sigma/d\Omega = \sigma_M \Big\{ F_{1p}^2(q^2) - (q^2/4M^2)[F_{1p}(q^2) + \varkappa F_{2p}(q')]^2 \tan^2(\theta/2) - \frac{q^2}{4M^2} \varkappa^2 F_{2p}^2(q^2) \Big\} , \tag{2.35}$$

where σ_m is the differential cross section for the scattering of electrons by a Coulomb field, $q^2 \approx -4EE' \sin^2(\theta/2)$; θ is the electron scattering angle in the laboratory frame, and $\varkappa$ is the anomalous magnetic moment of the proton.

Data on the form factors of the neutrons are found from experiments on elastic and inelastic scattering of electrons by deuterons, with a subsequent subtraction of terms associated with the elastic scattering of electrons by protons.

For the most part, experimental data are analyzed today not with the form factors $F_{1,2}^{s,v}$ but with some combinations of them which were introduced by Sachs:

$$G_{E_p} = F_1(q^2) + \tau \varkappa F_2(q^2); \; G_{M_p} = F_1(q^2) + \varkappa F_2(q^2) , \tag{2.36}$$

where $\tau = q^2/4M^2$. With the new Sachs form factors G_{E_p} and G_{M_p}, expression (2.35) can be rewritten in the standard form

$$d\sigma/d\Omega = \sigma_M\left[(G_E^2 - \tau G_M^2)/(1-\tau) - 2\tau G_M^2 \tan^2(\theta/2)\right]. \qquad (2.37)$$

A space-like vector is determined by the inequality $q^2 < 0$.

Expression (2.37) is frequently written in a form convenient for analyzing experimental data:

$$y = a - bx\,,$$

where

$$y = \frac{d\sigma}{d\Omega}\quad \sigma_M;\quad x = \tan^2\frac{\theta}{2};$$

$$a = \frac{G_E^2 - \tau G_M^2}{1-\tau};\quad b = 2\tau G_M^2\,.$$

By measuring y as a function of $\tan^2(\theta/2)$, we can determine the form factors G_E and G_M. The slope of the line depends on the value of $2\tau G^2{}_M$, while the segment $a = (G^2{}_E - \tau G^2{}_M)/(1-\tau)$ is determined at the point $\tan^2(\theta/2) = 0$ by extrapolating the experimental data to the point $\theta = 0$.

Another version of the Rosenbluth formula is often used:

$$d\sigma/d\Omega = A(\theta,q^2)G_E^2 - B(\theta,q^2)G_M^2\,, \qquad (2.38)$$

where

$$A(\theta,q^2) = \frac{\sigma_M}{1-\tau};\quad B(\theta,q^2) = \sigma_M\left(\frac{\tau}{1-\tau} + 2\tau\tan^2\frac{\theta}{2}\right).$$

This version of the formula stresses the absence of an interference term (in contrast with the term $[F_1(q^2) + 2MF_2(q^2)]^2$ in (2.35)); the effect is to reduce the correlation errors in a determination of the form factors from experimental data. In the discussion below we will use only the new form factors $G_E^{s,v}$ or $G_M^{s,v}$ (or $G_{E_{p,n}}$ or $G_{M_{p,n}}$).

From (2.36) we find

$$G_{E_p}(0) = F_{1p}(0) = 1;\quad G_{E_n}(0) = F_{1n}(0) = 0\,;$$

$$G_{M_p}(0) = F_{1p}(0) + \varkappa_p F_{2p}(0) = (1+\varkappa_p) = \mu_p;\quad \mu_p = 2.79\,;$$

$$G_{M_n}(0) = F_{1n}(0) + \varkappa_n F_{2n}(0) = \mu_n;\quad \mu_n = -1.91\,.$$

The form factor G_E is normalized by division by the charge and is accordingly called a "charge form factor"; G_M is normalized by division by the total magnetic moment of the nucleons and is a "magnetic form factor."

The experimental data indicate that the behavior of the neutron form factor G_{M_n} is similar to that of the form factors G_{E_p} and G_{M_p}. This similarity is confirmed by a test of the relation

$$G_{E_p} = G_{M_p}/\mu_p = G_{Mn}/\mu_n = (1 - q^2/0.71)^{-2} = G_D\,, \qquad (2.39)$$

where q^2 is expressed in units of $(\mathrm{GeV}/c)^2$.

The linear relationship of the form factors G_{E_p}, G_{M_p}/μ_p and $G_M{}^n/\mu_n$, which is described by (2.39), is called a "linear law" (scaling), while the dependence of these form factors on q^2 is called the "dipole model" and designat-

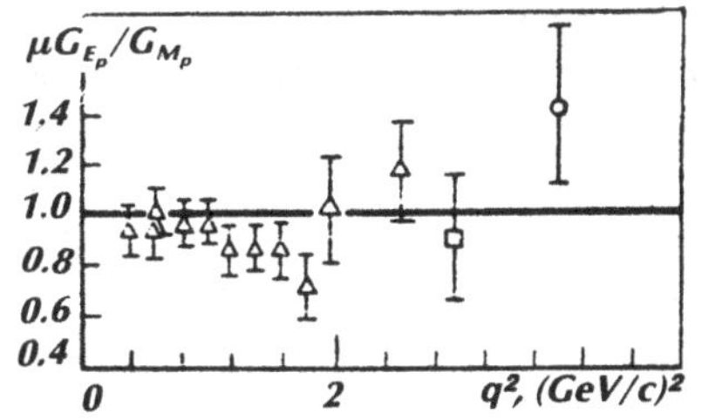

Figure 2.22. Test of the deviation from scaling (data obtained at SLAC, DESY, and Bonn).

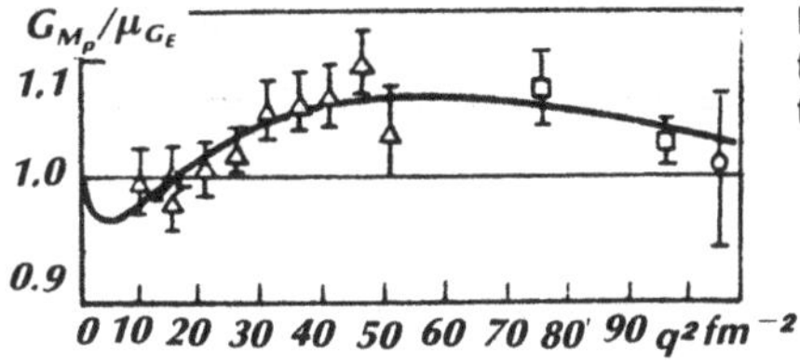

Figure 2.23. Comparison of the magnetic form factor of the proton with the dipole model (data from the same laboratories).

ed G_D. Relation (2.39) indicates the possiblity of describing the form factors with the help of a single-parameter model.

Data on the proton form factors of the G_{E_p} and G_{M_p} were obtained in 1969 at values up to $q^2 \leqslant 25$ (GeV/c)^2 (Ref. 78). In the region $1 \leqslant q^2 \leqslant 2$ (GeV/c)^2 a slight systematic deviation from the linear law $G_{E_p} = G_{M_p}/\mu_p$ was observed. Figure 2.22 compares the experimental data with the ratio $\mu G_{E_p}/G_{M_p} = 1$. We see that the deviation is about 10%. Figure 2.23 compares the magnetic form factor of the proton with the dipole model. The deviation of the ratio $G_{M_p}/\mu_p G_D$ from unity has yet to receive a satisfactory theoretical explanation. We write an equation[79] which gives a good description of this deviation in the region $0 \leqslant |q^2| \leqslant 3$ (GeV/c)^2:

$$G_{E_p} = (1 - v^2/c^2)^2 = (1 - q^2/2M^2)^{-2}, \qquad (2.40)$$

where v is the velocity of the recoil proton and q^2 is the square of the 4-momentum transfer of the proton. This expression may be interpreted as a dependence of the electromagnetic-interaction constant on the velocity of the particle—by analogy with QCD. Table 2.2 compares the predictions of (2.40) with experimental data.

Below we discuss a four-pole model which can successfully describe experimental data in both space-like and time-like regions of q^2.

Table 2.3 shows data on the electric form factor of the neutron at large values of q^2; we see that we have $G_{E_n} \approx 0$.

The experimental results indicate that the form factors G_{E_p}, G_{M_p}, and G_{M_n} fall off with increasing q^2 no more slowly than $1/q^2$. Consequently, if we wish to use a dispersion-relation method to describe the nucleon form factors, we need to take dispersion relations for the functions $G_{E,M}$ without subtractions. This statement means, however, that we also have to take the disper-

Table 2.2. Comparison of theoretical values [expression (2.40)] and experimental data.

$\lvert q^2\rvert$, (GeV/c)^2	$\frac{G_{E^p_{\text{theor}}}}{G_D}$	$\frac{G_{E^p_{\text{exp}}}}{G_D}$	$\lvert q^2\rvert$, (GeV/c)^2	$\frac{G_{E^p_{\text{theor}}}}{G_D}$	$\frac{G_{E^p_{\text{exp}}}}{G_D}$ [80]
0.670	1.05	1.05 ± 0.08	2.00	0.70	$0.81^{+0.13}_{-0.14}$
1.17	0.91	0.99 ± 0.08			
1.75	0.75	$0.78^{+0.13}_{-0.14}$	2.33	0.63	$0.7^{+0.18}_{-0.22}$
			3.00	0.52	$0.63^{+0.23}_{-0.33}$

Table 2.3. Experimental data on the function G_{E_n}.[81]

q^2, (GeV/c)^2	$\left(\frac{G_{E_n}}{e}\right)^2$	q^2, (GeV/c)^2	$\left(\frac{G_{E_n}}{e}\right)^2$
0.389	+0.026	1.17	−0.012
0.389	−0.034	1.75	−0.003
0.623	+0.007	2.92	0.102
0.857	+0.007	3.89	0.0066
1.17	−0.009		

sion relations for the form factors $F^{s,v}_{1,2}$ without subtractions [see (2.36)]. We see that the use of dispersion relations is convenient for a theoretical interpretation of the experimental data.

Before we write dispersion relations for the functions $G_{E,M}$, however, we would like to call attention to a particular circumstance. When we solve (2.36) for F_i, we find

$$F_1(t) = \frac{G_E(t) - \tau G_M(t)}{1-\tau}; \quad \varkappa F_2(t) = \frac{G_M(t) - G_E(t)}{1-\tau}. \tag{2.41}$$

We see that at the point $t = q^2 = 4M^2$, i.e., at the physical threshold of the annihilation channel, the denominator $1-\tau$ vanishes. The functions $F_{1,2}(t = 4M^2)$ thus becomes infinite unless we assume $G_M(4M^2) = G_E(4M^2)$. We assume that this inequality holds.

We thus write

$$G^v_{E,M}(t) = \frac{1}{\pi}\int_{4m_\pi^2}^{\infty} \frac{\operatorname{Im} G^v_{E,M}(q'^2)dq'^2}{q'^2 - q^2}. \tag{2.42}$$

Corresponding dispersion relations are written for the functions G^s_E and G^s_M, but in their case the lower limit is $9m_\pi^2$. The graphical representation of the unitarity condition for Im $G^{v,s}_{E,M}$ is the same as in Fig. 2.8. We assume that the spectral functions Im $G^{s,v}_{E,M}$ can be written as a sum of δ-function terms (in the preceding subsection, the spectral functions Im $F^{s,v}_{1,2}$ were approximated by a single δ function):

$$\text{Im } G_E^{s,v}(q^2) = \sum_i \alpha_i \pi\delta(q^2 - q_i^2);$$
$$\text{Im } G_M^{s,v}(q^2) = \sum_i \beta_i \pi\delta(q^2 - q_i^2), \tag{2.43}$$

where α_i and β_i are constants which are calculated theoretically and which should depend on the parameters of the interactions of the π and K mesons and other particles with nucleons; and $q_i^2 = M_i^2$ are the positions of the isovector or isoscalar mesons which contribute, by assumption, to the form factors $G_{E,M}$. Substituting relations (2.43) into (2.42), we find

$$G_E^{v,s}(q^2) = \sum_i \frac{\alpha_i}{1 - q^2/M_i^2};$$
$$G_M^{v,s}(q^2) = \sum_j \frac{\beta_j}{1 - q^2/M_j^2}. \tag{2.44}$$

All of the form factors (2.44) fall off in the limit $q^2 \to \infty$ approximately in accordance with q^{-2}, and the nucleon does not have a core. If any of the masses of the mesons turns out to be very large, the corresponding term in (2.44) will simulate the existence of a core of a nucleon.

We use the relations

$$G_{E_p} = G_E^s + G_E^v; \quad G_{M_p} = G_M^s + G_M^v;$$
$$G_{E_n} = G_E^s + G_E^v; \quad G_{M_n} = G_M^s - G_M^v,$$

and we write out the proton form factors explicitly:

$$G_{E_p} = \frac{\alpha_1}{1 + q^2/M_{1v}^2} + \frac{\alpha_2}{1 + q^2/M_{2v}^2} + \cdots + \frac{\alpha_3}{1 + q_2/M_{3s}^2} + \frac{\alpha_4}{1 + q^2/M_{4s}^2} + \ldots;$$
$$G_{M_p} = \frac{\beta_1}{1 + q^2/M_{1v}^2} + \frac{\beta_2}{1 + q^2/M_{2v}^2} + \cdots + \frac{\beta_3}{1 + q^2/M_{3s}^2} + \frac{\beta_4}{1 + q^2/M_{4s}^2} + \ldots. \tag{2.45}$$

The contributions of the isovector and isoscalar mesons in (2.45) are assigned subscripts v and s in (2.45) (M_s,M_v). The neutron form factors differ from those in (2.45) in that the signs on the contributions of the isovector mesons are changed.

For analyzing experimental data, we can choose a one-pole, two-pole, three-pole, etc., approximation. The following have been examined: (a) a dipole model, in accordance with (2.39); (b) a three-pole approximation without

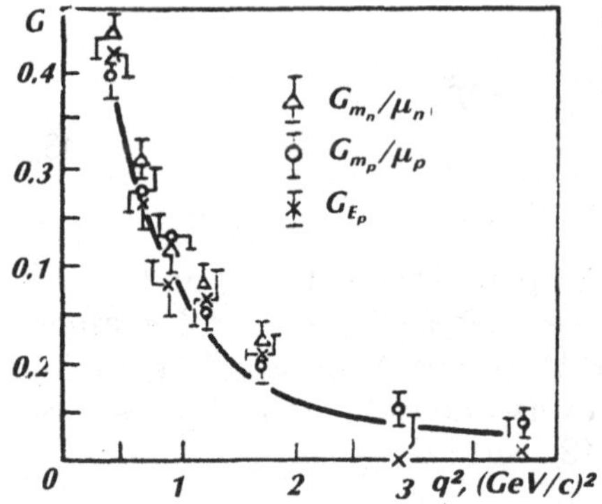

Figure 2.24. Comparison of the dipole model for the electromagnetic form factors of nucleons with experimental data.

a core; (c) a three-pole approximation with a core; and (d) a four-pole approximation. More complicated models could also be used.

The dipole approximation is exceptionally simple. It could also be interpreted as a two-pole approximation with resonances which are approximately equal in mass and which differ in the sign of the coupling constant. Figure 2.24 makes a comparison with experimental data for space-like values $q^2 < 0$. The region of time-like values, $q^2 > 0$, will be discussed below.

The three-pole approximation (with or without a core) is also capable of good agreement with experimental data. However, in this case it becomes necessary to choose values on the low side for the mass of the ρ meson.

Perhaps the most successful approach is to approximate the experimental data on elastic scattering by means of form factors described with the four-pole approximation.[82] In this case, the data obtained at various accelerators around the world (in the United States, France, West Germany, and Great Britain) reproduce very accurately the behavior of the differential cross section $d\sigma/d\Omega$ for the elastic scattering of electrons by protons over a broad range of the space-like 4-momentum transfer, $0 \leqslant |q^2| \leqslant 20$ $(\mathrm{GeV}/c)^2$.

Table 2.4 shows values of the square masses and the residues at the poles [see (2.45)]. In this parametrization, the physical meaning of certain poles and residues is not clear, and the parametrization should be regarded as empirical. The first pole with a small value of the mass provides the steep decay of the form factors at small values of q^2; the last pole describes the contribution of remote singularities (poles and cuts) to the dispersion integral and provides the necessary degree for the decay of the form factors at large values of q^2.

Data on the nucleon form factors in the time-like region of momentum transfer are rather scanty. The differential cross section for the process $e^+e^- \to p\bar{p}$ is described in the one-photon approximation by

$$d\sigma/d\Omega = (\beta\alpha^2/4s)\,[\,|G_E|^2(4M_p^2/s)\sin^2\theta + |G_M|^2(1+\cos^2\theta)]\,, \quad (2.46)$$

where β is the velocity of the proton, M_p is its mass, and θ is the angle at which

Table 2.4. Parameters of a four-pole fit of the nucleon form factors: M^2/residue.

Form factor	1st pole	2nd pole	3rd pole	4th pole
G_E	3.53/0.291	15.02/1.371	44.08/ − 0.634	154.20/0.044
G_M/μ_p	8.73/0.794	15.02/0.594	44.08/ − 0.393	355.40/0.005

Note. The masses are expressed in units of reciprocal femtometers.

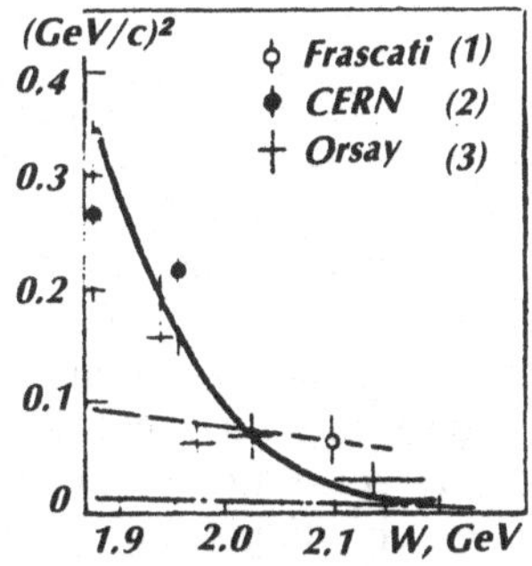

Figure 2.25. Square of the nucleon form factor found under the assumption $|G_E|^2 = |G_M|^2$. Solid line—with contributions from ρ, ω, φ, ρ′, and other mesons; dashed line—with contributions from only the ρ, ω, and φ mesons; dot-dashed line—calculated from the dipole model.

the proton is emitted with respect to the direction of the e^+e^- beams. The form factors here have been normalized in accordance with $G_E(0) = 1$, $G_M(0) = \mu_p$. If we set $G_E = G_M = 1$ in (2.46), we obtain (2.46) in a form completely identical to that of (1.53) ($\beta = p/E$, $4M^2_p/s = 1 - p^2/E^2$).

Because of the paucity of experimental data, we cannot separate G_E from G_M, as was done for the process $e + p \to e + p$, so that the extraction of form factors with the help of (2.46) is carried out under the assumption $G_E = G_M/\mu_p$.

Figure 2.25 compares the experimental data of Ref. 83 with calculations based on three models: the dipole model; the vector-dominance model with contributions from ρ, ω, and φ mesons; and the vector-dominance model[84] with contributions from ρ, ω, φ, ρ', and ρ'' mesons and other contributions. It turns out that the dipole formula leads to values for the total scattering cross section which are nearly an order of magnitude smaller than the experimental values (Table 2.5).

The vector-dominance model with contributions from only ρ, ω, and φ mesons give a good description of the region of space-like momentum transfer but a poor description of the region $q^2 > 0$. Only the vector-dominance model with contributions from ρ, ω, φ, ρ', and ρ'' mesons and from the family of J/ψ particles gives a satisfactory description of the region $q^2 > 0$.

Research on the form factors of elementary particles began in the mid-1950s. During the first five or ten years of this research it seemed that the problem was nearly solved. Today, however, certain aspects do not look as clear, and the problem of studying the form factors of elementary particles remains one of the most important in elementary-particle physics.

Table 2.5. Comparison of the values predicted for the total cross section for $e^+e^- \to p\bar{p}$ by the dipole model with experimental data ($\bar{\sigma}_{tot}$ is the mean value over the specified energy interval).

$\sqrt{s}$, GeV	$\bar{\sigma}_{tot\ expt}$ (10^{-30} cm^2)	σ_{dip} (10^{-30} cm^2)	$\sqrt{s}$, GeV	$\bar{\sigma}_{tot\ expt}$ (10^{-30} cm^2)	σ_{dip} (10^{-30} cm^2)
1.93 – 1.96	1.15 ± 0.46	0.14	2.00 – 2.06	0.75 ± 0.30	0.12
1.96 – 1.99	0.63 ± 0.36	0.14	2.10 – 2.18	0.63 ± 0.30	0.09

Quantum chromodynamics is not yet capable of describing the behavior of form factors at a small momentum transfer, since perturbation theory cannot be used at these values of q^2. Quantum chromodynamics predicts only the asymptotic behavior of the form factors. In QCD it is postulated that the bosons should consist of two valence quarks, and the baryons of three. If we do not know the quark interaction potential, however, we cannot derive an analytic functional dependence of the form factors on q^2.

Despite the exceptionally clear interpretation of the results which it provides, the dispersion method, which has been used so far, would seem to be obsolescent at a time when QCD is being used everywhere. As we have seen, however, for describing form factors the dispersion method remains the most convenient and most accurate method.

The form factors are known to be analytic functions of q^2. The problem of joining the form factors at the transition from space-like to time-like values of the square of the momentum transfer, q^2, however, is still hazy: Can this joining be accomplished for all of the form factors?

Few experimental results are available on the π and K form factors or on the nucleon form factors in the joining region and at time-like values $q^2 > 0$.

Research on the form factors of other mesons (other than π and K) is still in its infancy: It was only recently that the first results on the form factors of η and η' mesons were obtained.[85]

The kinematics of the elastic scattering by particles whose form factors are being studied in elementary-particle physics is surprisingly simple. The equations describing elastic scattering contain no unknown functions other than the form factors. Research on form factors has been under way since the mid-1950s. Despite its apparent simplicity, however, the problem of studying the form factors of elementary particles—a problem intimately related to research on the structure of elementary particles—has yet to be solved.

Chapter 3 Deep inelastic scattering of leptons by nucleons

1. Structure of the nucleon

Deep inelastic scattering of leptons (electrons, μ mesons, and neutrinos) by nucleons is a process in which a large 3-momentum is transferred from a lepton to a nucleon. On the curve of the cross section for this process the corresponding region of energies and momenta begins immediately after the resonances, after a plateau is reached (Fig. 3.1; the region of scattered-electron energies below 8 GeV).

Historically, research on deep inelastic scattering of electrons by protons and deuterons in experiments in the U.S.A. at the Stanford Linear Accelerator Center (SLAC) led initially to the development of the parton model (the theoretical papers by Feynman and Bjorken[86]) and then to the appearance of quantum chromodynamics (QCD).[87]

Research on deep inelastic scattering of leptons by nucleons provided powerful motivation for a more profound study of the asymptotic behavior of the cross section ("light-cone physics") in QED, the development of a non-Abelian gauge field theory, and the construction of several specific models of a nucleon.

Before the advent of the parton model, a nucleon model which was used had a definite physical meaning. The nucleon was represented as consisting of clouds of virtual pairs: e^+e^-, $\mu^+\mu^-$, $\pi^+\pi^-$, $K\bar{K}$, $N\bar{N}$, etc. (Fig. 3.2). The radius of each cloud was determined by the Compton wavelength of the corresponding particles:

$$r_e \sim \hbar/m_e c; \quad r_\pi \sim \hbar/m_\pi c\,,$$

etc.

In this model the nucleon cloud (the core of the nucleon) has dimensions on the order of the Compton wavelength of the nucleon, $r_N \sim 10^{-14}$ cm. The core of a nucleon can be observed in experiments in which the momentum transfer reaches values $|\Delta p| \gtrsim Mc \gg 1$ GeV/c. Such values of the momentum transfer have been observed for a long time in deep inelastic ep and μp scattering, but it has not been possible to detect deviations from a point-particle behavior of the nucleon.

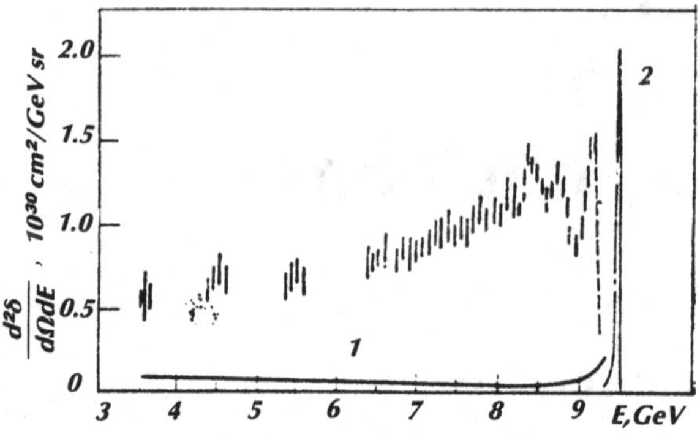

Figure 3.1. Spectrum of scattered electrons ($E = 10$ GeV, $\theta = 6°$). 1, radiative emission; 2, elastic-scattering peak.

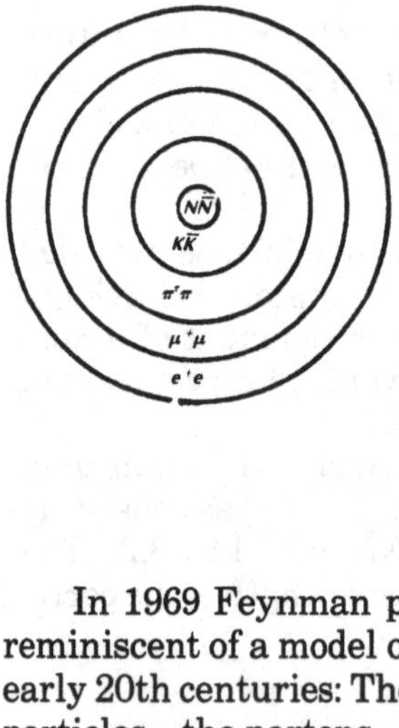

Figure 3.2. Model of the nucleon.

In 1969 Feynman proposed the parton model of the nucleon, which is reminiscent of a model of the electron which was extant in the late 19th and early 20th centuries: The nucleon was assumed to consist of extremely small particles—the partons—which fill the entire space within a nucleon. All the constituents of a nucleon are identical, as are their electric charges. This is the simplest parton model. It arose in study of the energy dependence of the cross section for deep inelastic scattering of electrons by nucleons. Experimental data indicated that the cross section behaved in the same way as the cross section for scattering by a point nucleon with a charge less than unity. This behavior of the cross section suggested that an electron was scattered not by the entire nucleon but by only part of it and that the responsible part has point-like dimensions. A corresponding suggestion had been offered earlier with respect to the interaction of neutrinos with nucleons.[88] If deep inelastic scattering is examined in a frame of reference in which the nucleon is moving at a velocity close to the velocity of light, and if the nucleon is assumed to consist of partons, then the motions of the constituents of the nucleon will be retarded in the relativistic time dilatation, and the distribution of the nucleon charge will become disk-shaped as a result of Lorentz contraction. Under these conditions the impinging lepton will be scattered essentially instantaneously and incoherently by an individual parton of the nucleon. This model is a good approximation of the real process of deep inelastic scattering of electrons by protons in the c.m. frame at high energies.

The more complex parton models differ from that described above in that they incorporate various other assumptions: that the partons come in different types, with different charges; that they may be nonpoint particles; that they may have a spin; etc. We will be discussing one such model of the nucleon (the Kuti–Weisskopf model[89]) below.

The parton model yields a natural explanation for the scaling (Sec. 2 of Chap. 3) of the cross section for deep inelastic scattering [this scaling should not be confused with the scaling of the form factors (Sec. 6 of Chap. 2)]. The representation that the constituents in the parton model have point-like dimensions makes it possible to retain the mathematical apparatus of perturbation theory. Together, these circumstances guaranteed the success of the parton model despite the fact that it contains certain fundamentally unsatisfactory postulates (there is no interaction among partons, and it is not possible to rigorously determine theoretically how many types of partons there are and their momentum distributions within a nucleon). Attempts to find the constituents of a nucleon in the free state have yet to meet with success. One possible reason is that the colliding-particle energies which can be reached at the existing accelerators are not high enough to eject a parton from a nucleon. We can thus understand the quest for ever higher energies at progressively more powerful accelerators—a desire extremely difficult to satisfy. It is particularly difficult to produce a high-energy beam of electrons because of their large radiative energy loss. It is safe to say at this point that protons will not be accelerated to energies above 5 TeV, or electrons above 100 GeV, before the year 2000.

The problem of studying the structure of a nucleon, which is intimately related to the philosophical problem of the structure of matter, has thus now evolved into one of the most important problems in the field of high-energy physics, both in connection with research on the nature of the new forces which act among the constituents of a nucleon and in connection with the need to construct powerful new accelerators capable of taking physicists close to the solution of this problem. Much hope is pinned on ring accelerators with colliding beams. Table 3.1 demonstrates the surprising progress in terms of the energy in the laboratory frame of reference which has been achieved by accelerators of this type. The next 20 or 30 years may shed new light on the problems of the structure of elementary particles, in particular, nucleons.

2. Kinematics and cross section of deep inelastic scattering; scaling of structure functions

The deep inelastic scattering of leptons by nucleons,

$$e(\mu) + N \to e(\mu) + X, \qquad (3.1)$$

where X is any set of particles allowed by the conservation laws, is described in the approximation of one-photon exchange by a single Feynman diagram (Fig. 3.3). Incorporating diagrams with two-photon exchange is a rather complicated matter. The contribution of such diagrams to the cross section for the process is proportional to α^4, and such diagrams are usually ignored. Invariant variables q^2 and ν are introduced to describe the process. The variable q^2 is defined as the difference between the 4-momenta of the lepton:

$$q^2 = (k - k')^2 = 2m^2 - 2EE' + 2|\mathbf{k}||\mathbf{k}'|\cos\theta. \qquad (3.2)$$

Here θ is the angle through which the lepton is scattered with respect to the direction of the impinging lepton beam, and m is the mass of the lepton. For an electron the relations $E, E' \gg m$ usually hold, so we can write

$$q^2 \approx -4EE'\sin^2(\theta/2). \qquad (3.3)$$

The variable ν is defined by

$$\nu = Pq/M, \qquad (3.4)$$

where P is the 4-momentum of the nucleon before the collision, and M is the mass of the nucleon.

We will also be using the variable W, which is defined as the invariant mass of all of the particles (except for the scattered lepton) in the final state (Fig. 3.2):

$$W^2 = M_X^2 = (P + q)^2. \qquad (3.5)$$

In the proper frame of the nucleon we have

$$\nu = E - E'; \quad W^2 = M^2 + 2M\nu + q^2.$$

In deep inelastic scattering, the 4-momentum transfer q^2 is a negative quantity, so for convenience one often introduces a quantity

$$Q^2 = -q^2. \qquad (3.6)$$

One then writes

$$W^2 = M^2 + 2M\nu - Q^2. \qquad (3.7)$$

For inelastic scattering we have $W^2 = M^2$ and $2M\nu = Q^2$.

The procedure for experimentally determining the differential cross section for the deep inelastic scattering of electrons (or μ mesons) by protons can be outlined as follows: A beam of leptons with a certain energy impinges on a

Figure 3.3. Feynman diagram describing deep inelastic scattering of an electron by a nucleon.

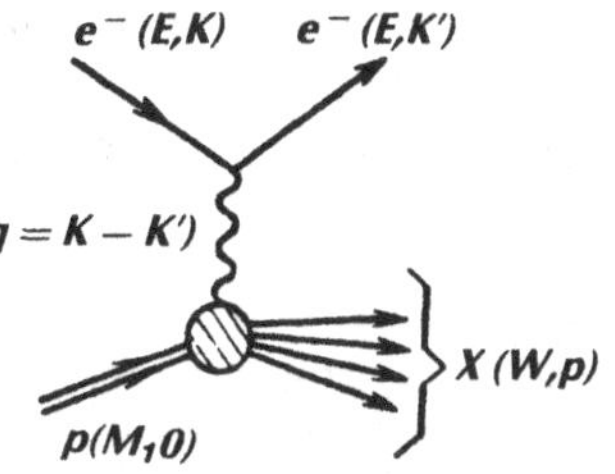

hydrogen target. The momenta of the scattered leptons are analyzed by a magnetic spectrometer. In the course of the measurements, particles other than the scattered leptons (π mesons, K mesons, protons, etc.) reach the magnetic spectrometer. The protons can be distinguished from leptons relatively easily because of their large mass; the K mesons decay rapidly (for the most part, into π mesons). The most complicated problem is distinguishing π mesons from leptons. For this purpose, threshold counters or several ordinary counters separated by material of some sort are installed, and the π mesons are identified on the basis of the energy loss per unit radiation length. Electrons are distinguished from the positrons which are produced in the decay of π^0 mesons into photons (followed by a conversion of the photons into e^+e^- pairs) by changing the direction of the magnetic field in the spectrometer. In this manner one measures the background from positrons and then uses this background to determine the contribution of π^0 mesons. It is also necessary to eliminate radiative corrections. This is a complicated problem: The radiative corrections depend on the geometry of the experiment. After all of these effects have been eliminated, one finds the cross section for the process under study.

Let us consider the scattering of electrons by protons. Figure 3.1 shows the standard shape of the scattered-electron spectrum in the process $e^+p \to e + X$. The highest peak, at an energy of about 9.5 GeV, corresponds to the elastic scattering $e + p \to e + p$. Between 8 and 9.5 GeV we see three peaks, which correspond to the excitation of nucleon resonances, $e + p \to e + \Delta$.

The peak closest to the elastic-scattering peak is associated with $M_\Delta = 1236$ MeV; the next one with $M_\Delta = 1520$ MeV; and the next with $M_\Delta = 1700$ MeV. At electron scattering energies below 8 GeV there is a broad plateau, which corresponds to deep inelastic scattering of leptons by nucleons: $e + p \to e + X$.

The spectrum in Fig. 3.1 should have contained peaks from all known low-lying nucleon resonances. However, because of the large widths of these resonances and their relatively low intensity, they merge and become indistinguishable.

If we assume that the one-photon diagram (Fig. 3.3) gives an adequate description of the process

$$e(\mu) + p \to e(\mu) + X ,$$

then we would expect to find that the cross section for deep inelastic scattering would contain all those resonances which are manifested in the inelastic interaction of real photons with protons:

$$\gamma + p \to X\,. \tag{3.8}$$

In fact, this is what we find. As the energy of the impinging electron and the scattering angle increase, i.e., as the "mass" of the virtual photon increases, the resonance behavior of the cross section gradually fades away.

The differential cross section for process (3.1) is determined in the one-photon approximation by the matrix element

$$M = e^2 \bar{u}(k')\gamma_\mu u(k)(g_{\mu\nu}/q^2)\langle x|J_\nu(0)|p\rangle\,. \tag{3.9}$$

If we are not interested in polarization effects, we should, after squaring matrix element (3.9), sum over the final spin states and take an average over the initial spin states. The expression for the differential cross section then becomes

$$d^2\sigma/dE'd\Omega = (8\pi\alpha^2/q^4)L_{\mu\nu}W_{\mu\nu}\,, \tag{3.10}$$

where $\alpha = e^2/\hbar c$,

$$\begin{aligned} L_{\mu\nu} &= \frac{m^2}{2}\sum_s\sum_{s'}[\bar{u}(k',s')\gamma_\mu u(k,s)]\,[\bar{u}(k,s)\gamma_\nu u(k',s')] \\ &= (1.2)[k_\mu k'_\nu + k'_\mu k_\nu + (1/2)q^2 g_{\mu\nu}]\,; \\ W_{\mu\nu} &= \frac{1}{M}\Big[\Big(p_\mu - \frac{(pq)}{q^2}q_\mu\Big)\Big(p_\nu - \frac{(pq)}{q^2}q_\nu\Big)W_2(\nu,q^2) \\ &\quad - M\Big(g_{\mu\nu} - \frac{q_\mu q_\nu}{q^2}\Big)W_1(\nu,q^2)\Big]\,. \end{aligned} \tag{3.11}$$

Here $\sum_s\sum_{s'}$ means a sum over the final spin states and an average over the initial spin states.

The tensor $W_{\mu\nu}$ contains only two independent functions: $W_1(\nu,q^2)$ and $W_2(\nu,q^2)$. The proof of this fact is similar to the proof of the assertion that only two form factors exist for the elastic scattering of electrons by protons.[90] There is the difference that while the form factors $F_1(q^2)$ and $F_2(q^2)$ for elastic scattering depend on only the single variable q^2 (since the other variable—the square of the 4-momentum of the proton in the final state, $p'^2 = M^2$—is a fixed quantity), in the case of inelastic scattering the functions W_1 and W_2 depend on two variables, since the quantity $W^2 = M^2{}_X$ is not fixed (it can be replaced by ν).

The tensor $W_{\mu\nu}$ satisfies the requirements of gradient invariance ($q_\mu W_{\mu\nu} = q_\nu W_{\mu\nu} = 0$) and that it be of a Hermitian nature. It follows from the latter requirement that the functions $W_1(\nu,q^2)$ and $W_2(\nu,q^2)$ are real.

Contracting the tensors $L_{\mu\nu}$ and $W_{\mu\nu}$, and using the approximations

$$m/E \ll 1;\quad m/E' \ll 1;\quad Q^2 \approx 4EE'\sin^2(\theta/2)\,,$$

we find

$$\frac{d^2\sigma}{dE'd\Omega} = \frac{4\alpha^2 E'^2\cos^2(\theta/2)}{Q^4}\left[W_2(\nu,q^2) + 2\tan^2\frac{\theta}{2}\,W_1(\nu,q^2)\right] \tag{3.12}$$

or

$$d^2\sigma/dE'd\Omega = \Gamma_t\,[\sigma_t(q^2,W) + \varepsilon\sigma_L(q^2,W)]\,, \tag{3.13}$$

where

$$\begin{aligned} &W_1 = \frac{K}{4\pi^2\alpha}\,\sigma_t; \quad W_2 = \frac{K}{4\pi^2\alpha}\,\frac{Q^2}{Q^2+\nu^2}\,(\sigma_t + \sigma_L);\\ &\Gamma_t = \frac{\alpha}{2\pi^2}\,\frac{K}{Q^2}\,\frac{E'}{E}\,\frac{1}{1-\varepsilon};\\ &\varepsilon = \frac{1}{1+2(1+\nu^2/Q^2)\tan^2(\theta/2)}; \quad K = \nu + \frac{Q^2}{2M}\,. \end{aligned} \tag{3.14}$$

The quantity $\sigma_t(q^2,W)$ in (3.13) describes the cross section for the absorption of virtual photons with transverse polarization; $\sigma_L(q^2,W)$ is the cross section for the absorption of virtual photons with longitudinal polarization [in the limit $q^2 \to 0$, expression (3.13) becomes the cross section for the photoabsorption of real photons]; Γ_t describes the spectrum of virtual photons; and ε is a polarization parameter.

Descriptions of deep inelastic scattering frequently make use of the variables x and y:

$$x = 1/\omega = Q^2/2M\nu; \quad y = \nu/E\,. \tag{3.15}$$

The variable x is called the "scaling variable." In terms of these variables, the expression for the differential cross section is

$$\frac{d^2\sigma}{dx\,dy} = \frac{8\pi\alpha^2}{Q^4}\,ME\left[\left(1 - y - \frac{Mxy}{2E}\right)\nu W_2 + \tfrac{1}{2}\,y^2(2Mx)W_1\right]. \tag{3.16}$$

The relationship between (3.12) and (3.16) is determined by the substitution

$$dx\,dy = (E'/2\pi M\nu)dE'd\Omega\,.$$

The functions $W_1(\nu,q^2)$ and $W_2(\nu,q^2)$ are sometimes called the "transition form factors of the inelastic scattering process." They are related to the form factors for elastic scattering by

$$\begin{aligned} &W_1(\nu,q^2) = \frac{Q^2}{4M^2}\,G_M^2\delta\left(\nu - \frac{Q^2}{2M}\right);\\ &W_2(\nu,q^2) = \frac{G_E^2 + (Q^2/4M^2)G_M^2}{1+Q^2/4M^2}\,\delta\left(\nu - \frac{Q^2}{2M}\right), \end{aligned} \tag{3.17}$$

where G_E and G_M are the electric and magnetic ("Sachs") form factors of the nucleon [see (2.36)].

In precisely the same way as in the determination of the nucleon form factors in the case of elastic scattering, the functions $W_1(\nu,q^2)$ and $W_2(\nu,q^2)$ can be determined, according to (3.12), from measurements of the cross section at various angles for the same values of ν and q^2. Knowing $W_1(\nu,q^2)$ and $W_2(\nu,q^2)$, we can find σ_t and σ_L, which we can in turn use to determine

$$R=\sigma_L/\sigma_t\,. \tag{3.18}$$

The five quantities R, σ_t, σ_L, W_1, and W_2, in (3.12)–(3.18) are the subjects of intense research, both experimental and theoretical. The functions $W_1(\nu,q^2)$ and $W_2(\nu,q^2)$ are usually separated by fixing the parameter R:

$$\begin{aligned} W_1&=\left(\frac{d^2\sigma}{dE'd\Omega}\right)_{\rm expt}\left(\frac{d\sigma}{d\Omega}\right)_M^{-1}\left[(1+R)\frac{q^2}{q^2+\nu}+2\tan^2\frac{\theta}{2}\right]^{-1};\\ W_2&=\left(\frac{d^2\sigma}{dE'd\Omega}\right)_{\rm expt}\left(\frac{d\sigma}{d\Omega}\right)_M^{-1}\left[1+2\left(\frac{1}{1+R}\right)\left(\frac{q^2+\nu}{q^2}\right)\tan^2\frac{\theta}{2}\right]^{-1}, \end{aligned} \tag{3.19}$$

where $(d^2\sigma/dE'd\Omega)_{\rm expt}$ is the experimental cross sections, and $(d\sigma/d\Omega)_M$ is the Mott scattering cross section,

$$\left(\frac{d\sigma}{d\Omega}\right)_M=\frac{4\alpha^2E'^2\cos^2(\theta/2)}{Q^4}=\frac{\alpha^2\cos^2(\theta/2)}{4E^2\sin^4(\theta/2)}\,. \tag{3.20}$$

The parameter R is determined from (3.14) and (3.18). The combination $\sigma_t+\varepsilon\sigma_L$ in (3.13) is first measured as a function of ε, and then the parameter R is found. When this approach is taken, there is a large error in the measurement of R, since the slope ε is difficult to determine near small values of q^2. The most probable values of R lie in the interval

$$0\lesssim R\lesssim 0.25\,. \tag{3.21}$$

An important result of the first measurements of the cross section for deep inelastic scattering was the observation of a weak q^2 dependence of the change in the ratio

$$\frac{d\sigma}{dE'd\Omega}\frac{1}{(d\sigma/d\Omega)_M}=W_2(\nu,q^2)+2\tan^2\left(\frac{\theta}{2}\right)W_1(\nu,q^2) \tag{3.22}$$

(Ref. 91; see Fig. 3.4 of the present book). In the first experiments the values of q^2 were small, and the function $W_1(\nu,q^2)$, which makes only a small contribution to the cross section for deep inelastic scattering, was measured only within large errors.

The smooth behavior of the functions $W_1(\nu,q^2)$ and $W_2(\nu,q^2)$ indicated that they adhered to the scaling which had been predicted earlier by Bjorken. Bjorken analyzed the behavior of the simultaneous commutators of the electromagnetic field in quantum field theory for large values of ν and q^2 and reached the conclusion that the functions $\nu W_2(\nu,q^2)$ and $W_1(\nu,q^2)$ should tend toward $F(\omega)$ in the limit $\nu\to\infty$, $q^2\to\infty$ under the condition $\omega=2M\nu/Q^2=\text{const}$:

$$\nu W_2(\nu,q^2)\to F(2M\nu/Q^2)\equiv F(\omega);\quad W_1(\nu,q^2)\to F(\omega)\,. \tag{3.23}$$

This behavior of the functions is called a "scaling."

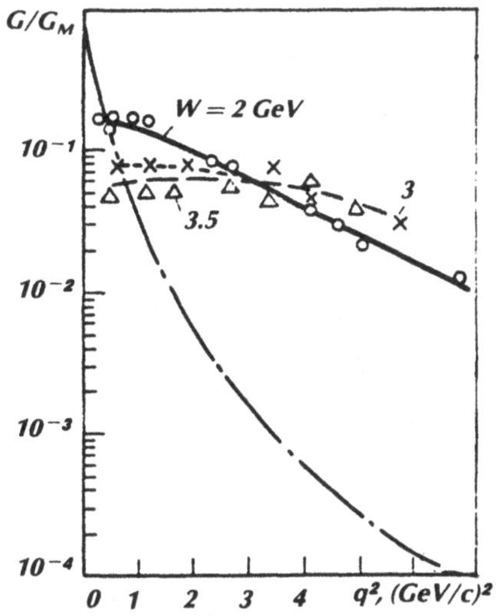

Figure 3.4. Cross-section ratio σ/σ_M as a function of Q^2. Shown for comparison is the ratio of the elastic cross section to σ_M for $\theta = 10°$ (the dot-dashed line).

The first experimental results did indeed indicate that the functions νW_2 and W_1 depend on only the ratio of invariant variables $\omega = 2M\nu/Q^2$ (Fig. 3.5). In these experiments, carried out with an error of about 10%, the scaling of the function νW_2 set in too early, at $\omega \gtrsim 4$ and at an energy of the impinging electrons below 18 GeV. In terms of the variables Q^2, the equivalent value is $W \approx 1.8$ GeV. This phenomenon became known as "precocious scaling." Small values of R ($R = 0$ in Fig. 3.5) meant that σ_L was small in comparison with σ_t; this circumstance led to the requirement that the function $W_1(\nu, q^2)$ must also exhibit a scaling at values of ν satisfying $\omega/2M < 1/\nu$ [see (3.14)].

There is experimental proof of a violation of the scaling behavior (Sec. 6 of Chap. 3; for simplicity we will say simply "scaling" in some places) at attainable impinging-electron energies (up to 20 GeV). The problem of precocious scaling is thus of no more than historical interest, and we will not discuss it further.

Together, these experimental results served as a foundation for the construction of several interesting theoretical models which were offered as candidates for describing the deep inelastic scattering of leptons by nucleons and also for describing the structure of a nucleon.

3. Relationships among structure functions in the simplest parton models; sum rules

In Sec. 1 of this chapter we discussed certain aspects of the simplest parton model of nucleons. Let us now assume the following: (1) All the partons are of spin 1/2. (2) In lowest-order perturbation theory, the interaction of a lepton with a nucleon occurs through the exchange of a single virtual photon, which

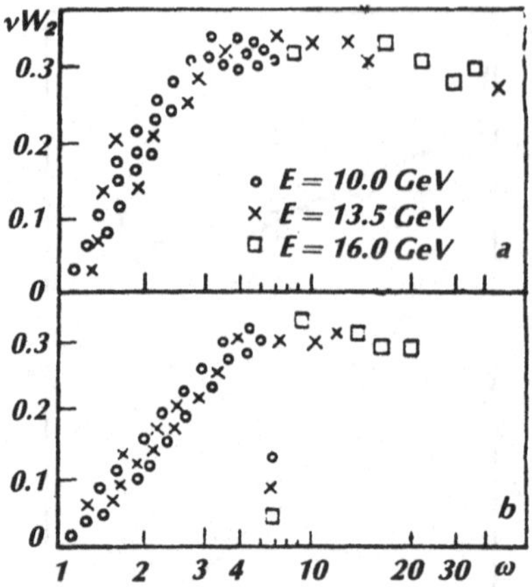

Figure 3.5. Function νW_2 vs ω. (a) $\theta = 6°$; (b) $\theta = 10°$. In both cases we have $R = 0$.

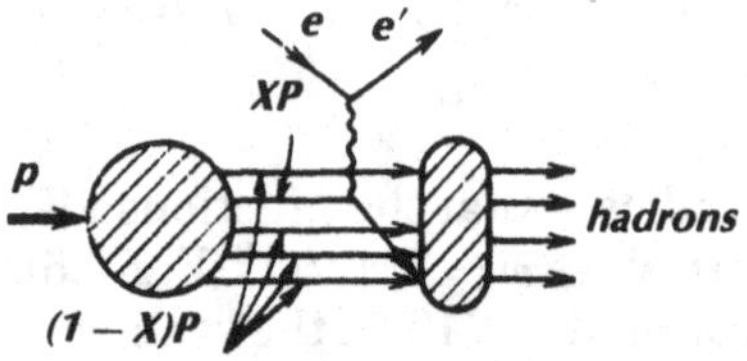

Figure 3.6. Lepton–nucleon interaction according to the simplest parton model for deep inelastic scattering.

interacts with one of the partons of the nucleon (Fig. 3.6). (3) The partons within a nucleon interact with each other (only under this condition is the momentum which is transferred to one of the partons transferred to the nucleon as a whole). Under these assumptions, as we mentioned earlier, there is no need to substantially change the mathematical apparatus of QED in describing the *ep* or μp interaction. All that we need to do is note that the lepton can interact with any parton of the nucleon, so that we need to take an average of the contributions from the interactions of the lepton with all the partons of the nucleon. The functions which arise when this averaging is carried out are identified with the functions $W_1(\nu,q^2)$ and $W_2(\nu,q^2)$. Under the rather natural assumptions which we just listed, a scaling arises in the functional dependence of these functions on the variable $\omega = 2M\nu/Q^2$.

How does a lepton interact with a nucleon in this version (Fig. 3.6) of the parton model? The total momentum of the nucleon, P, is the sum of the momenta of all the partons:

$$P = \sum p_i \,. \tag{3.24}$$

Each parton has a momentum

$$p_i = x_i\, P, \quad x_i = p_i / P \,.$$

We assume that all the partons have the same mass μ and that for each parton the following relation holds:

$$E_i^2 = p_i^2 + \mu^2 .$$

The momentum distribution of the partons in the nucleon is given by the function $f_n(x)$, where the index n corresponds to a certain configuration of the partons in the nucleon. The probability for observing parton configuration n in a nucleon is denoted by $\mathscr{P}(n)$. The index n is equal to the number of partons in the nucleon; $\left\langle \sum_i e_i^2 \right\rangle_n$ is equal to the expectation value of $\sum_i e_i^2$ in such configurations.

The functions $W_1(\nu,q^2)$ and $W_2(\nu,q^2)$ can now be calculated by the following method. The tensor $W_{\mu\nu}$, which is responsible for the interaction of a virtual photon with one of the partons of the nucleon, is written in the standard form:

$$(1/2E)(W_{\mu\nu})_{\text{part}} = (1/2E_1)|M|^2_{\text{part}}\,\delta[(p_i + q)^2 - \mu^2 - \Delta] . \tag{3.25}$$

Here p_i and q are 4-momenta; the factors $1/2E$ and $1/2E_1$ appear because of the particular choice of normalization of the quantities $(W_{\mu\nu})_{\text{part}}$ and $|M|^2_{\text{part}}$; and the quantity Δ is a measure of the unknown part of the interaction of partons with each other.[86] The matrix element M is, according to QED rules, proportional to the electric charge of a parton, e_i, and is written

$$M_{\text{part}} = e_i\,[\overline{\omega}(p_i')\gamma_\mu Y(p_i)] .$$

Its square is [see (3.11)]

$$|M|^2_{\text{part}} = e_i^2[p_{i\mu}\,p_{i\nu}' + p_{i\nu}\,p_{i\mu}' + (1/2)g_{\mu\nu}(pq)] . \tag{3.26}$$

Since the values of the transverse momentum transfer are small (an experimental fact which was observed in deep inelastic processes), we can write $p_i \approx p_i' \approx x_i P$ and therefore $(x_i \equiv \xi)$

$$|M|^2_{\text{part}} = e_i^2[2\xi^2 p_\mu p_\nu + (1/2)g_{\mu\nu}\xi(pq)] . \tag{3.27}$$

On the other hand, under the same assumption the tensor $W_{\mu\nu}$ for a point nucleon is

$$W_{\mu\nu} \approx (1/M)[W_2(\nu,q^2)p_\mu p_\nu + (1/2)M^2 W_1(\nu,q^2)g_{\mu\nu}] . \tag{3.28}$$

Taking an average of $W_{\mu\nu\,\text{part}}$ over all of the partons, and using the distribution function $f_n(x)$ and the probability for finding a given configuration of quarks, we find

$$W_{\mu\nu} = \sum_n \mathscr{P}(n)\left\langle \sum_i e_i^2 \right\rangle_n \int \frac{f_n(\xi)}{\xi}[2\xi^2 p_\mu p_\nu + \tfrac{1}{2}\,\xi g_{\mu\nu}(Pq)] \times \delta[(\xi P + q)^2 - \mu^2 - \Delta]\,d\xi , \tag{3.29}$$

where $\sum_n$ means a sum over all possible configurations of the partons in the nucleon. The factor of $1/\xi$ in the integral arises because of the relation

$E/E_1 \approx E/\xi E = 1/\xi$; the argument of the δ function, $(\xi P + q)^2 - \mu^2 - \Delta$, takes the following form in the laboratory frame of reference ($P = M$):

$$\xi^2 M^2 + 2\xi M(E - E') + q^2 - \mu^2 - \Delta\,. \tag{3.30}$$

In the Bjorken limit we can ignore $\xi^2 M^2$, μ^2, and Δ. We make use of the definition of the scaling variable $x = -q^2/2M\nu$. It then becomes a trivial matter to carry out the integration in (3.29). The δ function which appears in the integrand can be rewritten as

$$\delta(2M\xi\nu - 2Mx\nu) = (1/2M\nu)\delta(\xi - x)\,.$$

After integrating (3.29) we thus find

$$W_{\mu\nu} = \sum_n \mathscr{P}(n)\langle \Sigma e_i^2\rangle_n \frac{f_n(x)}{2M\nu}\left[2xp_\mu p_\nu + \tfrac{1}{2} g_{\mu\nu}(M\nu)\right]. \tag{3.31}$$

Equating (3.31) and (3.28), we find

$$\begin{aligned} \nu W_2(\nu,q^2) &= \sum_n \mathscr{P}(n)\left\langle \sum_i e_i^2\right\rangle f_n(x)x = x\,f(x)\tilde{e}^2\,; \\ 2MW_1(\nu,q^2) &= \sum_n \mathscr{P}(n)\left\langle \sum_i e_i^2\right\rangle_n f_n(x) = f(x)\tilde{e}^2\,. \end{aligned} \tag{3.32}$$

If there is only a single parton in configuration n, then we have $\mathscr{P}(n) = 1$, $\left\langle \sum_i e_i^2 \right\rangle = e^2$, where e^2 is the square charge of an individual parton, and

$$\nu W_2(\nu,q^2) = e^2 x f(x), \quad 2MW_1(\nu,q^2) = e^2 f(x)\,. \tag{3.33}$$

If we now assume that several parton species exist in the nucleon, we can write relations (3.33) in the more general form

$$\begin{aligned} \nu W_2(\nu,q^2) &= \sum_j e_j^2\, x f_j(x)\,; \\ 2MW_1(\nu,q^2) &= \sum_j e_j^2\, f_j(x)\,, \end{aligned} \tag{3.34}$$

where e_j^2 is the square charge of a parton of species j, and $f_j(x)$ is the momentum distribution of the partons of species j.

We will make frequent use of relations (3.34) in examining the quark model of the nucleon [see, e.g., (3.52), (3.53), and (3.55)]. From (3.34) we find

$$(\nu/2Mx)W_2(\nu,q^2) = W_1(\nu,q^2)\,. \tag{3.35}$$

Relation (3.35) is called the "Callan-Gross relation."[92] The functions $\nu W_2(\nu,q^2)$ and $2MW_1(\nu,q^2)$ thus depend on only the scaling variable

$$x = -q^2/2M\nu = Q^2/2M\nu \tag{3.36}$$

in the parton model in the Bjorken limit.

We wish to call attention to a property of the function $f_n(x)$. The probability for finding, in parton configuration n, one of the partons in a state with a certain value x_m of the longitudinal momentum is

$$f_n(x_m) = \int_0^1 dx_1 dx_2 \cdots dx_{m-1} dx_{m+1} \cdots dx_n$$

$$\times f_n(x_1, x_2, \ldots, x_m) \Sigma (1 - \sum_k x_k) ,$$

while the total probability for finding a system of partons with longitudinal-momentum values $x_1, \ldots, x_m$ is unity:

$$\int_0^1 dx_m \, f_i(x_m) = 1 . \tag{3.37}$$

It follows that $f_n(x_m)$ is a symmetric function of x_m. The first moment of the function $f_n(x_m)$ is therefore

$$\int_0^1 dx_m x_m \, f_n(x_m) = \frac{1}{n} . \tag{3.38}$$

It thus follows from (3.34) and (3.37) that we have

$$\int dx \sum_n \mathscr{P}(n) \langle \Sigma e_i^2 \rangle_n x f_n(x) = \frac{\sum_n \mathscr{P}(n) \langle \Sigma e_i^2 \rangle_n}{n} ; \tag{3.39}$$

i.e., the integral on the left-hand side is equal to the mean-square charge of an individual parton. Substituting (3.32) into (3.39) we find the well-known sum rule:

$$\int_0^1 dx \sum_n \mathscr{P}(n) \langle \Sigma e_i^2 \rangle_n x f_n(x) = \int \nu W_2(\nu, q^2) dx$$

$$= \frac{Q^2}{2M} \int_{\frac{Q^2}{2M}}^{\infty} W_2(\nu, q^2) \frac{d\nu}{\nu} = \frac{\sum_n \mathscr{P}(n) \langle \Sigma e_i^2 \rangle_n}{n} . \tag{3.40}$$

The values of $W_2(\nu, q^2)$ are known from experimental data, and the mean-square charge of a proton can be calculated numerically. For *ep* scattering the value is

$$\int_0^1 x f(x) dx \approx 0.17 ; \tag{3.41}$$

for *en* scattering it is

$$\int_0^1 x f(x) dx \approx 0.13 . \tag{3.42}$$

Returning to (3.12)–(3.14) and (3.18), we evaluate R, using (3.34) and (3.35):

$$\sigma_t = \frac{4\pi^2\alpha}{K} W_1(\nu,q^2); \quad \sigma_L = \frac{4\pi^2\alpha}{K}\left(\frac{Q^2+\nu^2}{Q^2} W_2 - W_1\right);$$

$$R = \sigma_L/\sigma_t = [(Q^2+\nu^2)/Q^2]\, W_2/W_1 - 1,$$

or

$$R = (Q^2+\nu^2)/\nu^2 - 1 = Q^2/\nu^2.$$

In the limit $\nu \to \infty, q^2 \to \infty$ (at fixed values of x) we have $R \to 0$.

Do assumptions (1) and (2) at the beginning of this subsection hold up in a comparison with experimental data? If partons were scalar particles, the term with $g_{\mu\nu}$ would be absent from (3.27), and $W_1(\nu,q^2)$ would be zero. We would then have $R = \infty$. If all the partons were spin-1/2 particles, then we would have

$$R = Q^2/\nu^2, \tag{3.43}$$

as we saw earlier. In the limit $\nu \to \infty$, $Q^2 \to \infty$ we find $R \to 0$. Finally, if the fraction of partons with a spin of 1/2 were $\gamma(x)$, then we would have $2MW_1 = \gamma(x) f(x)$ and

$$R = Q^2/\nu^2\gamma + [1-\gamma(x)]/\gamma(x). \tag{3.44}$$

In the Bjorken limit we find

$$R \to [1-\gamma(x)]/\gamma(x). \tag{3.45}$$

The experimental data yield a value $R \approx 0.14$; ignoring the first term in (3.44), we then find $\gamma(x) \approx 0.9$. In other words, we find an indication that the fraction of scalar partons in a nucleon is small. If $R = 0$, then we have $\gamma(x) = 1$, and there are no scalar partons in a nucleon.

We would like to know the behavior of the function $f(x)$ near $x = 0$ and $x = 1$ (the two extreme points). Experimental data indicate that the function $\nu W_2(\nu,q^2)$ behaves as a constant in the limit $x \to 0$, or we have

$$\lim_{x\to 0} f(x) \approx \frac{\text{const}}{x}; \tag{3.46}$$

i.e., the number of partons with small values of x is large. The behavior of the function $f(x)$ in the limit $x \to 1$ will be examined in Sec. 7 of Chap. 3. At this point we simply note that theoretical considerations yield

$$\lim_{x\to 1} f(x) \sim (1-x)^m, \tag{3.47}$$

where m is an integer.

3.1. Quarks in the role of partons

It might be suggested that quarks play the role of partons. Such a model allows the conversion of an individual quark into hadrons; such a conversion is possible, despite the fractional charge of quarks, since it is assumed that the quarks within a nucleon are not free and can undergo a redistribution in such a way that electric charge is conserved.

We will construct a quark model of the nucleon from quarks of three types: u (of charge $+2/3$ and strangeness 0), d (of charge $-1/3$ and strangeness 0), and s (of charge $-1/3$ and strangeness 1). The baryon charge of each of these quarks is 1/3.

Six functions are introduced to describe the distribution of quarks in a nucleon: $u(x)$, which is the number of u quarks with momentum x; $d(x)$, which is the number of d quarks with momentum x; $s(x)$, which is the number of s quarks with momentum x; and the functions $\bar{u}(x)$, $\bar{d}(x)$, and $\bar{s}(x)$, which describe the distributions of the antiquarks $\bar{u}$, $\bar{d}$, and $\bar{s}$, respectively, in the momentum x. The antiquarks have the same quantum numbers as the quarks but the opposite sign.

For a proton, with an electric charge of 1, we can write

$$1 = \tfrac{2}{3} \int^1 [u(x) - \bar{u}(x)]\,dx - \tfrac{1}{3} \int_0^1 [d(x) - \bar{d}(x)]\,dx - \tfrac{1}{3} \int_0^1 [s(x) - \bar{s}(x)]\,dx \,. \tag{3.48}$$

The strangeness of a proton is 0. We thus have yet another relation:

$$0 = \int_0^1 [\bar{s}(x) - s(x)]\,dx \,. \tag{3.49}$$

Finally, we can define a third projection of the isospin of the proton as follows:

$$\tfrac{1}{2} = \tfrac{1}{2} \int_0^1 [u(x) - \bar{u}(x)]\,dx - \tfrac{1}{2} \int_0^1 [d(x) - \bar{d}(x)]\,dx \tag{3.50}$$

(this relation is known in the literature as the Adler sum rule).

We thus have three equations for determining the numbers of quarks of species u, d, and s which comprise the nucleon. From (3.48)–(3.50) we find the solutions

$$\begin{gathered} \int_0^1 [u(x) - \bar{u}(x)]\,dx = 2; \quad \int_0^1 [d(x) - \bar{d}(x)]\,dx = 1\,; \\ \int_0^1 [s(x) - \bar{s}(x)]\,dx = 0\,. \end{gathered} \tag{3.51}$$

Equations (3.51) lead us to the conclusion that the proton consists of two u quarks and a single d quark. These three quarks are called "valence" quarks (since all of the quantum numbers of the nucleon are determined by these quarks). It might be suggested that a nucleon also has an infinite number of $u\bar{u}$, $d\bar{d}$, and $s\bar{s}$ quark pairs which do not change the quantum numbers of the proton. The set of these pairs of quarks is called the "sea."

Using the operation of charge conjugation (which reduces to replacing a u quark by a d quark), we can find the quark structure of a neutron: It consists of two d quarks and a single u quark. This is what the quark model of free nucleons looks like.

How are the structure functions $\nu W_2(\nu,q^2)$ and $2MW_1(\nu,q^2)$ related to the quark distribution functions $u(x)$, $d(x)$, and $s(x)$ and the corresponding antiquark functions?

For scattering by proton, the function $f(x)$ [see (3.34)] is determined in the form

$$\frac{\nu}{x}W_2 = 2MW_1 = f^{ep}(x) = \sum_i e_i^2 f_i(x) = \tfrac{4}{9}[u(x)+\bar{u}(x)] + \tfrac{1}{9}[d(x)+\bar{d}(x)] + \tfrac{1}{9}[s(x)+\bar{s}(x)]\,. \tag{3.52}$$

Using the operation of charge conjugation, we find from (3.52) a corresponding function for the neutron:

$$f^{en}(x) = \sum_i e_i^2 f_i(x) = \tfrac{1}{9}[u(x)+\bar{u}(x)] + \tfrac{4}{9}[d(x)+\bar{d}(x)] + \tfrac{1}{9}[s(x)+\bar{s}(x)]\,. \tag{3.53}$$

A very important point is that a profound relationship arises between the scattering of electrons and of muons by nucleons, on the one hand, and the scattering of neutrinos by nucleons, on the other, within the framework of the quark model. Specifically, the expression for the cross section for the scattering of neutrinos (or antineutrinos) by nucleons is (in terms of the variables x, $y=\nu/E$):

$$\frac{d^2\sigma^{\nu,\bar{\nu}}}{dx\,dy} = \frac{G^2ME}{\pi}\left[\left(1-y-\frac{Mxy}{2E}\right)\nu W_2 + \frac{y^2}{2}(2Mx)W_1 \mp y\left(1-\frac{y}{2}\right)x\nu W_3\right], \tag{3.54}$$

where the plus and minus signs correspond to the scattering of neutrinos and antineutrinos, respectively. Expression (3.54) is derived in detail in Chaps. 5 and 6 of Ref. 43.

Comparing (3.54) with (3.16), we see that an additional structure function νW_3 enters the cross section for the scattering of neutrinos (or antineutrinos) by a nucleon. This function arises because of the term with the γ_5 matrix in the weak-interaction Hamiltonian. Here are expressions for the neutrino structure functions in terms of the quark distribution functions for a reaction with a zero strangeness in the final hadronic state ($\Delta S=0$):

$$\begin{aligned}
\nu W_2^{\nu p} &= x f_2^{\nu p} = 2x[d(x)+\bar{u}(x)] = x f_2^{\bar{\nu} n}\,;\\
\nu W_2^{\bar{\nu} p} &= x f_2^{\bar{\nu} p} = 2x[\bar{d}(x)+u(x)] = x f_2^{\nu n}\,;\\
\nu W_3^{\nu p} &= f_3^{\nu p} = 2[\bar{u}(x)-d(x)] = f_3^{\bar{\nu} n}\,;\\
\nu W_3^{\bar{\nu} p} &= f_3^{\bar{\nu} p} = 2[\bar{d}(x)-u(x)] = f_3^{\nu n}\,.
\end{aligned} \tag{3.55}$$

The fraction of reactions with $\Delta S=1$ (the fraction of processes of deep inelastic scattering of neutrinos by nucleons in which the net strangeness of the hadrons in the final state is 1) is small. For $\Delta S=1$, we will therefore not consider the relationship between the functions $W_2^{\nu N}$ and $W_3^{\nu N}$, on the one hand, and the quark functions $u(x)$, $d(x)$, and $s(x)$, on the other. From (3.52), (3.53), and (3.55) we find some relations and sum rules:

$$6(f^{ep}-f^{en})=f_3^{\nu p}-f_3^{\nu n}\,; \tag{3.56}$$

$$f_2^{\nu n}+f_2^{\nu d}\leqslant(18/5)(f^{ep}+f^{en})\,. \tag{3.57}$$

Inequality (3.57) follows from the fact that all the distribution functions of the u, d, s, $\bar{u}$, $\bar{d}$, and $\bar{s}$ quarks are positive (by construction). We can also write several corollaries[93]:

$$1/4\leqslant f^{en}/f^{ep}\leqslant 4 \quad \text{for the } SU(2)\text{-symmetry quark model}\,; \tag{3.58}$$

$$1/4\leqslant f^{en}/f^{ep}\leqslant 3 \quad \text{for the } SU(3)\text{-symmetry quark model}\,; \tag{3.59}$$

$$\int_0^1 dx\, x[\bar{u}(x)+\bar{d}(x)+\bar{s}(x)+u(x)+d(x)+s(x)]\leqslant 1\,. \tag{3.60}$$

The last sum rule expresses conservation of momentum (of the longitudinal component of the momentum). The inequality sign arises when new particles—gluons—are introduced in the model of the nucleon, as we will be discussing below [see the discussion relating to (3.65)]. At a large momentum transfer (at small space-like intervals) a simplifying approximation is sometimes used: The contribution of antiquarks is ignored; i.e., we assume $\bar{u}(x)+\bar{d}(x)+\bar{s}(x)=0$. It then follows from (3.49) that we have $s(x)=0$. In this case, (3.57) becomes an exact equality:

$$f_2^{\nu n}+f_2^{\nu p}=(18/5)(f^{ep}+f^{en})\,. \tag{3.61}$$

It is sometimes written in integral form:

$$\int_0^1[f^{ep}(x)+f^{en}(x)]\,dx\Big/\int_0^1[f_2^{\nu p}(x)+f_2^{\nu n}(x)]\,dx=5/18\,. \tag{3.62}$$

We have presented here by no means all of the possible relations between deep inelastic eN and νN scattering. However, even from the relations which we have presented, we can see that there is an organic relationship.

Returning to the quark model of a nucleon, we note that the model which we have been discussing has a serious shortcoming. To see this we introduce the notation

$$U=\int_0^1 x[u(x+\bar{u}(x)]\,dx;\quad D=\int_0^1 x[d(x)+\bar{d}(x)]\,dx\,;$$
$$S=\int_0^1 x[s(x)+\bar{s}(x)]\,dx\,. \tag{3.63}$$

The quantities U, D, and S represent the fractions of the momentum of the nucleon which correspond to the u, d, and s quarks, respectively [see (3.60)].

From experiments we have the following values which were found as the areas under experimental curves; see (3.41) and (3.42):

$$\int_0^1 x f_p^{ep}(x)dx = 0.17; \quad \int_0^1 x f_n^{en}(x)dx = 0.13 . \tag{3.64}$$

The values of the functions f_p^{ep} and f_n^{en} at the lower and upper limits of the integration are not reached experimentally, so that one uses a reasonable extrapolation of the behavior of $f^{ep}(x)$ and $f^{en}(x)$ near these limiting points. Substituting (3.52), (3.53), and (3.63) into (3.64), we find

$$\begin{aligned} (4/9)U + (1/9)D + (1/9)S &= 0.17 ; \\ (1/9)U + (4/9)D + (1/9)S &= 0.13 ; \\ U + D + S &= 1 . \end{aligned} \tag{3.65}$$

The last equation in (3.65) means conservation of the momentum in a nucleon.

Solving system (3.65), we find the values $U = 0.19$, $D = 0.06$, and $S = 0.75$. These results are puzzling at best, since nucleons have a zero strangeness, while the fraction of the momentum corresponding to the strange quarks is very large. A way out of this difficulty was found by introducing the additional hypothesis of the existence of neutral vector massless partons (gluons) within a nucleon. Their role can be summarized by saying that they not only take on a significant fraction of the momentum corresponding to strange quarks [in this case, an inequality sign should appear in (3.60)] but also mediate an interaction between quarks. The quark model of the nucleon which arose in this manner can be described as follows. A nucleon consists of:

(1) three valence quarks, which are responsible for peripheral interactions and for all the quantum numbers of the nuclei;

(2) a sea of quark–antiquark pairs ($u\bar{u}$, $d\bar{d}$, $s\bar{s}$) [other quarks (c, b) are also incorporated in pairs ($c\bar{c}$, $b\bar{b}$) in the sea of the nucleon];

(3) gluons, which mediate an interaction among all of the quarks in the nucleon.

The interaction of gluons with quarks is described by a non-Abelian gauge-invariant Yang–Mills theory; QCD (Sec. 7 of Chap. 3). This theory is a natural extension of QED which is made through the introduction of the principle of local gauge invariance and the requirement that the strong-interaction theory be invariant under a local group of isotopic transformations.

The theory of gauge fields is now being used in attempts to construct a universal theory which describes strong, electromagnetic, and weak interactions from a common standpoint.

4. Automodality hypothesis (scale invariance)

Figure 3.4 shows a dependence of the function [see (3.12)]

$$W_2(\nu,q^2) + 2\tan^2(\theta/2)W_1(\nu,q^2)$$

from q^2 for three values of the energy: 2, 3, and 3.5 GeV. We see that this function remains essentially constant. This behavior of the structure functions W_1 and W_2 is reminiscent of the behavior of the automodal solutions of several problems in classical hydrodynamics, e.g., the problem of a powerful point explosion,[94] as was pointed some time ago by Bogolyubov. Methods of the theory of automodality and dimensionality are used in problems of this sort. The contents of the principle of automodality or scale invariance[95,96] when applied to deep inelastic scattering or the annihilation of leptons can be summarized as follows: The cross sections for these processes do not depend on fundamental physical scale values such as the masses of elementary particles, elementary lengths, etc.; they depend on only the energy and the momentum transfer.

If we sum over all the final states of the hadrons in the matrix element for deep inelastic scattering, we thus find the cross section for a process of the type $l + N \to l + N'$, where N' is a fictitious particle. The cross section for a process of the type $2 \to 2'$ depends on the variables s and q^2, the square of the invariant mass ($M_{N'}^2$), and also the masses of the lepton and the nucleon:

$$d^2\sigma/d\Omega\, dE' = \alpha^2 F(s,q^2,M_{N'}^2,m_l^2,m_N^2)\,.$$

If s, q^2, $M_{N'}^2 \gg m_l^2$, m_N^2, then

$$d^2\sigma/d\Omega\, dE' \simeq \alpha^2 F(s,q^2,M_{N'}^2)$$

is a good approximation. Now choosing the dimensionalities on the right and left, we can get an idea of the behavior of the cross section as a function of s, q^2, and $M_{N'}^2$ at high energies.

Let us consider some simple examples.

(1) The process

$$e^+ + e^- \to \mu^+ + \mu^- \tag{3.66}$$

is described in the one-photon approximation by a single diagram (Fig. 1.20). The total cross section for this process depends on only the one variable s, as can be seen directly from the diagram shown in Fig. 1.20; it is proportional to α^2. The cross section for the process, σ, is expressed in square centimeters; α is a dimensionless quantity. Working from dimensionality considerations, and ignoring terms of the type m^2/s at high energies, we thus find

$$\sigma = \text{const}\ \alpha^2/s\,. \tag{3.67}$$

The constant in (3.67) can be calculated by perturbation theory; it is $4\pi/3$. Hence

$$\sigma = (4\pi/3)\alpha^2/s\,. \tag{3.68}$$

Higher-order perturbation theories generate logarithmic corrections to (3.68).

(2) The differential cross section for the process $e+\mu\to e+\mu$ depends on the variables s and q^2. The differential cross section $d\sigma/dq^2$ is expressed in centimeters raised to the fourth power. In the approximation of one-photon exchange, the cross section for the process can be written in the general form

$$d\sigma/dq^2 = [\alpha^2/(q^2)^2]\, F(s,q^2)\,, \tag{3.69}$$

from which it follows that $F(s, q^2)$ must be a dimensionless function of its arguments. This case is possible if $F(s, q^2)$ depends on only the ratio of its arguments: $F(q^2/s)$. Indeed, theoretical calculations (in the approximation $s, q^2 \gg m^2$ lead to

$$\frac{d\sigma}{dq^2} = \frac{4\pi\alpha^2}{(q^2)^2}\left[1 - \frac{q^2}{s} + \frac{1}{2}\left(\frac{q^2}{s}\right)^2\right]. \tag{3.70}$$

(3) As was mentioned above, the differential cross section for the process $l+N\to l+X$ depends on three variables: s, q^2, and M_N^2. The expression for the differential cross section can thus be written in the form

$$d\sigma/dq^2 = [\alpha^2/(q^2)^2]\, F(s,q^2,M_N^2)\,. \tag{3.71}$$

We could introduce the variable $\omega = 2M\nu/Q^2$ in place of the variable M_N^2 (ω is a dimensionless quantity). The cross section $d\sigma/dq^2$ in (3.71) is expressed in units of centimeters raised to the fourth power, so that $F(s, q^2, M_N^2)$ must be a dimensionless function. We thus have

$$d\sigma/dq^2 = [\alpha^2/(q^2)^2]\, F(q^2/s,\omega)\,. \tag{3.72}$$

Theoretical calculations of the differential cross section in the one-photon approximation lead to the expression

$$\frac{d\sigma}{dq^2 d\omega} = \frac{4\pi\alpha^2}{(q^2)^2}\left[\left(\frac{1}{\omega} - \frac{q^2}{s}\right)\nu W_2 + \frac{q^2}{s}\, W_1\right]. \tag{3.73}$$

The form of this expression confirms the general form predicted for cross section (3.72) by the automodality principle. Note that in deriving (3.73) we need to carry out a summation over the hadronic states. If this is not done, the individual hadrons in the final state may not satisfy the requirement $s_i \gg m_i$, and the hypothesis of scale invariance in this variable will turn out to be inapplicable. In certain cases we cannot ignore the mass of the particle or its decay width. In the process $e^+ + e^- \to \rho^0$, for example, which is described by the Breit-Wigner formula, we can ignore neither the decay width of the ρ^0 meson (otherwise the cross section for the process would vanish) nor its mass (otherwise we would not find a resonance behavior of the cross section).

There is another formulation of the scale invariance principle. Under the scale transformations

$$q\to\lambda q,\ p\to\lambda p \tag{3.74}$$

the structure functions $W_1(\nu, q^2)$ and $W_2(\nu, q^2)$ must satisfy the relations

$$W_1(\lambda^2 q^2, \lambda^2 pq) = W_1(q^2,pq)\,; \tag{3.75}$$

$$W_2(\lambda^2 q^2, \lambda^2 pq) = W_2(q^2,pq)\,. \tag{3.76}$$

Relations (3.75) and (3.76) can evidently be satisfied for arbitrary q and p only if the functions W_1 and W_2 depend on the ratio q^2/pq, i.e., on the scaling variable $x = Q^2/2M\nu$.

Would it be possible to construct a scaling of the invariant functions $W_i(\nu, q^2)$ in an axiomatic quantum field theory? An affirmative answer to this question was given in the familiar paper by Bogolyubov *et al.*[97] On the basis of general principles of a local quantum field theory, a relationship was derived between singularity of the current commutator near the light cone and the asymptotic behavior of invariant functions.

5. High-energy physics on the light cone: Determination of the coefficients in Wilson's expansion near the light cone

The Bjorken limit ($q^2 \to \infty$, $\nu \to \infty$, $\omega = 2M\nu/q^2 = \text{const}$) for the amplitude for deep inelastic scattering means that the amplitude is considered in the region of small space-time intervals $x^2 \approx 0$, i.e., near the light cone.

This point can be demonstrated in the following way. The tensor $W_{\mu\nu}$ is expressed in terms of a product of currents [see (3.9) and (3.10)]. The functions W_1 and W_2 are thus also expressed in terms of a product of currents. For simplicity we restrict the discussion to the case of scalar currents. We consider an arbitrary function

$$W(q,...) = \int d^4x \exp(iqx)\langle f|A(x),B(-x)|i\rangle\,, \tag{3.77}$$

where $A(x)$ and $B(-x)$ are scalar currents. We introduce the variable $\nu_i = qp_i$, $i = 1, 2, 3$, where p_i are 4-momenta. We assume that at least one of the momenta satisfies $p_i \neq 0$ and that the nonvanishing momenta are linearly independent. We define the energy variables $s_{ij} = p_i p_j$. In terms of these new variables, the Bjorken limit, in which we are interested, means that we have

$$q^2 \to \infty, \quad \text{while} \quad q^2/\nu \quad \text{and} \quad q^2/s_{ij} \tag{3.78}$$

are fixed.

We can show that in limit (3.78) the most important part of the integration range in (3.77) is that near the light cone, i.e., the region $x^2 \approx 0$. We make use of the identity

$$W(q,...) \equiv |J(q^2,\nu_1,\nu_2,\nu_3)|^{-1} \times \int d^4k\, \theta(k_0)\delta(q^2 - k^2)\prod_i \delta(\nu_i - kp_i)W(k_1,...)\,, \tag{3.79}$$

in which the Jacobian $J(q^2, \nu_1, \nu_2, \nu_3)$ arises in the transformation from the integration variables q to the integration variables k:

$$J(q^2,\nu_1,\nu_2,\nu_3) = \partial(k^2,kp_1,kp_2,kp_3)/\partial(k_0,k_1,k_2,k_3)\,. \tag{3.80}$$

Substituting (3.77) into (3.79), we find

$$W(q,...) = |J|^{-1} \int d^4x \langle f|A(x)B(-x)|i\rangle$$
$$\times \int d^4k\, \theta(k_0)\delta(k^2 - q^2) \prod_i \delta(\nu_i - kp_i)\exp(ikx)$$
$$= \int d^4x\, D(q^2,\nu_1,\nu_2,\nu_3;x)\langle f|A(x)B(-x)|i\rangle , \tag{3.81}$$

where

$$D(q_2,\nu_1,\nu_2,\nu_3;x) = |J|^{-1} \int d^4k\, \theta(k_0)\delta(k^2 - q^2)\exp(ikx)$$
$$\times \int \prod_i \frac{da_i}{2\pi} \exp[i\, \Sigma a_i(\nu_i - kp_i)] . \tag{3.82}$$

In writing (3.82) we made use of the definition

$$\prod_i \delta(\nu_i - kp_i) = \int \prod \frac{da_i}{2\pi} \exp[i\, \Sigma a_i(\nu_i - kp_i)] .$$

The positive-frequency part of the Pauli-Jordan function is[33]

$$D+(x) = \frac{1}{2\pi i} \int \exp(ikx)\theta(k_0)\delta(k^2 - m^2)d^4k .$$

Using it, we can rewrite expression (3.82) as

$$D(q^2,\nu_1,\nu_2,\nu_3;x) = |J|^{-1} \int \left(\prod_i \frac{da_i}{2\pi}\right)\exp(i\, \Sigma a_i\nu_i)iD^+[(x - \Sigma a_ip_i)^2;q^2] , \tag{3.83}$$

where

$$iD+[(a - \Sigma a_ip_i)^2;q^2] = \int d^4k\, \theta(k_0)\delta(k^2 - q^2)\exp[ik(x - \Sigma a_ip_i)] .$$

Here the role of a mass is being played by the square momentum q^2. It follows from the definition of the D function in (3.83) that this function is concentrated in a region bounded by the inequalities

$$a_i \lesssim 1/\nu_i \quad \text{and} \quad |(x - \Sigma a_ip_i)^2| \lesssim 1/q^2 . \tag{3.84}$$

In the limit in which we are interested, (3.78), we have $a_i \to 0$, and when we square the second inequality, we need consider only the first power of a_i. In the limit (3.78) we thus find from (3.84)

$$a_i \lesssim 1/\nu_i, \quad |x^2 - 2(\Sigma a_ip_i)x| \lesssim 1/q^2 .$$

Since Σaip_ix is a bounded quantity, we have

$$|x^2| < \max(1/q^2, 1/\nu_i) ,$$

as was asserted at the beginning of this subsection.

As was mentioned earlier, the tensor $W_{\mu\nu}$ is expressed in terms of a product of currents. It turns out that in the region $x^2 \approx 0$ the product of two currents can be written in a relatively simple form. For theories which are invariant under a change of scaling variable, the product of operators $A(x)B(x)$ can be expanded in a series near the light cone in such a way that the basic singularities of this product are determined by rules of simple dimensionality.[98] A theory which is invariant under the change of scaling variable means that a scale transformation $x \to Sx$ is represented by a unitary transformation in a Hilbert space, so that each local field operator $\mathscr{P}(x)$ transforms in accordance with the rule

$$U^{-1}(s)\mathscr{P}(x)U(s) = s^{d(\mathscr{P})}\mathscr{P}(sx)\,,$$

where $d(\mathscr{P})$ is the "dimensionality" of the operator $\mathscr{P}(x)$.

Two forms of the expansion of a product of operators at small distances are encountered (this expansion is frequently called "Wilson's expansion"):

$$P(A(x)B(0)) \underset{x_\mu \to 0}{\sim} \sum_i C_i(x)E_i(0) \tag{3.85}$$

and

$$P(A(x)B(0)) \underset{x^2 \to 0}{\sim} \sum_i \tilde{C}_i(x^2)\tilde{E}_i(x,0)\,, \tag{3.86}$$

where P means some action which is performed on the product of the operators $A(x)$ and $B(0)$ (for example, P might represent a commutation relation); $A(x)$, $B(0)$, and $E_i(0)$ are local operators; $\tilde{E}_i(x,0)$ are bilocal operators which are not singular in the limit $x \to 0$; and the functions $C_i(x)$ and $\tilde{C}(x^2)$, the "Wilson coefficients," are C-number singular functions which are generally not the same.

The main point in expansions (3.85) and (3.86) is that all the singularities of the currents $A(x)B(0)$ on the light cone are specified by the C-number functions: the Wilson coefficients.

Expansion (3.86) exists in the theory of free and super-renormalizable fields. This expansion is also written in the theory of renormalizable interacting fields, but in a more complicated form. Let us attempt to reproduce expansions (3.85) and (3.86) in a theory of free scalar fields. We define the current operator $J(x)$ by

$$J(x) = :\Phi^2(x): = :\Phi(x)\Phi(x): = :\Phi^+\Phi^+ + 2\Phi^+\Phi^- + \Phi^-\Phi^-:\,,$$

where Φ^+, Φ^- are positive- and negative-frequency parts of the scalar fields, and the colons mean that the operator is written in normal form.[33]

We understand the operators $A(x)$ and $B(0)$ in (3.85) and (3.86) to mean the current operator $J(x)$. In the expression

$$T(A(x)B(0)) \equiv T(J(x)J(0))$$

we thus find terms which are not of normal form. If we introduce a normal ordering in this expression, causal functions $D^c(x, m^2)$ appear upon permutations of Φ^+ and Φ^-, and we finally find

$$T(J(x)J(0)) = -2[D^c(x,m^2)]^2 + 4iD^c(x,m^2):\Phi(x)\Phi(0): + :\Phi^2(x)\Phi^2(0):\,.$$

The normal-product sign means that the bilocal operators do not have singularities in the limit $x \to 0$.

Retaining only the leading singularities in the functions $D^c(x, m^2)$,

$$D^c(x,m^2) \underset{x_2 \to 0}{\sim} \frac{1}{4\pi^2} \frac{1}{-x^2 + i\alpha} = \frac{1}{4\pi} \delta(x^2) - \frac{i}{4\pi^2 x^2},$$

we find an expansion of the T product of the currents $J(x)$ and $J(0)$ in the form

$$T\{J(x)J(0)\} \underset{x_2 \to 0}{\approx} -\frac{I}{8\pi^4(x^2 - i\alpha)^2} - i\frac{:\Phi(x)\Phi(0):}{\pi^2(x^2 - i\alpha)} + :\Phi^2(x)\Phi^2(0):,$$

which is of the same form as (3.86). The bilocal operator $:\Phi(x)\Phi(0):$ can be replaced by a local operator if we expand it around the point $x = 0$:

$$:\Phi(x)\Phi(0): = :\Phi(0)\Phi(0): + x_\mu :\Phi(0)\partial_\mu \Phi(0): + \cdots .$$

In this case we find an expansion in the form of (3.85):

$$T\{J(x)J(0)\} \approx -\frac{I}{8\pi^4(x^2 - i\alpha)^2} - i\frac{J(0)}{\pi^2(x^2 - i\alpha)} - i\frac{x_\mu J^\mu(0)}{\pi^2(x^2 - i\alpha)} + \cdots,$$

where $J^\mu(0) = :\Phi(0)\partial_\mu \Phi(0):$. The singularities of the C-number functions $C_i(x)$ and $C_i(x^2)$ are determined here by the rules of simple dimensionality. For example, the coefficients $C_i(x^2)$ are functions with singularities of the type

$$(1/x^2)^{\frac{1}{2}(d_A + d_B - d_E)}, \tag{3.87}$$

where d_A, d_B, and d_E are real numbers—the "dimensionalities" of the operators A, B, and E, expressed in mass units.

In a renormalizable perturbation theory, Wilson's expansion of the product of two or more local operators has singularities, which can be counted by rule (3.87), in which the values of d_A, d_B, and d_E are the same as the physical dimensionality expressed in mass units ("with a naive dimensionality," as it is sometimes said). For a scalar field we have $d = 1$; for a spinor field we have $d = 3/2$ and $d(\partial_\mu) = 1$; for a current we have $d = 3$; etc.

In Thirring's model, however, the dimensionality of the basic field is a function of the coupling constant. Such a dimensionality has been called "dynamic," in contrast with a "canonical" dimensionality, which does not depend on the coupling constant (and which is the same as the "naive" dimensionality).

The presence of two dimensionality hypotheses leads to singularities of different types, and it is not possible to theoretically resolve which of these hypotheses is correct. However, they do have different experimental consequences, and the experimental data are described better by the hypothesis of a canonical dimensionality.

When applied to the description of deep inelastic scattering, the expansion of a product of two currents can be utilized to derive several consequences. We will discuss one of them here: the behavior of the structure functions $W_i(\nu, q^2)$ as $Q^2 \to \infty$. When we square matrix element (3.9), we find the tensor $W_{\mu\nu}$ which appears in (3.10):

$$W_{\mu\nu} \sim \sum_{\sigma,\sigma'} \int \langle p'\sigma' | [J_\mu(x), J_\nu(0)] | p\sigma \rangle \exp(iqx) dx . \tag{3.88}$$

Tensor (3.88) is related to the imaginary part of the amplitude $T_{\mu\nu}$ for the Compton scattering of a virtual photon, $\gamma^* N \to \gamma^* N$, by

$$W_{\mu\nu} = (1/2\pi) \mathrm{Im} T_{\mu\nu} . \tag{3.89}$$

This relation is illustrated in Fig. 3.7. The amplitude $T_{\mu\nu}$ can be written in the following form by means of Wilson's expansion:

$$\begin{aligned} T_{\mu\nu}(\nu,Q^2) &= \int d^4z \exp(iqz) \langle p | i\Theta(z_0) [J_\mu^+(z), J_\nu(0)] | p \rangle \\ &= \sum_{i,n;z^2\to 0} \int d^4z \exp(iqz) C_i^n(z^2) z_{\alpha_1} z_{\alpha_2} \cdots z_{\mu_n} \langle p | O_{i\mu\nu\alpha_1\alpha_2\cdots\alpha_n}(0) | p \rangle \\ &= \sum_{i,n} \left(\frac{Q^2}{2} \right)^{-n} q_{\alpha_1} \cdots q_{\alpha_n} C_n^i(Q^2) \langle p | O_{i\mu\nu\alpha_1\cdots\alpha_n}(0) | p \rangle . \end{aligned} \tag{3.90}$$

The matrix element in (3.90) can be written in the general form

$$\langle p | O_{i\mu\nu\alpha_1\cdots\alpha_n} | p \rangle = A_n^i (p_\mu \cdots p_{\alpha_n} - g_{\mu\nu} M^2 p_{\alpha_1} \cdots + \cdots) , \tag{3.91}$$

where the terms proportional to $g_{\mu\nu}$ are a symmetric tensor of rank n, found by permuting the indices, and the coefficient A_i^n carries information on the properties of the target. The terms proportional to $g_{\mu\nu}$ arise when we use traces; their contribution is proportional to $(M^2/Q^2)^k$, where M is the target mass and k is a positive number. At large values of Q^2 the contribution of these terms is thus suppressed in comparison with that of the first term in (3.90). A contribution proportional to M^2/Q^2 could also result from higher-order twist terms with $\tau > 2$ (a definition of the twist terms will be given below). Twist effects cannot yet be calculated in QCD; they are usually incorporated in a phenomenological way.

Wilson's expansion for amplitude (3.90) with (3.91) can be written (for simplicity, we are omitting indices)

$$T(\nu,Q^2) = \sum_{i,n} C_n^i(Q^2) x^{-n} A_n^i , \quad x = \frac{Q^2}{2M\nu} . \tag{3.92}$$

Making use of the optical theorem [(3.89)], recalling that $W_{\mu\nu}$ is expressed in terms of the structure functions W_1 and W_2, and integrating (3.92) over the variable x in the interval $0 < x < 1$, we find a relationship between the structure functions for deep inelastic scattering, $W_i(x, Q^2)$, and the coefficients in Wilson's expansion. This relationship is usually written in the form of moments $M_n(Q^2)$:

$$M_n(Q^2) = \int_0^1 dx\, x^{n-2} F(x,Q^2) = \sum_i C_n^i(Q^2) A_n^i , \tag{3.93}$$

where $F = xMW_1$ or $F = W_2(x,Q^2)$ or $F = \nu W_3(x,Q^2)$.

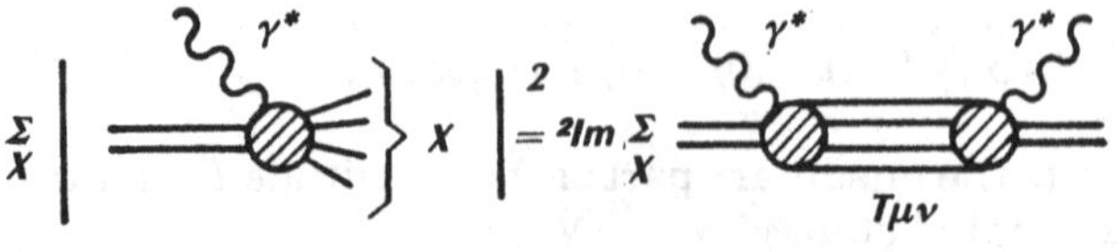

Figure 3.7. Graphic representation of relation (3.89) (optical theorem).

For interacting fields with a spin, the singularities in the Wilson's coefficients $C_n(z^2)$ in expression (3.87) are determined by an expression similar to (3.87):

$$(1/x^2)^{\frac{1}{2}[d_A + d_B - (d_E - x)]} .$$

We see that the terms which are the most singular arise for the smallest values of the twist τ, defined by $\tau = d_E - x$, where $\varkappa$ is the spin of the field which forms the current (the twist is equal to the difference between the dimensionality and the spin).

Simple nontrivial examples of local operators $O_{\mu\nu\alpha_1\alpha_2 \dots \alpha_n}(0)$ which arise in expansions of the type in (3.90) have the following twist values:

$$\tau = 1\text{:scalar field } \Phi, \partial_\mu \Phi, \partial_\mu \partial_\mu \Phi \dots ;$$
$$\tau = 2\text{:} \bar{\psi}\gamma_\mu \psi, \bar{\psi}\gamma_\mu d_\nu \psi, \Phi d_\mu \Phi \dots .$$

The operators with a twist $\tau = 1$ contribute vanishing diagonal matrix elements. The dominant contributions come from the operators with $\tau = 2$. The operators used in (3.90) correspond to a current $\bar{\psi}\gamma_\mu \psi$ and have smallest values $\tau = 2$. All those operators $O_{i\mu\nu\alpha_1 \dots \alpha_n}$ from (3.90) which have the smallest twist value ($\tau = 2$) contribute to relation (3.93). The dimensionality $\tau = 2$ is equivalent to the circumstance that the coefficients $C_n(Q^2)$ of the corresponding operators have δ-function singularities; i.e., $C_n(x^2) \sim \delta(x^2)\varepsilon(x_0)$.

As follows from (3.93), the behavior of the moments in the limit $Q^2 \to \infty$ is determined by the behavior of the coefficients $C_n(Q^2)$, which in turn depends on the behavior of the currents near the light cone. Ultimately, the functional dependence $M_n(Q^2)$ in the limit $Q^2 \to \infty$ is determined by the nature of the theory. Exact Bjorken scaling is found if we have $C_n(Q^2) \to \text{const.}$ as $Q^2 \to \infty$.

In QCD, the Q^2 dependence of the moments is found from the solution of the renormalization-group equation for the coefficients $C_n^i(Q^2)$.

In non-Abelian gauge theories (including QCD), a dependence of the coefficients $C_n^i(Q^2)$ on the coupling constant (g) of the interacting fields arises. The invariant charge $\bar{g}^2$ found by the renormalization-group technique in the approximation of the lowest (second) order of the QCD perturbation theory [see (3.126)] depends on Q^2 and appears in the expression for the coefficients $C_n(Q^2, g)$ in the form

$$C_n(Q^2, g) = C_n(1, \bar{g}) \exp\left\{ - \int_0^\infty \gamma_n(\bar{g}, y) \partial y \right\}, \tag{3.94}$$

where $\gamma_n(\bar{g},y)$ is the anomalous dimensionality (not determined by the physical dimensionality, mentioned earlier) of the operator $O_{i\mu\nu\mu\ldots\mu_n}$ [see Sec. 7 of Chap. 3 for a definition of the anomalous dimensionality $\gamma_n(\bar{g},y)$].

For asymptotic-freedom theories (including QCD) we have $\bar{g}\to 0$ in the limits $Q^2\to\infty$ and $\gamma_n(\bar{g},y)\to 0$ at the upper limit of the integral in (3.94). However, $\gamma_n(\bar{g},y)$ does not vanish rapidly enough to cause the integral in the exponential function in (3.94) to vanish; as a result, logarithmic corrections to the coefficients $C_n(Q^2)$ appear in the limit $Q^2\to\infty$:

$$C_n(Q^2) \underset{Q^2\to\infty}{\to} C_n\left(\frac{1}{\ln Q^2}\right)^{\gamma_n b}, \tag{3.95}$$

where b is a constant known from the theory. Substituting (3.95) into (3.93), we find the standard representation for the evolution of the moments in Q^2:

$$M_n(Q^2) = \sum_i C_n^i(Q_0^2)A_n^i[\ln Q^2]^{-\gamma_n b} \to M_n(Q_0^2)[\ln Q^2]^{-\gamma_n b}, \tag{3.96}$$

where Q_0^2 is a fixed value. In QCD the scaling is thus violated in accordance with a logarithmic law. The violation of scaling is discussed in more detail in Secs. 6–8 of Chap. 3.

6. Experimental evidence for a violation of scaling

The scaling hypothesis, which appeared correct as long as the error in the experimental data was about 10%, did not hold up under a more thorough experimental test. Some more-accurate SLAC experiments on the scattering of electrons by nucleons showed that a scaling in the variable $\omega = 2M\nu/Q^2$ is not correct. If there is a deviation from scaling in ω, the implication may be that (a) either the value of R was not determined accurately and is perhaps not constant, or (b) the scaling variable ω has a different form (if a scaling nevertheless does exist).

In particular, various scaling variables have been proposed in various studies:

$$\omega_a = \frac{2M\nu + a}{Q^2}; \quad \omega_a = \frac{s}{Q^2}; \quad \omega_r = \frac{2M\nu + M^2}{Q^2 + \Lambda^2}. \tag{3.97}$$

A variable which is used fairly often in theoretical work is

$$1/\xi = 4\nu/[Q^2 + \sqrt{Q^4 + 4M^2Q^2}]. \tag{3.98}$$

The variable ω_a has perhaps been the most successful. In some early papers the disruption of scaling was explained by introducing this variable with a parameter value $a = M^2$. It was suggested that the variable ω_a be used to continue the scaling behavior of the functions νW_2 and $2MW_1$ into the low-energy region: from $W = 2.6$ GeV to $W = 1.8$ GeV.

Some new experimental indications[99] of a possible violation of scaling soon appeared. For example, it follows from the quark models describing scaling in deep inelastic scattering that the cross-section ratio

$$R = \sigma(e^+e^- \to \text{hadrons})/\sigma(e^+e^- \to \mu^+\mu^-) \tag{3.99}$$

should be a constant (this quantity should not be confused with $R = \sigma_L/\sigma_t$).

The experimental data disagree with the theoretical predictions and indicate an increase in R with increasing energy of the e^+e^- pair; this is a consequence of a violation of scaling. Indications of a violation of scaling also emerged from experiments on the scattering of μ mesons.[99]

A further experimental test of the scaling hypothesis thus required improving the accuracy of the experimental data and also increasing the energy of the impinging leptons and the momentum transfer.

New data on deep inelastic scattering of electrons by protons and deuterons were obtained[100] in 1974–1975 over broad regions of angle (6–60°) and energy, up to the maximum energies obtainable at electron accelerators (4.5–19.5 GeV). Figure 3.8 shows 75 points which were used to determine the functions νW_2 and $2MW_1$, and the quantity $R(Q^2, W^2) = \sigma_L/\sigma_t$ for the deep inelastic scattering of electrons by protons. The kinematics ($W^2 = 2M\nu + M^2 + Q^2$) was such that the data spanned a broad range of the quantity $x = Q^2/2M\nu$ ($0.1 < x < 0.8$). The values of $\sigma_L(Q^2, W^2)$, $\sigma_t(Q^2, W^2)$, and $R = \sigma_L/\sigma_t$ were measured [see (3.13) and (3.14)]. Figure 3.9 shows the results of the measurements of R. The lines with the long dashes show results of a fit at various values of x under the condition $R = \text{const}$. The best value turns out to be $R = 0.14 \pm 0.011$. The lines with short dashes show a fit of R by the function

$$R = \alpha^2(x)/\ln(Q^2/\beta^2), \tag{3.100}$$

where $\alpha^2(x)$ is some function of x, and β^2 is a parameter. The solid lines are fits of the experimental data by the function

$$R = C(x)Q^2/(Q^2 + d^2)^2.$$

Equations (3.14) have been used to distinguish the functions νW_2 and $2MW_1$ for various values of x (Figs. 3.10 and 3.11). The experimental data indicate that in the scaling variable $x = Q^2/2M\nu$, neither νW_2 nor $2MW_1$ is a constant. An attempt to introduce a new scaling variable of the type in (3.97),

$$\omega_a = \omega + a/Q^2, \tag{3.101}$$

was also unsuccessful. It was suggested that scaling was violated. For the function $\nu W_2(Q^2, W^2)$ the following behavior was chosen:

$$\nu W_2(Q^2, W^2) = F_2(\omega_a)(1 + b_2 Q^2), \tag{3.102}$$

where b_2 is a parameter, and $F_2(\omega_a)$ is an adjustable function of the form

$$F_2(\omega_a) = \sum_{i=3}^{7} \alpha_i (1 - x_a)^i, \quad x_a = \frac{1}{\omega_a}. \tag{3.103}$$

It was found that with

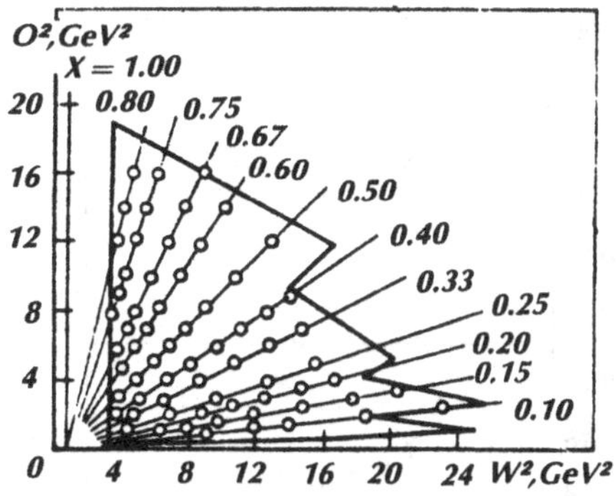

Figure 3.8. Kinematic region in the (W^2, Q^2) plane and points used in finding the structure functions νW^2 and $2MW_1$ in the reactions $e + p(d) \to e + X$.

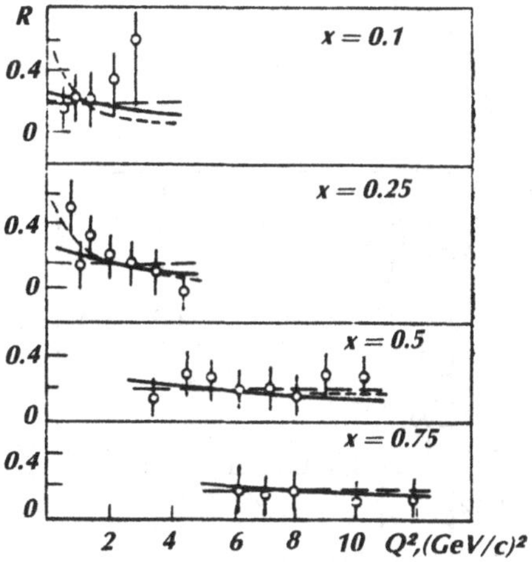

Figure 3.9. Values of R (for protons) for the points marked in Fig. 3.8. (Dashed line) best fit of data ($R = \text{const}$); (solid line) the function $R = C(x)Q^2/(Q^2 + d^2)^2$; (points) the curve $R = \alpha^2(x)/\ln(Q^2/\beta^2)$, where α and β are constants. Here $C(x)$ and $\alpha^2(x)$ are arbitrary functions.

$$b_2 = -(0.011 \pm 0.001)\text{GeV}^{-2} \tag{3.104}$$

it was possible to generate a relatively satisfactory description of the experimental data (except at very small values $x < 0.03$).

A corresponding analysis was carried out in a study of the scaling of the function $W_1(Q^2, W^2)$ which was written in the form

$$W_1(Q^2, W^2) = F_1(\omega_a)(1 + b_1 Q^2)\,; \tag{3.105}$$

$$F_1(\omega_a) = \sum_{i=3}^{7} \beta_i (1 - x_a)^i\,. \tag{3.106}$$

This analysis yielded

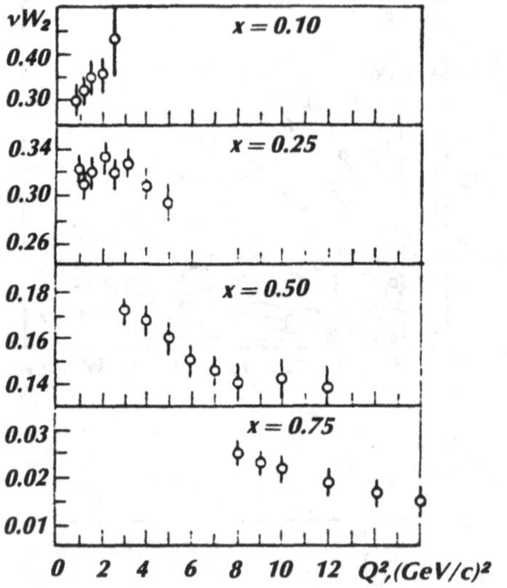

Figure 3.10. Structure function νW_2 for various values of x. The function νW_2 turns out not to be a constant; there is accordingly a deviation from the scaling.

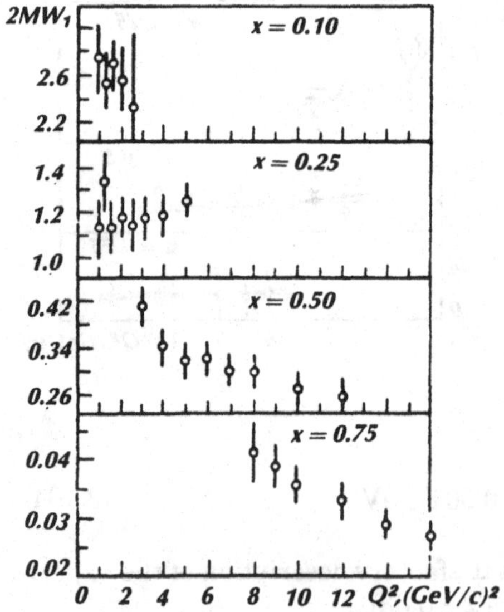

Figure 3.11. Scaling function $2MW_1$ for various values of x. For this function, as for νW_2, there is a deviation from scaling.

$$b_1 = -(0.009 \pm 0.004)\mathrm{GeV}^{-2}, \tag{3.107}$$

in good agreement with the value in (3.104).

This analysis thus indicated that scaling was violated slightly for the functions νW_2 and W_1 (in the variables $1/x_a = \omega_a = \omega + M^2/Q^2$). The functions νW_2 and W_1 decrease by about 1% as Q^2 is changed by 1 $(\mathrm{GeV}/c)^2$.

There exists the variable

$$\omega_s = 1/x_s = \omega + 1.5/Q^2, \tag{3.108}$$

whose introduction in place of ω and ω_a results in a scaling of both ωW_2 and W_1 in the region $\varepsilon \leqslant 0.15$, $W^2 \geqslant 4$ GeV^2 with $R = 0.14$. However, whether this variable must be introduced requires an experimental test through an expansion of the range of applicability in ε and W^2.

The choice of the function $F^2(\omega_a)$ in the form in (3.103) is justified in the following way. If it is assumed that the structure function νW_2 for inelastic *ep* scattering has a power-law behavior near the threshold, $\omega = 2M\nu/Q^2 \to 1$, i.e.,

$$\nu W_2 \sim (1 - x)^p, \tag{3.109}$$

then it turns out that the form factor of the proton, F_1, for elastic *ep* scattering has the following behavior in the limit $Q^2 \to \infty$:

$$F_1(Q^2) \sim (1/Q^2)^{(p+1)/2}. \tag{3.110}$$

Relations (3.109) and (3.110) are called the "Drell–Yan–West relations."[101] Experimental data indicate $F_1(Q^2) \sim 1/Q^4$, i.e., $p + 1 = 4$. We thus have $p = 3$ in (3.109), so that the series in (3.103) should begin at $i = 3$.

In the derivation of (3.110) it was assumed that the form factor of the proton is dominated by the diagrams in Fig. 3.12 and that the coupling of the π-meson field with the nucleon is a pseudoscalar coupling. It can also be shown that the ratio G_M/G_E should fall off with increasing Q^2:

$$\frac{G_M(Q^2)}{G_E(Q^2)} \equiv \frac{F_1 + kF_2}{F_1 + (q^2/4M^2)kF_2} \sim \frac{1}{Q^2}; \tag{3.111}$$

this result also indicates a violation of the scaling of the form factors for elastic *ep* scattering.

A possibility which has not been ruled out is that the proton form factors fall off more rapidly than $1/Q^4$ with increasing Q^2. In this case the relation $p + 1 > 4$ should hold, and the series in (3.103) can begin with an exponent higher than $i = 3$.

In general, the exponent in (3.109) is equal to $p + \delta$, where δ is an anomalous dimensionality, and the Drell-Yan-West relation may be violated.

Figure 3.13 compares experimental data on $2MW_1(Q^2)$ with results calculated from (3.105) and (3.106). We see that the behavior $2MW_1 \sim (1 - x_s)^3$ is explicitly at odds with the experimental data and that the function W_1 is apparently closer to $(1 - x_s)^4$ in behavior. However, it would be premature to assume that the function $(1 - x_s)^3$ does not describe the experimental data correctly. A new experimental test of (3.102)–(3.106) should be carried out.

Important to reaching an understanding of the violation of scaling are experiments on the scattering of electrons by deuterons carried out under the same kinematic conditions as for the scattering of electrons by protons (data on the functions νW_2^d and $2MW_1^d$ were obtained for the same points as are shown in Fig. 3.8).

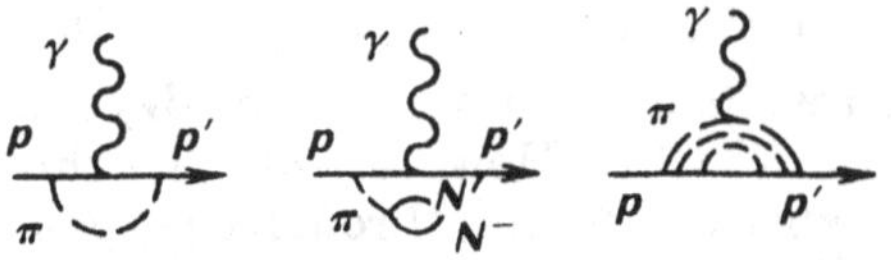

Figure 3.12. Types of diagrams considered in the derivation of expression (3.110).

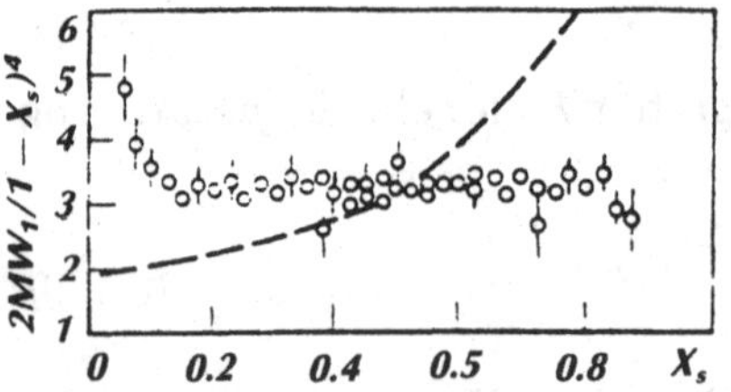

Figure 3.13. Comparison of experimental and theoretical data on the function $2MW_1$. The experimental points show that the function $2MW_1$ is described by expressions (3.105) and (3.106) beginning at $i = 4$; the dashed line corresponds to the ratio $2MW_1/(1 - x_s)^3$, with $x_s = 0.5$.

The following values were found for R_d (under the assumption that R_d is constant):

$$R_d = 0.175\begin{cases} \pm 0.09 \text{ (statistical error)} \\ \pm 0.060 \text{ (possible systematic errors)}. \end{cases}$$

Within the errors, this result agrees with $R_p = 0.14$. If we assume $R_d = R_p$, we find $R_p = R_n$. However, the use of (3.102), (3.103), and (3.105) to describe νW_2^d and W_1^d leads to significant deviations from the experimental data. The circumstance can be seen clearly from Table 3.2 (Ref. 100), since the values of b_1 are different for W_1^p and W_1^n. Use of the variable $\omega_s = 1/x_s = \omega + 1.5/Q^2$ also fails to explain the data on the proton and the neutron simultaneously.

The nature of the deviations from scaling for the neutron structure functions is thus different from that for the proton structure functions.

Some very interesting results, but with a less obvious physical interpretation, have emerged from a study of the scattering of μ mesons by iron nuclei.[102] The scaling hypothesis means that if the quantities ν and Q^2 are multiplied by the same number, then the ratio $x = Q^2 M\nu$ will remain unchanged, so that the cross section for inelastic scattering will remain unchanged under a scaling transformation of this sort. We thus introduce the transformation

$$E_0 \to \lambda E; \quad E' \to \lambda E'; \quad \theta \to \theta/\sqrt{\lambda}. \tag{3.112}$$

Here we have $\nu \to \lambda\nu$, $Q^2 \to \lambda Q^2$, and $x = \text{const}$.

An expression for the differential cross section for the inelastic scattering of μ mesons at small angles can be written

$$E\frac{d^2\sigma}{dx\,dy} = \frac{4\pi\alpha^2}{2M}\frac{(\nu W_2)}{x^2y^2}\left[1 - y + \frac{y^2}{2(1+R)}\right], \tag{3.113}$$

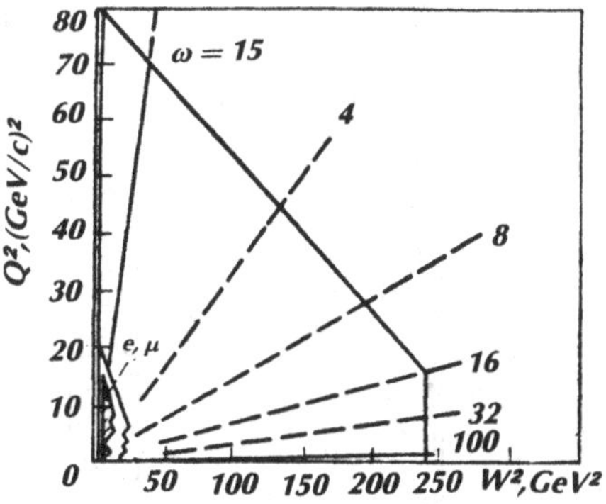

Figure 3.14. Kinematic region in an experiment on the scattering of μ mesons by iron nuclei. (e, μ) region of earlier experiments.

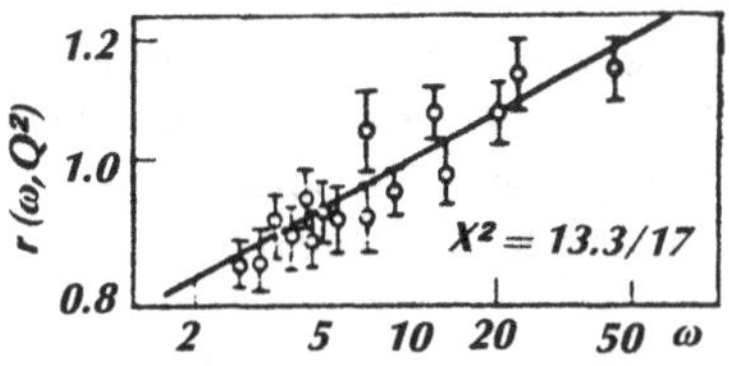

Figure 3.15. Experimental data on the function $r(\omega, Q^2)$. The solid line is the function $r = (\omega/\omega_0)^n$, where $n = 0.096 \pm 0.028$ and $\omega_0 = 6.1^{+8.9}_{-3.6}$.

where $y = \nu/E$. The ratio of cross sections

$$r(\omega, Q^2) = \frac{(E\, d^2\sigma/dx\, dy)_{E=\lambda E_0}}{(E\, d^2\sigma/dx\, dy)_{E=E_0}}$$

should then remain constant. Measurements were carried out at two energies: 150 and 56 GeV ($\lambda = 150/56 = 8/3$).

The kinematic region corresponding to these energies is shown in Fig. 3.14. Figure 3.15 shows $r(\omega, Q^2)$ as a function of ω.

A violation of scaling is thus manifested again in this case, but we easily see that it will be a far more complicated matter to interpret the results. The violation of scaling can be regarded as an established fact only when the nucleus is treated as a point object with a charge $Z = 26$. However, there is the possibility that scaling is not violated and that the deviation from scaling in the case of scattering by the iron nucleus arises because of the complex structure of this nucleus.

The scaling of the functions νW_2 and $2MW_1$ was subsequently tested in experiments on the scattering of μ mesons by protons and deuterium.[102] These experiments were carried out in beams of μ mesons with energies of 100 and 150 GeV. Figure 3.16 shows the kinematic region covered in these experiments. The hatched region has a small statistical base. As usual, radiative corrections were made, and in the case of scattering by deuterium Glauber

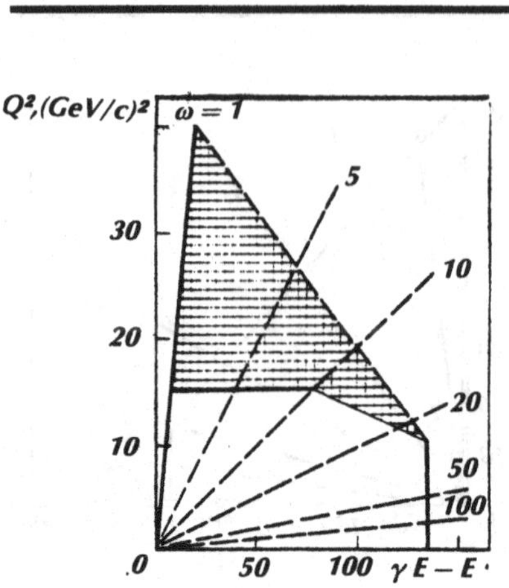

Figure 3.16. Kinematic region in experiments on the scattering of μ mesons by protons and deuterons. The data from the hatched region were not used in the analyses because of the small statistical base.

corrections (nuclear shadowing effects) and corrections for the Fermi motion of the nucleons in the nucleus were also made (the two latter corrections were small in these experiments).

The scattering $\mu + p \to \mu + X$ was analyzed on the basis of the formula[103]

$$\frac{d^2\sigma}{d\nu\, dQ^2} = \frac{\pi}{pp'}\frac{2\alpha^2}{q^4}\left(\frac{p'}{p}\right)\left[\left(2EE' - \frac{Q^2}{2}\right)W_2(\nu,Q^2) + (Q^2 - 2m_\mu^2)W_1(\nu,Q^2)\right], \tag{3.114}$$

$$Q^2 = 2(EE' - pp'\cos\theta - m_\mu^2).$$

With $m_\mu^2 \approx 0$, $E \approx p$, and $E' \approx p'$, we find the usual expression for $d^2\sigma/dE'd\Omega$. Formula (3.114) can also be written [cf. (3.113)]

$$d^2\sigma/d\nu\, dQ^2 = \Gamma(E,E',\theta)[\sigma_t + (\varepsilon + \delta)\sigma_L],$$

$$\delta = \frac{2m_\mu^2(1-\varepsilon)}{Q^2}; \quad \varepsilon = \left[1 + \frac{2(Q^2+\nu)\tan^2(\theta/2)}{Q^2(1 - Q^2_{\min}/Q^2)}\right]^{-1}.$$

This formula was used to analyze the ω dependence of R:

$W^2 = 2M\nu - Q^2 + M^2$, GeV^2	Q^2, (GeV/c)^2	$\langle\omega\rangle$	$R = \sigma_L/\sigma_T$
100 − 144	1 − 2	80	-0.10 ± 0.27
	2 − 6	30	0.02 ± 0.30

It can be seen from these results that there is no indication of an increase in the parameter R with increasing ω. With regard to the deviation of the behavior of the function $\nu W_2(x,Q^2)$ from the scaling for deep inelastic scattering of electrons and μ mesons by protons, we note that it is of the same nature at given values of x (Ref. 103; Fig. 3.17 of the present book).

New pieces of evidence for violations of scaling at high values of Q^2 were

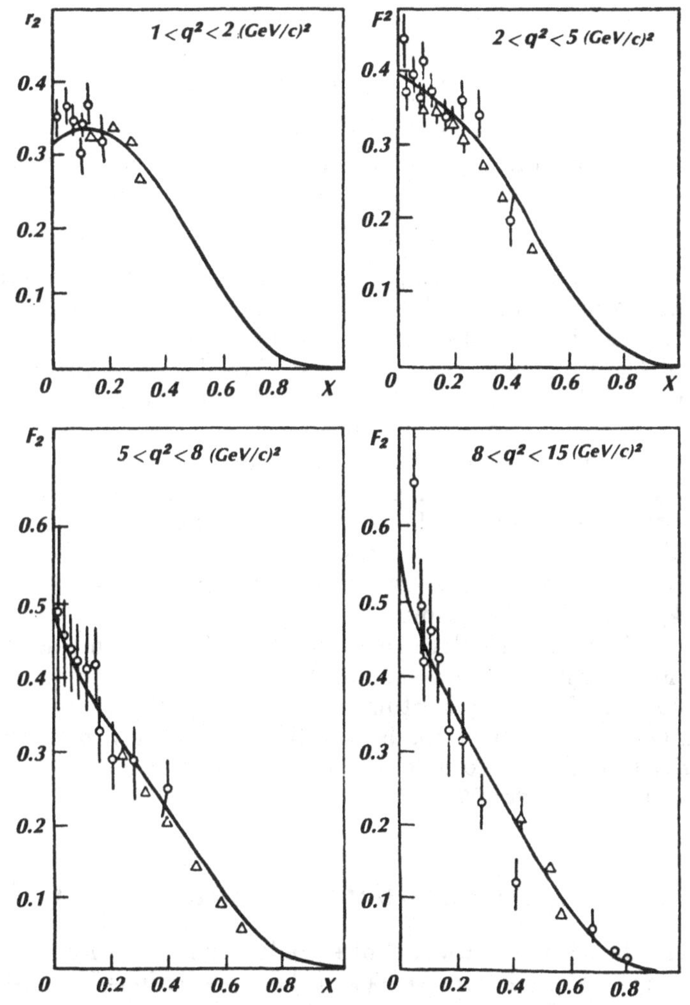

Figure 3.17. $F_2 = \nu W^p_2$ **vs** x **in various** q^2 **intervals for inelastic scattering of (triangles) electrons and (circles)** μ **mesons. The solid line is a fit of the combined experimental data.**

obtained in experiments on the scattering of μ mesons by nucleons and nuclei at the SPS accelerator at CERN (the BCDMS and EMC[1] groups)[104] and at the Fermilab accelerator.[105] In the experiments carried out by the BCDMS group, the maximum energy of the μ beam was 200 GeV, and the maximum value of Q^2 was 200 $(\mathrm{GeV}/c)^2$. The value of x ranged from 0.3 to 0.7; the μ mesons were scattered by carbon nuclei. Analysis of the experimental data revealed that the dependence of the function $W_2(x, Q^2)$ on Q^2 is weak.

An experiment carried out by the EMC group showed that the behavior of the structure function $W_2(x, Q^2)$ as a function of x and Q^2 is the same as the corresponding behavior which had been found previously in experiments by the CDHS group (CERN) on deep inelastic scattering of neutrinos:

$$W_2(x,Q^2)_{\mathrm{EMC}} / W_2(x,Q^2)_{\mathrm{CDHS}} \approx 1, \quad 0.05 \leqslant x \leqslant 0.65 .$$

The data obtained by the EMC group in an experiment on μH_2 scattering agree with data on eH_2 scattering obtained at the SPEAR accelerator (SLAC) within the normalization (the error is about 10%):

$$W_2(x,Q^2)_{\mathrm{EMC}} / W_2(x,Q^2)_{\mathrm{SLAC}} \approx 0.9 .$$

Analyses of experimental data on deep inelastic scattering of electrons, muons, and neutrinos by nucleons and nuclei which have been carried out by various experimental groups have used various values of R: $R = 0.21 \pm 0.10$ (SLAC, 1978) and $R = -0.13 \pm 0.19$ (EMC, μFe scattering, 1981). Different methods have been used to incorporate radiative corrections. The Fermi motion of the nucleons in the nuclei has been taken into account in some cases, but not all. The accelerator beams have been normalized in different ways. Consequently, a combined quantitative analysis of the experimental data obtained in different laboratories is not always effective, and it must allow for all of the necessary reservations. It would seem to be time to develop some worldwide standards regarding the extraction of structure functions $W_i(x, Q^2)$ from experimental data.

Summarizing all the results obtained on deep inelastic *ep*, *ed*, μp, μC, and μFe scattering, we can conclude that scaling is violated over the entire ranges of E and of the 4-momentum transfer Q^2 which have been studied. We should stress that the functions W_1 and W_2 are relatively insensitive to the choice of R. At small values of x, for example, a variation in R from 0.15 to 0.35 changes W_2 by only about 5%.

7. About quantum chromodynamics

The theory offers quite definite predictions regarding the nature of the violation of scaling. Before we look at these predictions, however, we would like to summarize the basic positions of quantum chromodynamics. Quantum chromodynamics (QCD), which describes the interaction of quarks and gluons, is presently the only candidate for the role of a theory of strong interactions. In QCD the interaction constant α_s [see (3.126)] depends on the square of the 4-momentum transfer Q^2 (in contrast with the constant charge e in QED) in such a way that as Q^2 decreases, the value of $\alpha_s(Q^2)$ increases, becoming so large at $Q^2 \lesssim 1\ (\mathrm{GeV}/c)^2$ that perturbation theory cannot be used within the framework of QCD. In the limit $Q \to \infty$ we have $\alpha_s(Q^2) \to 0$, and a regime of "asymptotic freedom" sets in.

In the interaction Lagrangian in QCD it is necessary to allow for the fact that the quark species (or flavors) introduced earlier (u, d, s, c, b) have an additional quantum number: color.

One of the well-known (and earliest) arguments in favor of the introduction of a new quantum number—color—is the construction of the wave function for the Δ^{++} resonance in a state with an angular momentum $J = 3/2$. The quark wave function should have the form

$$|\Delta^{++},\ J_{3/2}(u\uparrow, u\uparrow, u\uparrow)\rangle,$$

and all three of the u quarks must be in the same state in order to provide the spin $J = 3/2$ (an electric charge of 2). In this case, however, the wave function of the Δ^{++} resonance turns out to be a symmetric function of its arguments, contradicting Fermi statistics for quarks. In an effort to eliminate this contradiction it was suggested[106] that a quark of each species can be in three different states or, in today's terminology, three different color states. In this case the wave function of the Δ^{++} resonance can be rendered antisymmetric by constructing it from u quarks taken in different states (colored quarks):

$$|\Delta^{++},\ J_{3/2}\rangle = (1/\sqrt{6})\varepsilon^{ijk}|u_i\uparrow, u_j\uparrow, u_k\uparrow\rangle,$$

where i, j, and k are color indices. In terms of the other variables (the spatial coordinates, the spin, and the quark species), this function is symmetric. The introduction of color[106] gives rise to a new symmetry group: the colored group $SU(3)_c$.[107]

The principle for constructing hadrons from colored quarks can also be applied to mesons. Since colored baryons and colored mesons have yet to be observed experimentally in a free state, a "color prison" confinement rule has been postulated: All physical observables (hadrons, currents, etc.) are colorless.

The pieces of experimental evidence indicating that quarks have color are based on a comparison of results calculated from theoretical equations (in which color is taken into consideration) with experimental data. For example, the ratio of total cross sections [see (3.99)]

$$R = \sigma(e^+e^- \to \text{hadrons})/\sigma(e^+e^- \to \mu^+\mu^-)$$

agrees with experimental data (Chap. 4) if quarks are assumed to be colored entities.

It is thus proposed that each quark is a tricolor entity, so that additional indices appear on the quark fields in the Lagrangian to correspond to the introduction of the $SU(3)_c$ color group in the theory. Along with the colored quarks, the Lagrangian must incorporate colored photons: gluons. In a theory which is constructed systematically and which is invariant under local gauge transformations and the local group of isotopic transformations, vertices describing the interaction of three and four gluons appear in addition to the vertices describing the interaction of quarks with gluons.

The interaction Lagrangian in QCD has the form:

$$L = -(1/4)G^a_{\mu\nu}G^a_{\mu\nu} + i\bar{\psi}_\alpha(\gamma^\mu D^\mu_{\alpha\beta} + i\,m\delta_{\alpha\beta})\psi_\beta, \tag{3.115}$$

where $a = 1, \ldots, 8$; and $\alpha, \beta = 1, 2, 3$. The tensor $G^a_{\mu\nu}$ describes the gluon field:

$$G^a_{\mu\nu} = \partial_\mu G^a_\nu - \partial_\nu G^a_\mu + g f^{abc} G^b_\mu G^c_\nu$$

[g is the gluon-quark interaction constant, and f^{abc} are the structure constants of the $SU(3)_c$ group]. Here ψ_α is a spinor describing a quark of color α,

$D^{\mu}_{\alpha\beta} = \delta_{\alpha\beta}\partial_{\mu} - ig\lambda^{a}_{\alpha\beta}G^{a}_{\mu}$ is a covarient derivative, γ^{μ} are Dirac matrices, m is the mass of the quark, and the matrices λ^{a} obey the commutation relations

$$[\lambda^{a}, \lambda^{b}] = i f^{abc}\lambda^{c}.$$

A complete determination of the Lagrangian requires choosing a gauge. In QCD, the gauge is determined by an auxiliary parameter. In the 't Hooft-Feynman gauge, we need to add to Lagrangian (3.115) some additional terms ("Faddeev-Popov ghosts"), which we will not write out here. It then becomes possible to introduce rules for calculating the QCD diagrams—rules analogous to the Feynman rules in QED:

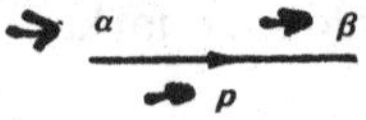

—is a quark propagator,

$$\frac{i\delta_{\alpha\beta}}{\hat{p} - m + i\varepsilon};$$

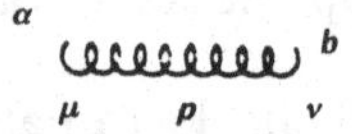

—is a gluon propagator,

$$-i\left[g_{\mu\nu} - (1-\alpha)\frac{p_{\mu}p_{\nu}}{p^2 + i\varepsilon}\right]\frac{\delta_{ab}}{p^2 + i\varepsilon};$$

—is a ghost propagator,

$$\frac{i\delta_{ab}}{p^2 + i\varepsilon};$$

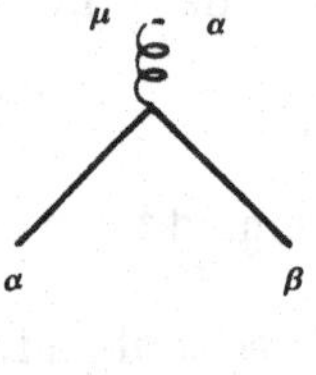

—is a gluon vertex of a quark,

$$ig\gamma_{\mu}\lambda^{a}_{\alpha\beta};$$

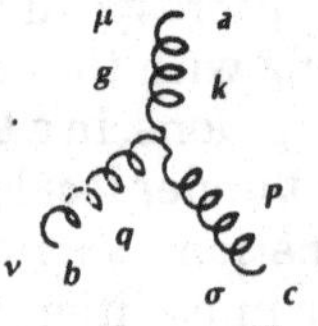

—is a three-gluon vertex,

$$g f_{abc}\left[g_{\mu\nu}(k-q)_{\sigma} + g_{\nu\sigma}(q-p)_{\mu} + g_{\sigma\mu}(p-k)_{\nu}\right];$$

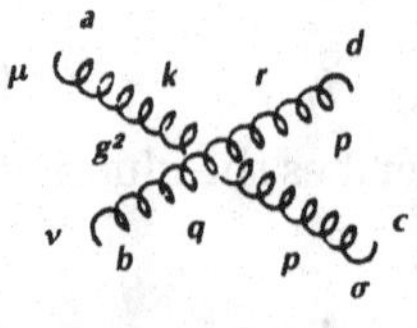

—is a four-gluon vertex ($\sim g^2$),

$$-ig^2\left[f_{abe}f_{cde}(g_{\mu\sigma}g_{\nu\rho} - g_{\mu\rho}g_{\nu\sigma}) + f_{ace}f_{bde}(g_{\mu\nu}g_{\sigma\rho} - g_{\mu\rho}g_{\nu\sigma}) + f_{ade}f_{cbe}(g_{\mu\sigma}g_{\nu\rho} - g_{\mu\nu}g_{\sigma\rho})\right];$$

and

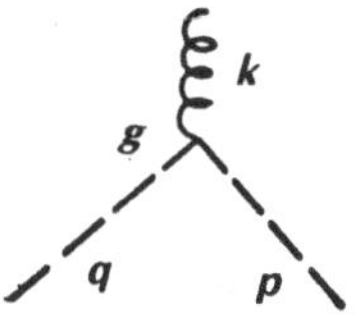

is a gluon vertex of a ghost,

$g f_{abc} p_\mu$.

Although we will not be carrying out perturbation-theory calculations anywhere below, we will make systematic use of diagrams to illustrate the arguments. When we write the quark propagator in the form $\sim(\hat{p} - m + i\varepsilon)^{-1}$, we are saying that the propagator has a pole at the point $\hat{p} = m$; i.e., the quark is on the mass shell. This circumstance might be understood as the existence of a quark in a free state, but we know that quarks do not leave nucleons: A quark confinement ("quark prison") prevails.

Many of the QCD diagrams, as in QED, diverge at infinitely large momenta; also as in QED, it is necessary to introduce a procedure for eliminating divergences: a renormalization. This procedure consists of two steps: a regularization and a renormalization. Let us use the specific example of calculating the lowest-order correction to the eigenenergy of a quark, $\Sigma(p^2)$, to examine the fundamental aspects of this two-step renormalization procedure. Working from the rules established above, we write the diagram in Fig. 3.18 as

$$\left.\begin{aligned} &\Sigma(\hat{p}) \sim g^2 \int \frac{dk}{(2\pi)} \frac{\gamma_\mu(\hat{k} - k^2)\gamma_\mu}{(k-p)^2}, \\ \text{or} \quad &\Sigma(\hat{p}) = \hat{p}\,\Sigma(p^2). \end{aligned}\right\} \tag{3.116}$$

An evaluation of integral (3.116) by the method of dimensional regularization leads to the expression

$$\Sigma(p^2) = \text{const}(g^2/16\pi^2)[2/\varepsilon - \ln(p^2/\mu^2) + 1 + \ln 4\pi - \gamma_E], \tag{3.117}$$

where ε is the parameter of the dimensional regularization ($\varepsilon \to 0$), μ is an arbitrary constant with the dimensionality of a mass, and γ_E is Euler's constant. The divergence of the integral is incorporated in the expression $2/\varepsilon$. A divergence-removal operation which is completely analogous to the removal of divergences in QED can be expressed with the help of counterterms[33] introduced in Lagrangian (3.115) in such a way that the fields and coupling constants are renormalized in accordance with the rules

$$(G^a_\mu)^0 = Z_G^{1/2} G^a_\mu; \quad \psi^0_\alpha = Z_\psi^{1/2}\psi_\alpha; \quad g^0 = Z_g \mu^{\varepsilon/2} g, \tag{3.118}$$

where the superscript 0 specifies an unrenormalized quantity; Z_G, Z_ψ, and Z_g are renormalization constants, which tend toward infinity in the limit $\varepsilon \to 0$; and μ is a constant with the dimensionality of a mass which makes the coupling constant g dimensionless. The operation of eliminating divergences reduces to deleting the term $2/\varepsilon$ from expression (3.118); this procedure is called a "minimal subtraction scheme." It would be possible to discard from (3.117) any other combination of quantities which does not depend on the momentum p^2, e.g., $2/\varepsilon + 1 + \ln 4\pi - \gamma_E$. In any case, the quantities $Z_G^{1/2}$, $Z_\psi^{1/2}$ and Z_g

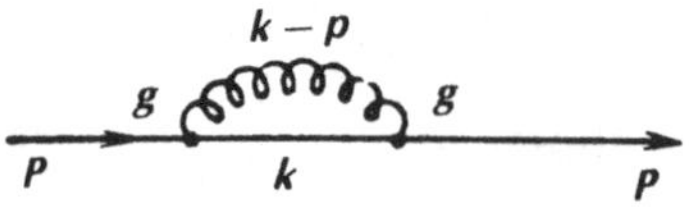

Figure 3.18. Diagram corresponding to the contribution proportional to g^2 to the self-energy of a quark.

are known if the explicit form of the counterterms in the Lagrangian has been determined. In an expression regularized in this manner, i.e., in an expression without divergences, we are left with yet one more undetermined quantity, μ^2, on which $\Sigma(p^2)$ of course depends. A renormalization is introduced in order to make the calculated expression $\Sigma(p^2)$ independent of the choice of normalization mass μ^2 and of the regularization method. The quantity μ was introduced in (3.118) in order to render the coupling constant in the vertex parts (Fig. 3.18) dimensionless. The renormalization process thus reduces to a renormalization of a vertex function, denoted by $\Gamma(p, q_0)$. In our case the renormalized vertex is

$$\Gamma_R(p,g,\varepsilon,\mu) = Z_\psi^{1/2} Z_G^{1/2} \Gamma(p,g_0)_g \tag{3.119}$$

and there must exist a limit

$$\Gamma_R(p,g,\varepsilon,\mu) = \lim_{\varepsilon\to 0} \Gamma(p,g,\varepsilon,\mu) .$$

The independence of the vertex function $\Gamma(p, g_0)$ from the choice of μ can be expressed as the requirement that the following derivative vanish:

$$\partial\Gamma(p,g_0)/\partial\mu = 0 . \tag{3.120}$$

From (3.120) and (3.119) we find a renormalization-group equation. We will write it out here not for the particular case which we have been discussing but for the general case:

$$[\mu\,\partial/\partial\mu + \beta(g)\,\partial/\partial g - N_\psi\gamma_\psi(g) - N_G\gamma_G(g)]\,\Gamma_R^{(N_\psi,N_G)} = 0 , \tag{3.121}$$

where $\Gamma_R^{(N_\psi,N_G)}$ is a complicated vertex function which contains N_ψ external quark lines and N_G gluon lines. In (3.121) we have introduced the notation

$$\begin{aligned} \beta(g) &= \mu\,\partial g/\partial\mu ; \\ \gamma_\psi(g) &= (1/2)\mu(\partial/\partial\mu)\ln Z_\psi ; \\ \gamma_G(g) &= (1/2)\mu(\partial/\partial\mu)\ln Z_G ; \end{aligned} \tag{3.122}$$

the quantities γ_ψ, γ_G, and β depend on only the interaction constant g (γ_ψ and γ_G are called the "anomalous dimensionalities" of the quark and the gluon fields, respectively).

It is important that we recognize that (3.121) relates a vertex function which depends on the momentum p to the same vertex function but with a momentum $\exp t \times p$. For deep inelastic scattering we would have $t = \ln(Q^2/\mu^2)$ and

$$\Gamma_R(\exp tp,g) = \Gamma_R(p,\bar g(t))\exp\left[\tilde D t - \int_g^{\bar g(t)} dg' \frac{\gamma_\psi(g') + \gamma_G(g')}{\beta(g')}\right] , \tag{3.123}$$

where $\bar g(t)$ is an effective coupling constant which satisfies the equation

$$d\bar{g}^2(t)/dt = \bar{g}\beta(\bar{g}); \quad \bar{g}(t=0) = g; \tag{3.124}$$

and $\tilde{D}$ is the dimensionality of the vertex function Γ_R.

Knowing γ_ψ, γ_G, $\beta(g)$, and Γ_R at a single point, e.g., $t = 0$, we can thus use (3.123) to calculate Γ_R for any value of the momentum exp tp at $t \neq 0$. The renormalized vertex function Γ_R and the coupling constant $g(q^2)$ are chosen in each perturbation-theory order in such a way that the result of the elimination of the divergence turns out to be independent of the choice of the parameter μ^2. Using perturbation theory, we can calculate the function $\beta(g)$ in any order in g^2.

An expansion of $\beta(g)$ in a series in the coupling constant g^2 takes the form

$$\beta(g) = -\beta_0 g^3/16\pi^2 - \beta_1 g^5/(16\pi^2)^2 + \dots, \tag{3.125}$$

where $\beta_0 = 11 - (2/3)f$, and $\beta_1 = 102 - (38/3)f$ (f is the number of quark flavors). Retaining only the first term in the expansion in (3.125), and substituting this expression into (3.124), we find an approximate expression for the effective coupling constant $\bar{g}^2(Q^2)$:

$$\alpha_s(Q^2) = \bar{g}^2(Q^2)/4\pi = 12\pi/(33 - 2f)\ln(Q^2/\Lambda^2). \tag{3.126}$$

This is an expression for $\alpha_s(Q^2)$ in the leading order (the leading-log approximation). Incorporating the next term in expansion (3.125) leads to the following expression for $\alpha_s(Q^2)$:

$$\frac{\bar{g}^2(Q^2)}{4\pi} = \frac{12\pi}{(33-2f)\ln(Q^2/\Lambda^2)}\left[1 - \frac{\beta_1}{\beta_0}\frac{\ln\ln(Q^2/\Lambda^2)}{\ln(Q^2/\Lambda^2)}\right]. \tag{3.127}$$

In (3.126) and (3.127), Λ is an adjustable parameter of the theory, which is to be determined through a comparison with experimental data. The relationship between this parameter and μ in the leading order in $\alpha_s(Q^2)$ is

$$\Lambda^2 = \mu^2 \exp\{-16\pi^2/[11 - (2/3)f]g^2\}.$$

The parameter Λ determines the numerical value of the effective coupling constant which appears in all the expressions calculated in QCD, for any physical process. It is one of the fundamental parameters of the theory. The existence in QCD of a coupling constant $\alpha_s(Q^2)$ which depends on the square of the 4-momentum transfer Q^2 automatically leads to a violation of scaling.

We turn now to the QCD predictions regarding deep inelastic scattering. Using Wilson's expansion of the product of current operators, we can relate the structure functions to the Wilson coefficients $C_n(Q^2)$ [see (3.93)]. In writing this relationship we retain the necessary indices and arguments of the functions [see (3.133) and (3.134)], but we first introduce the notation (cf. Sec. 3 of Chap. 3):

$$\begin{aligned} &q(x) = u(x) + d(x) + s(x) + c(x) = \sum_i q_i(x); \quad \bar{q}(x) = \sum_i \bar{q}_i(x); \\ &\Sigma(x) = q(x) + \bar{q}(x); \quad \Delta_{ij}(x) = q_i(x) - q_j(x); \\ &V(x) = q(x) - \bar{q}(x); \quad \overline{\Delta}_{ij} = \bar{q}_i(x) - \bar{q}_j(x); \quad S(x),\ G(x), \end{aligned} \tag{3.128}$$

where $q(x)$ is the momentum distribution of all the quarks (u, d, s, c) in the nucleon; $\bar{q}(x)$ is the momentum distribution of all the antiquarks in the nucleon; the difference $q(x) - \bar{q}(x) = 2u + d$ for a proton is equal to the distribution of the valence quarks alone, $V(x)$; for a neutron, it is $V(x) = q(x) - \bar{q}(x) = u + 2d$; $S(x)$ is the momentum distribution of the sea quarks; and $G(x)$ is the momentum distribution of the gluons. The combination $\Sigma(x)$ is called a "singlet combination" with respect to $SU(4)$ symmetry, i.e., with respect to the replacement of quarks of one species by others; and the distribution $G(x)$ is also a singlet combination. The distributions Δ_{ij}, $\bar{\Delta}_{ij}$, and $V(x)$ are nonsinglet distributions with respect to $SU(4)$ symmetry.

In terms of the new definitions in (3.128), the relationship between the structure functions $W_i(x)$ and distributions (3.128) for the deep inelastic scattering of electrons by protons is described by [see also (3.34)]

$$\nu W_2^{ep}(x) = F_2^{ep}(x) = \sum_i e_i^2 x[q(x) + \bar{q}(x)] \, . \tag{3.129}$$

The Callan-Gross relation [see (3.35)] takes the form

$$2x\, F_1^{ep}(x) = F_2^{ep}(x)(MW_1(x) = F_1(x)) \, . \tag{3.130}$$

The function $F_2(x)$ can be rewritten as

$$\begin{aligned} F_2^{ep}(x) &= (4/9)x[u(x) + \bar{u}(x)] + (1/9)x[d(x) + \bar{d}(x)] + (1/9)x[s(x) + \bar{s}(x)] \\ &\quad + (4/9)x[c(x) + \bar{c}(x)] = (5/18)x\Sigma(x) + (1/6)x\Delta^{ep}(x) \, , \end{aligned} \tag{3.131}$$

where $\Delta^{ep}(x) = (u + \bar{u}) - (d + \bar{d}) - (s + \bar{s}) + (c + \bar{c})$. It follows from (3.131) that $F_2^{ep}(x)$ contains a singlet part (Σ) and a nonsinglet part (Δ).

For electron scattering by an isoscalar target N (i.e., a target in which the number of protons is equal to the number of neutrons, e.g., the carbon nucleus ${}^{12}_{6}C$), the expression for F_2^{eN} becomes

$$F_2^{eN}(x) = (5/18)x\Sigma(x) + (1/6)x\Delta^{eN}(x) \, ,$$

where $\Delta^{eN} = (c + \bar{c}) - (s + \bar{s})$.

In QCD, the distribution functions of all of the quarks and gluons in the elementary particles depend on not only x but also the 4-momentum transfer $q^2 = -Q^2$. At the beginning of Sec. 9 of Chap. 3 we offer a qualitative interpretation of the dependence of the quark and gluon distribution functions on the variable Q^2. Below we will use the notation customary in QCD: $q(x, Q^2)$ and $\bar{q}(x, Q^2)$.

The scattering of neutrinos by nucleons is described by three structure functions [see (3.54) and (3.55)]. We write the relationship between the structure functions $F_2(x, Q^2)$ and $F_3(x, Q^2)$ and the distribution functions as follows:

$$\begin{aligned} F_2(x,Q^2) &= 2x\, F_1(x,Q^2) = x[q(x,Q^2) + \bar{q}(x,Q^2)] \, ; \\ F_3(x,Q^2) &= q(x,Q^2) - \bar{q}(x,Q^2) \, . \end{aligned} \tag{3.132}$$

The function $F_3(x, Q^2)$ is a nonsinglet combination with respect to the colored $SU(4)$ transformation group and is the same as the distribution of valence quarks.

The distinction between singlet and nonsinglet combinations has been drawn for both electromagnetic and weak interactions because the relationships between these combinations and the coefficients in Wilson's expansion are different, and the expressions for the moments of these functions are written differently.

In the case of nonsinglet combinations, the expressions for the moments of the distribution functions are given in the leading order by[108]

$$M_n^{(2)NS}(Q^2) = \int_0^1 dx\, x^{n-2} F_2^{NS}(x,Q^2) = \delta_2^{NS} A_n^{NS} \left[\ln \frac{Q^2}{\Lambda^2}\right]^{-\gamma_n^{NS}};$$
$$M_n^{(3)NS}(Q^2) = \int_0^1 dx\, x^{n-1} F_3^{NS}(x,Q^2) = \delta_3^{NS} A_n^{NS} \left[\ln \frac{Q^2}{\Lambda^2}\right]^{-\gamma_n^{NS}}. \tag{3.133}$$

Here we have $\delta_2^{NS} = 1/6$ for F_2^{ep} and F_2^{eN} and $\delta_3^{NS} = 1$ for $F_3^{\nu,\bar{\nu}}$ (charged currents). For neutral currents, δ_3^{NS} has a different value. Values of the anomalous dimensionalities γ_n^{NS} are given in Table 3.3.

For singlet structure functions, the expressions for the moments in the leading order are

$$M_n^{(2)S}(Q^2) = \int_0^1 dx\, x^{n-2} F_2^S(x,Q^2)$$
$$= \delta_2^{NS} A_n^- \left[\ln \frac{Q^2}{\Lambda^2}\right]^{-\gamma_n^-} + \delta_2^S A_n^+ \left[\ln \frac{Q^2}{\Lambda^2}\right]^{-\gamma_n^+}. \tag{3.134}$$

Here we have $\delta_2^S = 5/18$ for F_2^{ep} and F_2^{eN}, and we have $\delta_2^S = 1$ for $F_2^{\nu,\bar{\nu}}(x, Q^2)$ (charged currents). For neutral currents, δ_2^S takes on a different value. The quantities γ_n^- and γ_n^+ can be calculated in QCD. They determine the evolution of the moments as functions of Q^2; numerical values of these quantities for various values of f and n are given in Table 3.3.

The coefficients A_n^{NS} and $A_n^{\pm}$ are found by comparing moments (3.133) and (3.134) with experimental data for some value Q_0^2; they remain constant for all other values of Q^2.

It follows from (3.133) and (3.134) that the moments of the structure functions in QCD evolve in a logarithmic way as Q^2 is varied [see also (3.96)]. The moments for the nonsinglet combinations evolve in a simpler way (they depend on the single exponent γ_n^{NS}).

Using the expression for $\alpha_s(Q^2)$ [see (3.126)], we can describe the evolution of moments (3.133) and (3.134) in terms of an effective coupling constant. For example, the expression for nonsinglet moments can be written

$$M_n^{NS}(Q^2) = M_n^{NS}(Q_0^2)[\alpha_s(Q_0^2)/\alpha_s(Q^2)]^{-\gamma_n^{NS}}, \tag{3.135}$$

where

$$\delta^{NS} A_n^{NS} = M_n^{NS}(Q_0^2)[\ln(Q_0^2/\Lambda^2)]^{\gamma_n^{NS}}$$

(Q_0^2 is an arbitrary fixed value of the square of the 4-momentum transfer at which the evolution of the moments as functions of Q^2 begins).

Finally, expression (3.133) can be rewritten as

$$\ln M_n(Q^2) = \ln(\delta_2^{NS} A_n^{NS}) - \gamma_n^{NS} \ln \ln(Q^2/\Lambda^2),$$

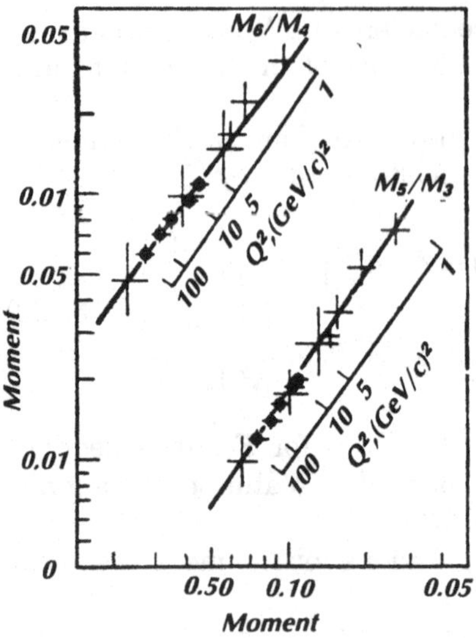

Figure 3.19. Ratios of moments of the structure function xF_3 (neutrino-scattering data obtained by CERN experimental groups).

or

$$\frac{d[\ln M_n(Q^2)]}{d[\ln (Q^2/\Lambda^2)]} = -\gamma_n^{NS}\frac{1}{\ln(Q^2/\Lambda^2)} = -\gamma_n^{NS}\frac{\alpha_s}{2\pi}\frac{33-2f}{3} = \frac{\alpha_s}{2\pi}a_n^{NS}, \quad (3.136)$$

where

$$a_n^{NS} = -\frac{33-2f}{6}\gamma_n^{NS} = -\frac{2}{3}\left(1-\frac{2}{n(n+1)}+4\sum_{j=2}^{n}\frac{1}{j}\right).$$

If we substitute expression (3.126) into (3.136), we find functional dependence (3.133).

From (3.136) we find

$$d[\ln M_k(Q^2)]/d[\ln M_n(Q^2)] = \gamma_k^{NS}/\gamma_n^{NS}; \quad (3.137)$$

i.e., a plot of $\ln M_k(Q^2)$ versus $\ln M_n(Q^2)$ reveals a straight line, whose slope is determined by the anomalous dimensionalities γ_n, which in turn depend on the gluon spin. Expression (3.137) thus presents an opportunity for a graphic test of the positions of QCD: to determine the spin of the gluon, the evolution as a function of Q^2, the value of Λ, and the form of the structure functions $F_i(x, Q^2)$.

Comparison with experimental data shows convincingly that several of the basic positions of QCD are valid (Fig. 3.19).

Up to this point we have been working in the leading-log approximation. Incorporating the following orders leads to more-complicated expressions for the moments. The relationship between the structure functions and the quark and gluon distribution functions becomes complicated and multivalued; a supplemental definition of this relationship becomes necessary. We will not take up these questions here.

A test of the QCD predictions consisted of comparing the evolution as a function of Q^2 for moments (3.133) and (3.134) with experimental values of the moments as functions of $F_2(x, Q^2)$ and $F_3(x, Q^2)$ or combinations thereof. However, a test of this sort has two substantial shortcomings.

In the first place, the x interval in which the structure functions $F_i(x, Q^2)$ are measured does not reach the limiting values $x = 0$ and $x = 1$ in the actual experiments. The experimental data must therefore be extrapolated to these points, at the risk of significant errors. Second, uavoidable correlations arise between the moments calculated from the same structure function.

The most direct test of a violation of scaling in deep inelastic scattering of leptons by nucleons can be carried out with the help of the Lipatov-Altarelli-Parisi (LAP) equations[109] rather than by the moment method. These equations determine the evolution of the distribution functions of the valence quarks, of the sea quarks, and of the gluons as functions of the 4-momentum transfer Q^2; they are called the evolutionary equations of QCD. The solutions of the LAP equations describe the functional dependence of the structure functions on Q^2 and x over the entire x interval ($0 < x < 1$). The gluon distribution function does not have to be known in order to determine the violation of scaling by the function $xF_3(x, Q^2)$, which appears in the cross section for the scattering of neutrions by a nucleon. In the leading order in the constant $\alpha_s(Q^2)$, the LAP equation for the distribution function of the valence quarks is

$$\frac{dq_v(x,t)}{dt} = \frac{\alpha_s(t)}{2\pi}\int_x^1 \frac{dy}{y} K_{q\to q'}\left(\frac{x}{y}\right) q_v(y,t)\,, \tag{3.138}$$

where $t = \ln(Q^2/\mu^2)$, and $K_{q\to q'}(x/y, t)$ is the kernel of the equation which describes the process $q \to q' + g$ (Fig. 3.20a), in which a gluon carries off a fraction $q = 1 - z$ of the momentum of the quark, while the quark q' has a fraction z of the momentum of the primary quark. The kernel $K_{q\to q'}$ is calculated from the rules of QCD perturbation theory; $\alpha_s(Q^2)$ is a known coupling constant. In the lowest order in $\alpha_s(Q^2)$, the kernel $K_{q\to q'}$ is independent of the variable t (Ref. 110):

$$K_{q\to q'}(z) = (4/3)[(1 + z^2)/(1 - z)_+ + (3/2)\delta(1 - z)]\,. \tag{3.139}$$

Here $1/(1 - z)_+$ means that the integral of the function $f(x/z)$, which is regular over the entire range of integration, including the end points, must be used in accordance with the rule

$$\int_x^1 dz\, z\,\frac{f(x/z)}{(1 - z)_+} = f(x)\ln(1 - x) + \int_x^1 \frac{dx}{1 - x}\left(zf\left(\frac{x}{z}\right) - f(x)\right).$$

Equation (3.138) can be rewritten as

$$q_v(x,t) + dq_v(x,t) = \int_0^1 dy \int_0^1 dz\, \delta(zy - x) q_v(y,t)\left[\delta(1 - z) + \frac{\alpha_s(t)}{2\pi} K(z) dt\right]. \tag{3.140}$$

Expression (3.140) is another version of (3.138), which allows a clear physical interpretation: There is a certain probability that a quark with a momen-

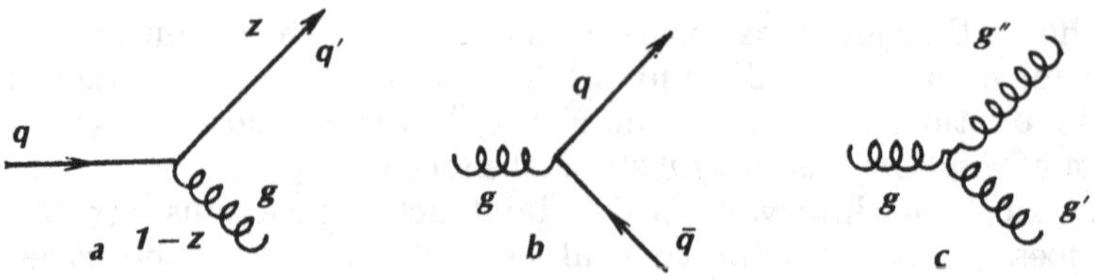

Figure 3.20. Graphical representation of the kernels of the LAP equations. (a) $q \to q' + g$; (b) $g \to q + \bar{q}$; (c) $g \to g'g''$.

tum y will emit a gluon g and go into a state with momentum x. This probability can be calculated by a QCD perturbation theory.

The quantity

$$K_{qq} + dK_{qq} = \delta(1-z) + [\alpha_s(t)/2\pi]K(z)dt$$

may be interpreted as the probability density for finding a quark in a final state with a fraction z of the momentum of the primary quark ($zy = x$). The change in this probability density as t changes leads to a variation in the momentum distribution of the quark. The quantity $[\alpha_s(t)/2\pi]K(z)$ is thus a measure of the change per unit t in the probability density for finding within the initial quark another quark, with a fraction z of the momentum of the initial quark, in the QCD perturbation theory of lowest order in α_s.

The LAP equations for singlet functions can be interpreted in a corresponding way. These equations can be written

$$\frac{d\Sigma(x,t)}{dt} = \frac{\alpha_s(t)}{2\pi}\int_x^1 \frac{dy}{y}\left[\Sigma(y,t)K_{q\to q'}\left(\frac{x}{y}\right) + 2fG(y,t)K_{q\to g}\left(\frac{x}{y}\right)\right];$$
$$\frac{dG(x,t)}{dt} = \frac{\alpha_s(t)}{2\pi}\int_x^1 \frac{dy}{y}\left[\Sigma(y,t)K_{g\to q}\left(\frac{x}{y}\right) + G(y,t)K_{g\to g'}\left(\frac{x}{y}\right)\right]. \tag{3.141}$$

In (3.141), the kernels $K_{q\to g}$, $K_{g\to q}$ and $K_{g\to g'}$ do not depend on the variable t in the lowest order in $\alpha_s(Q^2)$ and can be written

$$K_{q\to g}(z) = (1/2)[z^2 + (1-z^2)];$$
$$K_{g\to q}(z) = (4/3)[1 + (1-z)^2]/z; \tag{3.142}$$
$$K_{g\to g'}(z) = 6\left[\frac{z}{(1-z)_+} + \frac{(1-z)}{z} + z(1-z) + \left(\frac{11}{12} - \frac{f}{18}\right)\delta(1-z)\right].$$

The kernels $K_{q\to g}$, $K_{g\to q}$, and $K_{g\to g'}$ are illustrated by the diagrams in Fig. 3.20, b and c. The t dependence in the kernels $K(x/y, t)$ arises in higher orders in $\alpha_s(Q^2)$. Incorporating the next order in $\alpha_s(Q^2)$ which follows the leading order results in a more complicated form of the evolutionary equations. In order to solve the LAP equation, it is necessary to specify an *a priori* functional dependence of the distribution functions on the variable x over the entire range of this variable ($0<x<1$). The x dependence cannot be found within the framework of QCD. It is usually assumed that at small values of $x=Q^2/2M\nu$ one can use the predictions of the Regge phenomenology with regard to the behavior of the amplitudes for deep inelastic scattering (the Regge limit: $\nu\to\infty$; Q^2 is fixed; and $x\to 0$). With regard to the region $x\to 1$, we note that on occasion use is made of a generalization of the quark counting rules which had been proposed earlier for describing the behavior of the cross sections for various processes at high energies and at large values of the momentum transfer Q^2 (Ref. 55). This new generalized rule states that in the region $x\to 1$ the x distributions of the quarks and gluons should be of the form

$$q(x,Q_0^2)\sim(1-x)^{2n-3}, \tag{3.143}$$

where the values of Q_0^2 are chosen in such a way that the simplest parton model has a physical meaning [$Q_0^2\approx 2$–5 (GeV/$c)^2$], and n is the minimum number of "hard" (large-momentum) constituents in the hadron. The distribution of valence quarks, for example, is determined by the number of hard valence quarks, $n=3$: $q(x,Q_0^2)\sim(1-x)^3$. The gluon distribution is determined by the number $n=4$ (three valence quarks plus a single hard gluon, emitted by one of the valence quarks): $G(x,Q_0^2)\sim(1-x)^5$. The distribution of the sea quarks is determined by the number $n=5$, since it is assumed that in addition to the three valence quarks there is a single hard $q\bar{q}$ pair, which is formed by one hard gluon: $q_s(x,Q_0^2)\sim(1-x)^7$.

Rule (3.143) looks so contrived that in practice one uses a very simple parametrization of the distribution functions:

$$q(x,Q_0^2)\sim x^a(1-x)^b, \tag{3.144}$$

where a and b are arbitrary numbers. Such a parametrization, however, gives a poor description of the x distribution of the quarks and gluons in the intermediate region $0<x<1$. The choice of distribution functions of the quarks and gluons in x will be discussed in Sec. 8 of Chap. 3.

We have one last, brief comment regarding the incorporation in QCD of corrections to the leading-log approximation, which is used to describe deep inelastic scattering. Quantum chromodynamics begins where the emission of gluons is taken into consideration. Incorporating the gluon corrections to deep inelastic scattering reduces in the simplest cases to incorporating the diagrams in Fig. 3.21, b–e. Particularly remarkable is the diagram in Fig. 3.21e, which describes the exchange of a gluon between a quark which has collided with a lepton and a quark which is in a bound state in the nucleon. This contribution is proportional to Q^{-2} and is called the "contribution of higher twist states." It cannot be calculated by perturbation theory; it becomes more significant as Q^2 decreases [the region $Q^2\lesssim 5(\text{GeV}/c)^2$] and as x increases. The contributions of the higher twist states are taken into account phenomenologically by adding terms of the type $[1/Q^2(1-x)^m]^n$ ($m, n=1,2,3,\ldots$ are integers) to the structure functions $F_i(x,Q^2)$.

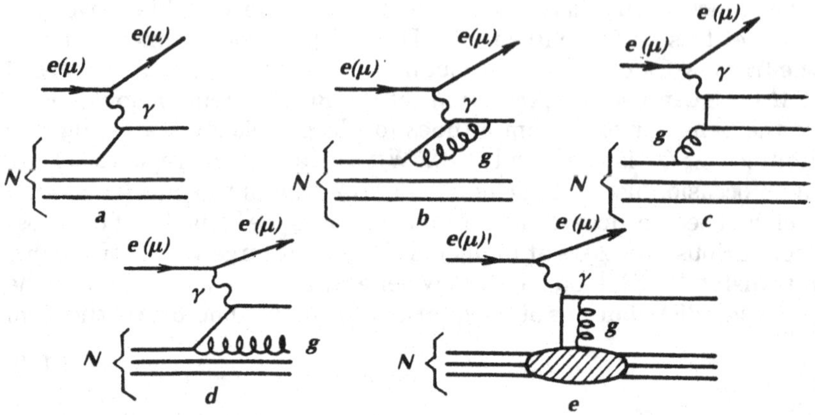

Figure 3.21. Diagrams describing deep inelastic scattering of a lepton by a nucleon. (a) In a simple quark–parton model; (b–d) with radiative corrections for the emission of a gluon in the leading order in α_s; (e) with the contribution of higher twist states.

When we substitute theoretically grounded structure functions $F_i(x, Q^2)$ into the expressions for the cross sections for various types of deep inelastic scattering of leptons by nucleons, we find predictions for these processes within the framework of QCD.

8. The Kuti–Weisskopf model

Theoretical aspects of deep inelastic scattering associated with the behavior of the cross section at high energies and large momentum transfer were examined in Secs. 4, 5, and 7 of Chap. 3. Almost nothing was said about the structure of nucleons; information on this structure can be found from what is known about the structure functions $W_1(\nu, Q^2)$ and $W_2(\nu, Q^2)$. While the Q^2 dependence of the functions $W_i(\nu, Q^2)$ can be predicted in QCD, their dependence on x ($x = Q^2/2M\nu$) is almost completely unknown. All that we have are some experimental and theoretical indications regarding the x distribution of the quarks and gluons within a nucleon in the limits $x \to 0$ and $x \to 1$.

In the discussion below, when we speak of studying the quark-gluon structure of nucleons (or hadrons) we mean searching for the momentum distribution functions of the valence quarks, the sea quarks, and the gluons within the hadrons or, equivalently, determining the x dependence of the structure functions $W_i(\nu, Q^2)$. The momentum distributions of the quarks and gluons are usually specified on the basis of some model, whose parameters are determined from a comparison with experimental data.

We will discuss in more detail the Kuti–Weisskopf model,[89] because of its comparative simplicity and methodological importance. In this model it is assumed that a nucleon consists of three valence quarks and a sea of partons (quarks and gluons).

The valence quarks determine all the quantum numbers of the nucleon and are responsible primarily for peripheral collisions. The sea of partons constitutes the interior part of a nucleon and is responsible primarily for processes involving a large momentum transfer.

The total momentum of the nucleon, P, is the sum of the momenta of the constituents of the nucleon:

$$\sum_{i=1}^{\infty} p_i = P . \tag{3.145}$$

If we introduce the relative momentum of each parton, $y_i = p_i/P$, we can instead write

$$\sum_i y_i = 1 . \tag{3.146}$$

It is assumed that the parton satisfies the condition

$$p_i^2 + \mu_i^2 = E_i^2 ,$$

so that we have

$$\sum_i E_i = E = \sqrt{P^2 + M^2} , \tag{3.147}$$

where μ_i and E_i are the mass and energy of parton i, and M is the mass of the nucleon. These assumptions underlie all models of the nucleon.

The next step in the construction of a model is to specify initial (seed) momentum distribution functions of the constituents of the nucleon. For the quark-antiquark pairs of the sea, the distribution function is written in a form similar to that of an ordinary phase-space distribution:

$$\tilde{g}_s = (1/3)g/\sqrt{y^2 + \mu^2/P^2} . \tag{3.148}$$

That for gluons is written in the same form:

$$\widetilde{G}(y) = g'/\sqrt{y^2 + \mu^2/P^2} . \tag{3.149}$$

The constants $g/3$ and g' are responsible for the interaction of the sea quarks and gluons. For valence quarks the distribution is chosen in the form

$$\tilde{q}_v(y) = sy^{1-\alpha(0)}/\sqrt{y^2 + \mu^2/P} , \tag{3.150}$$

where s is a proportionality factor, and $\alpha(0)$ is the Regge trajectory of the A_2 meson, which is responsible for the exchange in the deep inelastic lepton-nucleon scattering. The correct choice of $\alpha(0)$ leads to the appropriate asymptotic behavior of the cross section for the process under consideration.

Several arguments can be offered in favor of model distributions (3.148)–(3.150). If the masses of the valence quarks, the sea quarks, and the gluons are set equal to zero, the distribution functions take the forms

$$\tilde{q}_s(y) \sim y^{-1}; \tag{3.148a}$$

$$\widetilde{G}(y) \sim y^{-1}; \tag{3.149a}$$

$$\tilde{q}_v(y) \sim y^{-\alpha(0)}. \tag{3.150a}$$

These spectra have a definite physical meaning.

For example, the gluon spectrum $\widetilde{G}(y)$ is reminiscent of the spectrum of bremsstrahlung photons in the equivalent-photon method (Chap. 5). Analysis of deep inelastic scattering in the Regge phenomenology leads to the conclusion that in the Regge limit ($Q^2 = \text{const}, \nu \to \infty, x = Q^2/2M\nu \to 0$) the structure functions W_i take the form

$$\nu W_2(\nu,Q^2) \sim \nu^{\alpha-1} f_2(Q^2); \quad W_1(\nu,Q^2) \sim \nu^{\alpha} f_1(Q^2). \tag{3.151}$$

If the functions $W_i(\nu, Q^2)$ are to conform to a scaling, it is necessary to stipulate that the following relations hold in the limit $Q^2 \to \infty$:

$$f_1(Q^2) \sim (Q^2)^{-\alpha}; \quad f_2(Q^2) \sim (Q^2)^{1-\alpha}.$$

In the Bjorken limit ($Q^2 \to \infty$, $\nu \to \infty$, $Q^2/\nu \to \text{const}$) we would then have

$$\begin{aligned} &\sum_j e_j^2 x f_j(x) = \nu W_2(\nu,Q^2) \sim \nu^{\alpha-1}/(Q^2)^{\alpha-1} \to \text{const}; \\ &\sum_j e_j^2 f_j(x) = W_1(\nu,Q^2) \sim \nu^{\alpha}/(Q^2)^{\alpha} \to \text{const}. \end{aligned} \tag{3.152}$$

Relations (3.152) can be written as

$$\nu W_2(\nu,Q^2) \sim (1/x)^{\alpha-1} \sim x^{1-\alpha}; \quad W_1 \sim x^{-\alpha}. \tag{3.153}$$

Experimental data indicate that in the limit $x \to 0$ the most probable behavior of the functions is $\nu W_2 \to \text{const}$, $W_1 \to x^{-1}$. This behavior is possible if the exponent in (3.153) is $\alpha = 1$; in the language of Regge phenomenology, this value corresponds to the exchange of a pomeron in the t channel of deep inelastic scattering (Fig. 3.22). The exchange of a pomeron is the exchange of a particle with the quantum numbers of vacuum. Pomeron exchange can occur between particles (or between systems of particles) which do not have color degrees of freedom. In a nucleon, therefore, pomeron exchange occurs only with a $q\bar{q}$ pair of the sea, since only such a pair constitutes a color singlet in a nucleon. In other words, pomeron exchange serves as an analyzer of the momentum distribution of the quark-antiquark pairs. It follows from these considerations that the sea quarks should have the distribution described by (3.148a). As for the valence quarks, which are responsible for peripheral collisions, we note that they form a nonsinglet part of a nucleon, and it is natural to suggest that the Regge trajectory which is responsible for interactions with valence quarks is extremely close to a pomeron trajectory, i.e., that with a value $\alpha(0) = 1/2$ [see (3.150a)]. An exchange corresponding to this trajectory corresponds in deep inelastic scattering to the exchange of ρ, ω, A_2, and f mesons in the t channel of the processes in which we are interested here (Fig. 3.22).

An n-quark-gluon state of a nucleon is written as the product of one-particle distribution functions (3.148)–(3.150):

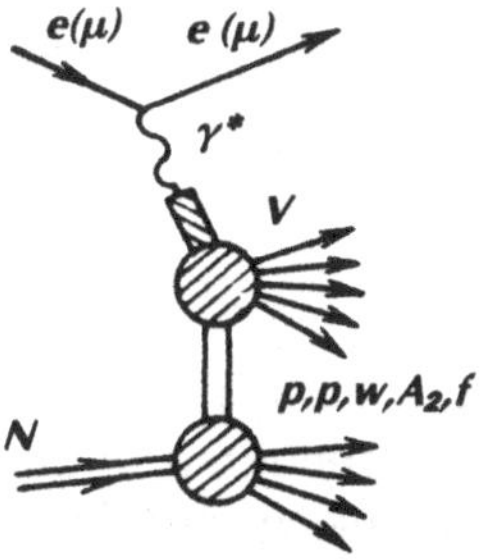

Figure 3.22. Model for deep inelastic scattering in the Regge phenomenology (*V*) Vector meson: (*P*) pomeron.

$$dP_n(y_1,\ldots,y_n) = Z\frac{(g/3)^{k_1}(g/3)^{k_2}(g/3)^{k_3}(g')^l}{k_1!k_2!k_3!l!}$$

$$\times\prod_{j=1}^{n}\frac{dy_j}{(y_j^2+\mu_j^2/P^2)^{1/2}}\prod_{i=1}^{3}y_i^{1-\alpha(0)}\delta\left(1-\sum_{j=1}^{n}y_j\right), \tag{3.154}$$

where Z is a normalization constant which is found from the condition that the total probability is 1; the positive constants k_1, k_2, and k_3 correspond to the numbers of u, d, and s quarks (these numbers must be even if we wish to avoid changing the quantum numbers of the nucleons); and $l = 0, 1, 2, \ldots$ is the number of gluons in the n-particle state ($n = 3 + k_1 + k_2 + k_3 + l$). The 3 which appears in this expression for n is the number of valence quarks. The δ function is responsible for momentum conservation in the nucleon and also expresses the condition of primitive confinement of quarks and gluons.

We introduce the functions $G_i^{p,n}(y)$, where $i = 1, 2, 3, 4$ corresponds to the species of quark (u, d, s) and the gluon; the superscripts p and n correspond to the proton and the neutron. The functions $G_i^p(y)$ and $G_i^n(y)$ are defined as the probabilities for finding a parton of species i with momentum y in the proton or neutron. It is assumed in this model that $SU(3)$ symmetry holds, so that we have the following relations between G_i^p and G_i^n

$$\begin{aligned} G_1^p(y) &= G_2^n(y); \quad G_2^p(y) = G_1^n(y); \\ G_3^p(y) &= G_3^n(y); \quad G_4^p(y) = G_4^n(y). \end{aligned} \tag{3.155}$$

For convenience in the calculations, the functions G_i are broken up into two parts:

$$G_i^{p,n} = G_{iv}^{p,n} + G_{ic}^{p,n}, \tag{3.156}$$

where G_{iv} is responsible for the contribution of valence quarks, and G_{ic} for that of the sea quarks.

To find the momentum distribution of one of the quarks in the nucleon, we need to integrate the function $dP_n(y_1,\ldots,y_n)$ over the momentum of all of the quarks except that in which we are interested. As a result, we find the functions

$$G_{1v}^p = 2G_{2v}^p = 2\frac{\Gamma[\gamma + 3(1-\alpha(0))]}{\Gamma[1-\alpha(0)]\Gamma[2(1-\alpha(0))+\gamma]} \times y^{-\alpha(0)}(1-y)^{-1+\gamma+2[1-\alpha(0)]}; \quad (3.157)$$

$$G_{1c}^p = G_{2c}^p = G_{3c}^p = \frac{g}{3g'}G_{4c}^p = \frac{1}{3}g\frac{1}{y}(1-y)^{-1+\gamma+3[1-\alpha(0)]},$$

where $\gamma = g + g'$.

Although it would appear that there is much latitude in the choice of the initial (seed) distribution functions $\tilde{q}_s(y)$, $\tilde{q}_v(y)$, and $\tilde{G}(y)$, there are several conditions which restrict this choice. These functions must be of such a nature that the functions (3.157) found with their help satisfy the requirements

$$G_i(y) \underset{y\to 1}{\rightarrow} (1-y)^n; \quad (3.158a)$$

$$G_i(y) \underset{y\to 0}{\rightarrow} 1/y; \quad (3.158b)$$

$$\int_0^1 G_v(y)dy = \text{const}. \quad (3.158c)$$

The asymptotic behavior in (3.158a) is dominated by the part $G_i^{p,n}$, which is associated with valence quarks; that in (3.158b) is dominated by the function $G_{ic}^{p,n}$; and the integral in (3.158c) is associated with the charge and can be normalized to the charge of the nucleon. To conditions (3.158) we must add the condition that the scaling functions νW_2 and $2MW_1$ be positive definite.

It remains to determine the relationship between the functions $G_i^{p,n}$ on the one hand, and νW_2 and $2MW_1$ on the other. We can do this in the following way. The scattering of an electron by a nucleon is treated as a scattering by one of the constituent quarks of the nucleon. The quark is assumed to be point-like, and it is assigned a Mott cross section:

$$d\sigma_M/dQ^2 = 4\pi\alpha^2/Q^4. \quad (3.159)$$

The cross section for the scattering by one quark thus takes the form

$$d^2\sigma/dQ^2d\nu = (d\sigma_M/dQ^2)e_i^2(y_i/\nu)\delta(x-y_i), \quad (3.160)$$

where e_i is the charge of a quark, expressed in units of the electric charge; $\delta(x-y_i)$ ensures that the quark momentum fraction y_i is identical to x; and $x = Q^2/2M\nu$ is an invariant scaling variable. The cross section for the scattering by the nucleon is the sum of the cross sections for the scattering by the three valence quarks:

$$\frac{d^2\sigma^{p,n}}{dQ^2d\nu} = \frac{d\sigma_M}{dQ^2}\sum_{i=1}^{3} e_i^2\frac{y_i}{\nu}G_i^{p,n}(x). \quad (3.161)$$

Comparison of cross section (3.161), expressed in terms of the functions $G_i^{p,n}(x)$, which depend on only the scaling variable x, with cross section (3.12) leads to

$$\nu W_2^{p,n}(q^2,\nu) = x\sum_{i=1}^{3} e_i^2 G_i^{p,n}(x) \equiv x f^{p,n}(x) \equiv F^{p,n}(x). \quad (3.162)$$

By analogy with (3.156), we break up the function $F^{p,n}(x)$ into two parts:

$$F^{p,n}(x) = F_v^{p,n}(x) + F_c^{p,n}(x)\,. \qquad (3.163)$$

Using (3.157), we find

$$F_v^p(x) = \tfrac{3}{2}\, F_v^n(x) = \frac{\Gamma[\gamma + 3(1-\alpha(0))]}{\Gamma[1-\alpha(0)]\,\Gamma\{\gamma + 2[1-\alpha(0)]\}} \times x^{1-\alpha(0)}(1-x)^{-1+\gamma+2[1-\alpha(0)]}\,; \qquad (3.164)$$

$$F_c(x) = (2/9)g(1-x)^{-1+\gamma+3[1-\alpha(0)]}\,.$$

In the limit $x \to 1$, the contribution of the valence quarks are dominant, since $F_c(x)$ falls off more rapidly as $x \to 1$. In the limit $x \to 1$ we have

$$F^n/F^p = 2/3\,. \qquad (3.165)$$

The function $W_1(\nu,q^2)$ is related to the functions G_i by

$$2MW_1^{p,n}(\nu,q^2) = \sum_{i=1}^{3} e_i^2 G_i^{p,n}(x)\,. \qquad (3.166)$$

Using (3.162) and (3.166), we find [see (3.43)]

$$R = Q^2/\nu^2 = 2Mx/\nu\,. \qquad (3.167)$$

In particular, in the limit $\nu \to \infty$ under the condition $x = \text{const}$, we find $R \to 0$. This is the essence of the Kuti-Weisskopf model. This model gives a qualitatively satisfactory description of the behavior of the function νW_2, correctly describes the behavior of the structure functions in the limit $x \to 0$, and reflects the basic features of the cross section for deep inelastic scattering of leptons by nucleons. On the other hand, this model suffers from several disadvantages: (1) It gives a statistically unconvincing description of deep inelastic *ep* scattering (according to the χ^2 test); (2) it does not accurately predict the behavior of the difference $F_2^p - F_2^n$ near small values of x; and (3) it does not give the structure functions a dependence on Q^2 (there are also other disadvantages). Since we know that scaling is violated, we need to modify the Kuti-Weisskopf model in such a way that the structure functions W_i contain the Q^2 dependence predicted by QCD. We do this in the following subsection.

9. Models describing the scaling violation of the structure functions

The deviation from scaling can be seen at accessible energies. We do not know how the structure functions behave at very high energies. Strictly speaking, scaling should be violated (if for no other reason) because there are mass terms in the interaction Lagrangians. The masses of elementary particles are associated with a length scale which is manifested in the structure functions of deep inelastic scattering as a violation of scaling (Sec. 4 of Chap. 3).

In the simplest parton model, a violation of scaling can be achieved by assuming that the partons have a structure and are described by some form factor.

In QCD one can introduce the following physical interpretation of the scaling violation of the structure functions. In the interaction of a virtual photon of momentum Q^2 with one of the quarks of the nucleon, some momentum distribution function of the quarks within the nucleon is manifested. If the square of the 4-momentum of a virtual photon, Q^2, is increased to Q'^2 ($Q'^2 > Q^2$), we can imagine a case in which a quark which has undergone an interaction emits a gluon, and an additional gluon with a momentum $k \lesssim Q' - Q$ appears in the nucleon. It can be assumed that the emitted gluon will have an energy high enough to convert into a $q\bar{q}$ pair; the addition of this pair to the sea of the nucleon changes the original momentum distribution of the sea quarks. In QCD, in which the emission of gluons and their conversion into $q\bar{q}$ pairs are taken into account, the distribution functions of the valence quarks, the sea quarks, and the gluons thus acquire a Q^2 dependence.

When the perturbation-theory series in QCD are summed, the deviation from scaling arises in a natural way,[111] as a result of the Q^2 dependence of the coupling constant $\alpha_s(Q^2)$. A deviation from scaling can be introduced in a model. Below we will discuss the construction of structure functions $W_i(\nu, Q^2)$ in the model of Ref. 112, which meets all the requirements of QCD and which gives a good description of a large body of experimental evidence.

Quantum chromodynamics predicts two equivalent forms of the evolution of the distribution functions as functions of Q^2: as an evolution of moments (3.133) and (3.134) and as a system of Lipatov–Altarelli–Parisi integro-differential equations (3.138) and (3.141). In Sec. 7 of Chap. 3 we mentioned some disadvantages of describing the experimental data by means of the moments of structure functions, so we would like to have predictions of the scaling violation in QCD directly for the quark and gluon distribution functions. In the model[112] discussed below, an explicit analytic expression is found for the dependence of the quark and gluon distribution functions on the variables x and Q^2. The distribution functions themselves satisfy the QCD evolutionary equations highly accurately both in the leading-log approximation and in the next higher order of a perturbation theory in the effective coupling constant $\alpha_s(Q^2)$.

The model structure functions which are found may thus be regarded as approximate solutions of the QCD evolutionary equations. We already mentioned in Secs. 7 and 8 of Chap. 3 that the initial conditions (the x distributions of the quarks and gluons at a fixed value of Q_0^2) cannot be calculated within the framework of QCD. According to the model which we are considering here, the initial conditions are calculated by a procedure of reconstructing the distribution functions over the entire range of x ($0 \leqslant x \leqslant 1$) on the basis of their limiting behavior in the region at the point $x \approx 0$ (Sec. 8 of Chap. 3).

We choose initial (seed) distribution functions for the valence quarks ($\tilde{q}_v$), the sea quarks ($\tilde{q}_s$), and the gluons ($\tilde{G}$) in the following form [cf. (3.148a)–(3.150a)]:

$$\begin{aligned} \tilde{q}_v(x,Q^2) &= a_v(Q^2)x^{-\alpha(0)}; \\ \tilde{q}_s(x,Q^2) &= a_s(Q^2)x^{-1}; \\ \tilde{G}(x,Q^2) &= a_g(Q^2)x^{-1}\exp(-\beta x), \end{aligned} \tag{3.168}$$

where $a_v(Q^2) = a_s(Q^2)$ are functions which describe the deviation from scaling (these functions will be refined below); $\alpha(0) = 1/2$ corresponds to the choice of the A_2-meson Regge trajectory, which ensures the correct asymptotic behavior of the cross section for deep inelastic scattering; and the factor $\exp(-\beta x)$ has the meaning of a Boltzmann exponential function. In other words, the gluon distribution within a nucleon is interpreted in a statistical way. Writing an arbitrary n-particle quark-gluon state in the form in (3.154), and integrating over the momenta of all the quarks and gluons except the one in which we are interested, we find the distribution functions

$$q_v(x,Q^2) = x^{-1/2} \frac{(1-x)^\tau}{B(1/2,\tau+1)} \frac{\Phi(a_g,\tau-1;-\beta_q(1-x))}{\Phi(a_g,\tau+3/2;-\beta_q)};$$

$$q_s(x,Q^2) = \frac{a_s(Q^2)}{8x}(1-x)^{\tau+1/2} \frac{\Phi(a_g,\tau+3/2;-\beta_q(1-x))}{\Phi(a_g,\tau+3/2;-\beta_q)}; \tag{3.169}$$

$$G(x,Q^2) = \frac{a_g(Q^2)(1-x)^{\tau+1/2}}{x} \frac{\Phi(a_g,\tau+3/2;-\beta_g(1-x))}{\Phi(a_g,\tau-3/2;-\beta_g)} \exp(-\beta_g x),$$

where $\tau(Q^2) = a_s(Q^2) + a_g(Q^2)$, $\Phi(a, b; z)$ is the confluent hypergeometric function, and β_q and β_g are the numbers which determine the evolution of the quarks and the gluons in different ways. The integration is carried out with the help of the δ function $\delta(1-\Sigma y_i)$, which expresses conservation of 4-momentum; this circumstance corresponds to primitive confinement of quarks and gluons. The divergences which arise in the course of the integration are removed by introducing a normalization constant Z [see (3.154)].

The x dependence in (3.169) is determined by the choice of the functions $\tau(Q^2)$, $\beta_q(Q^2)$, and $\beta_g(Q^2)$. The QCD evolutionary equations impose some extremely severe restrictions on the analytic form of the x dependence of the distribution functions $q_v(x, Q^2)$, $q_s(x, Q^2)$, and $G(x, Q^2)$. It is by no means true that just any prespecified x dependence will permit a satisfactory solution of this problem. It is necessary to choose Q^2 dependences of the functions τ, β_q, and β_g which will satisfy the Lipatov-Altareli-Parisi equations after functions (3.169) are substituted into them. For this purpose it is sufficient to choose these functions in the form

$$\begin{aligned} \tau(Q^2) &= \tau_0 + \tau_1 s; & a_v(Q^2) &= a_0 + a_1 s; \\ \beta_g(Q^2) &= \beta_0 + \beta_1^g s; & \beta_q(Q^2) &= \beta_0 + \beta_1^q s, \end{aligned} \tag{3.170}$$

where the numbers τ_0, a_0, and β_0 are determined at the fixed value of $Q^2 = Q_0^2$ (in this case we have $s = 0$ and $\beta_q = \beta_g$); $\tau_1 = 16/25$ is a parameter which is defined in QCD [in the leading-log approximation, it is $\tau_1 = 16/(33-2f)$, where f is the number of quark flavors; for $f = 4$, we would have $\tau_1 = 16/25$]; and $s = \ln[\ln(Q^2/\Lambda^2)/\ln(Q_0^2/\Lambda^2)]$ is an evolutionary variable.

Setting $Q^2 = Q_0^2$, we recognize expressions (3.170) as the initial conditions on the evolutionary equations. If we use a set of experimental data for

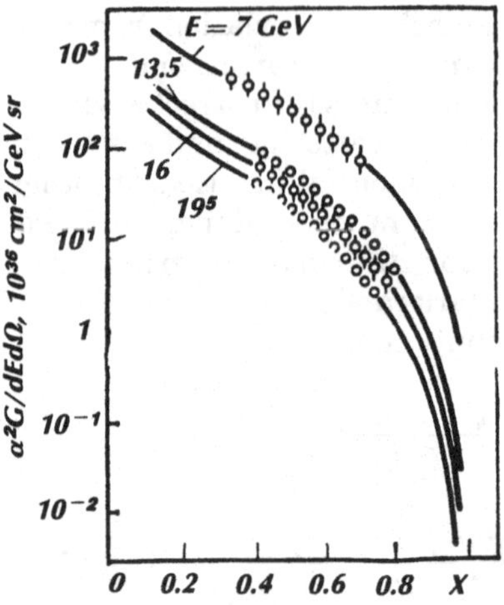

Figure 3.23. Differential cross section for deep inelastic *ep* scattering. $\theta = 50°$. The solid lines are the predictions of the model of Ref. 112.

the functions $F_2(x, Q^2)$ and $F_3(x, Q^2)$ at a certain value of Q_0^2 and for various values of the variable x, we can determine the values of τ_0, a_0, and β_0. After we substitute the numerical values of these parameters into (3.169), we find distribution functions (3.169) which depend on only the three adjustable numerical parameters a_1, β_1^q, and β_1^g. The parameters a_1, β_1^q, and β_1^g are determined from the requirement that the evolution as a function of Q^2, which is predicted by the QCD equations, and the evolution as a function of Q^2, which is determined by expressions (3.169), must be the same for all values of x in the interval $0 \leqslant x \leqslant 1$ and for $Q^2 \geqslant Q_0^2$. This problem reduces to one of minimizing a certain functional by numerical calculation.

It turns out that a model with the three parameters a_1, β_1^q, and β_1^g reproduces all the moments up to $n = 20$ highly accurately. The deviation of the Q^2 evolution of the model moments from that of the QCD moments does not exceed 8–10% up to $Q^2 \leqslant 2 \times 10^4$ $(\mathrm{GeV}/c)^2$ and the deviation at $Q^2 \leqslant 1000$ $(\mathrm{GeV}/c)^2$ does not exceed 2–3%. Incorporating the corrections of next higher order in $\alpha_s(Q^2)$ in the QCD evolutionary equations improves the agreement with experimental data.

Here are the parameter values for the approximate solutions of the QCD evolutionary equations in the leading-log approximation:

τ_0	τ_1	a_0	a_1	β_0	β_1^q	β_1^g
1.97	16/25	0.992	0.619	– 2.32	– 0.995	5.95

Figures 3.23–3.26 demonstrate the accuracy with which the experimental data are reproduced by the model of Ref. 112.

Distribution functions (3.169) have been used to calculate the electromagnetic and weak form factors of the nucleon and also for calculations on the production of the C^{++} (1232) resonance and the C^{++} (2260) and C^{++} (2420) charmed resonances in neutrino interactions. In all cases, they give a good description of the experimental data.

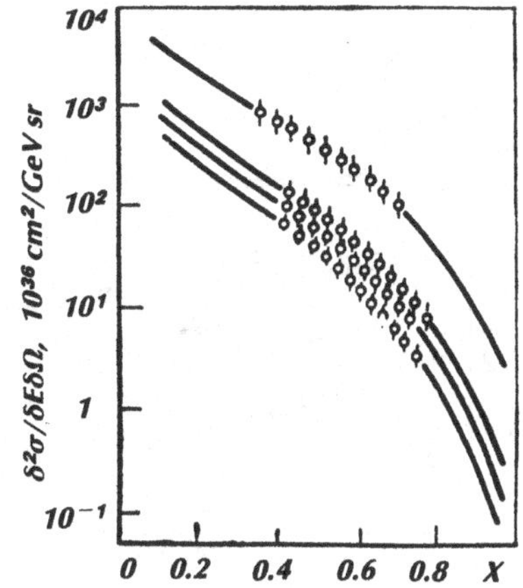

Figure 3.24. Differential cross section for deep inelastic *ed* scattering. The notation is the same as in Fig. 3.23.

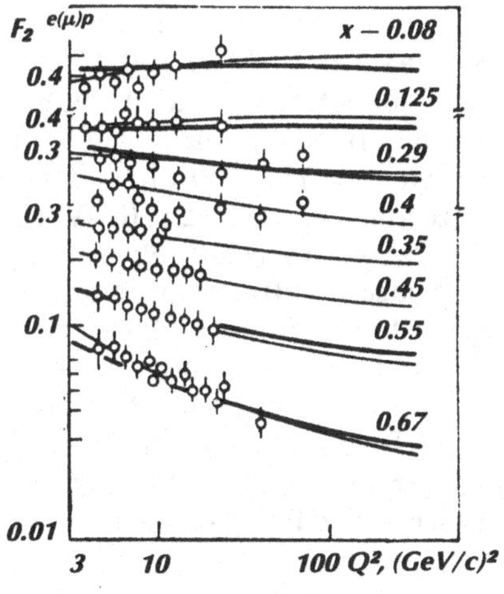

Figure 3.25. Structure function $F_2^{e(\mu)p}$ for the deep inelastic scattering $e(\mu) + p \to e(\mu) + X$. (Heavy lines) leading-log approximation; (light lines) predictions incorporating the order in α_s following the logarithmic order. $\Lambda = 0.3$ GeV (SLAC and Fermilab data).

Functions (3.169) can thus be regarded as approximate solutions of the QCD evolutionary equations. They are universal functions in the sense that they can be used for calculations on various physical processes.

We have already emphasized, in Sec. 7 of Chap. 3, the fundamental importance of determining the value of Λ which is a parameter of QCD. In a description of deep inelastic scattering, this parameter is an organic part of all the relations derived by QCD perturbation theory, since it appears in the definition of the effective coupling constant. In the model of Ref. 112 it enters the functions $\tau(Q^2)$, $a(Q^2)$, and $\beta(Q^2)$ through the variable s.

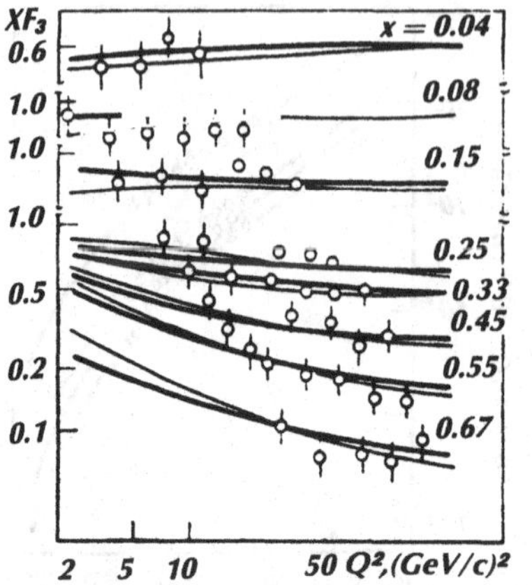

Figure 3.26. Structure function xF_3 for the deep inelastic scattering $\nu(\bar{\nu})N \rightarrow \mu^-(\mu^+) + X$ by an isoscalar target. The notation is the same as in Fig. 3.25. $\Lambda = 0.3$ GeV (data from the CDHS group at CERN).

The values of the parameter Λ have been decreasing continually, from 750 MeV (in 1975) to 150 MeV (1981). It follows from an analysis of deep inelastic scattering that the value of Λ lies in the interval 150–400 MeV, and it follows from an analysis of bound $c\bar{c}$ of charmed particles that it lies in the interval $\Lambda = 150$–100 MeV. The following picture is seen in deep inelastic scattering: At $Q^2 \lesssim 25$ $(\text{GeV}/c)^2$, the structure functions $F_2(x, Q^2)$ and $xF_3(x, Q^2)$ vary relatively rapidly with Q^2, and this behavior is described more successfully by large values of Λ ($\Lambda > 400$ MeV). At $Q^2 < 25$ $(\text{GeV}/c)^2$, higher twist values $\tau > 2$ can contribute significantly to the structure functions and can thus change the conclusions regarding the value of Λ. From the theoretical standpoint, the twist corrections to the structure functions $F_i(x, Q^2)$ should be of the form $h_k^2 f_k(x)/Q^{2k}$, where h_k^2 are positive constants, and $f_k(x)$ $(k = 1, 2, \ldots)$ are unknown functions of x. The general expression for the structure functions is thus written

$$F_i(x,Q^2) = F_i^{\text{QCD}}(x,Q^2)\left[1 + \sum_k \frac{h_k^2 f_k x}{Q^{2k}}\right],$$

where $F_i^{\text{QCD}}(x, Q^2)$ are structure functions which satisfy the LAP equations. In perturbation theory, the twist contributions cannot be calculated accurately because perturbation theory is not applicable in QCD at small values of Q^2, where these contributions can become significant (since they are inversely proportional to Q^{2k}). Approximate calculations, however, can be carried out. Further difficulties confront attempts to determine the twist corrections at small values of Q^2, because over a small interval of Q^2 it is difficult to distinguish the functional behavior of the twist terms, $1/Q^{2k}$, from the QCD behavior of the structure functions $F_i^{\text{QCD}}(x, Q^2)$, which have a logarithmic dependence on Q^2. At our present level of understanding of QCD, we are obliged to resort to models for the functions $f_i(x)$. They are usually chosen in the form $f(x) \sim P(x)/(1-x)^r$, where $P(x)$ is a simple polynomial $a + bx$ or $a + bx + cx^2$, and r is 1 or 2. At $Q^2 > 50$ (GeV/c)^2 the structure functions $F_2(x, Q^2)$ and $xF_3(x, Q^2)$ vary slowly with increasing Q^2, and in this case it is more convenient to use small values of Λ ($\Lambda < 200$ MeV). However, Λ also depends on the number of quark flavors in the expression for the coupling constant $\alpha_s(Q^2)$ [see (3.126) and (3.127)]. At $Q^2 < 5$ (GeV/c)^2, we should use the value $f = 3$; in the interval $5 \lesssim Q^2 \lesssim 100$ (GeV/c)^2 we should use $f = 4$; and in the interval $100 \lesssim Q^2 \lesssim 1000$ (GeV/c)^2 we should use $f = 5$.

The value $f = 4$ is ordinarily used in analyzing experimental data on the deep inelastic scattering of leptons by nucleons and nuclei. This approach generally reduces the value of Λ if data from the region $Q^2 > 100$ (GeV/c)^2 are included in the analysis. For deep inelastic scattering of leptons by nuclei it is necessary to consider nuclear-structure effects, which lead to an increase in Λ. Both corrections (choosing a correct value of f and allowing for the nuclear structure) increase the value of Λ; we can expect that the value of Λ will exceed 200 MeV in the deep inelastic scattering of leptons by nucleons and nuclei.

Since the asymptotic behavior of the cross section for deep inelastic scattering is determined by the exponent in the valence-quark distribution [see the discussion of (3.152) and (3.153)], it appears that at $Q^2 \gg 50$ (GeV/c)^2 the structure functions can be described by means of solely nonsinglet combinations of the quark distribution. Specifically, analysis of the experimental data reveals that the contribution of the singlet combinations of the quark and gluon distribution does not exceed 10–15% at $Q^2 > 100$ (GeV/c)^2. Despite the fact that the singlet combinations contribute little, their incorporation in an analysis of experimental data can raise the value of Λ by a factor of 1.5–2. The value of Λ is sensitive to the incorporation of the next higher order in $\alpha_s(Q^2)$ in the QCD evolutionary equations and to the evolution of the moments with Q^2.

All these brief comments show how unreliable the determination of Λ is. Since Λ appears in the argument of a logarithm in the effective coupling constant, it would seem that the interval of Q^2 values accessible to physicists is still too narrow for a relatively accurate determination of Λ. Even in the experiments by the BCDMS and EMC groups (Sec. 6 of Chap. 3), where the value of Q^2 reached 300 (GeV/c)^2, the parameter Λ was determined only within a large uncertainty, and its value varied significantly with the particular choice of scheme for analyzing the experimental data. The problem of determining reliable values of Λ will apparently be resolved at the accelerators of the next generation, with particle energies of 2–5 GeV.

In Chaps. 4 and 5 we will return repeatedly to the determination of the parameter Λ from other processes.

10. Inclusive spectra in deep inelastic scattering of leptons by nucleons

Logunov *et al.*[113] were the first to introduce the concept of inclusive processes. Study of the inclusive spectra of the hadrons produced in the deep inelastic scattering of leptons is of major interest both for learning about the structure of the nucleon and for studying the mechanism underlying scaling.

Let us consider the process (Fig. 3.27)

$$l + N \to l' + h + \text{anything}\,, \tag{3.171}$$

where the particles l' and h are detected in coincidence (h represents the inclusive hadron which has been singled out). We examine process (3.171) in accordance with the following chain:

$$l \to l'\gamma\,; \tag{3.172}$$

$$\gamma + N \to \text{hadrons} \tag{3.173}$$

or

$$\gamma + N \to h + \text{hadrons}\,. \tag{3.174}$$

The virtual photon in reactions (3.172)–(3.174) is characterized by a 4-momentum transfer q^2 and a photon polarization parameter ε [see (3.14)]. We introduce yet another variable: s, which is the square of the total energy of the $\gamma + N$ system in its c.m. frame.

The relationship between the differential cross section for the reaction $l + N \to l' +$ hadrons, characterized by definite values of q^2, ε, and s, and the total cross section for reaction (3.173), σ_{total}, for the same values of q^2 and s, is described by

$$\sigma_{\text{total}}(q^2,s) = (1/\Gamma)d^2\sigma/dq^2ds\,, \tag{3.175}$$

where Γ is the flux of virtual photons in reaction (3.172). Let us consider reaction (3.174), which corresponds to the inclusive production of a hadron h (a π meson) by a virtual photon, which is characterized (as mentioned above) by an invariant value of q^2 and ε in the c.m. frame of the $\gamma + N$ system. The kinematics in this system is simple. The 4-momenta of the photon and nucleon are denoted by $q(q_0, \mathbf{q})$ and $P(E_N, \mathbf{P})$, respectively; that of the hadron h is $p(E_h, \mathbf{p})$; μ is the mass of the hadron; the projections of the vector $\mathbf{p}$ which are, respectively, longitudinal and transverse with respect to $\mathbf{q}$ are denoted by $p_\parallel$ and $p_\perp$; and M_X is the invariant mass of the final state of the hadrons (without hadron h). By definition we have

$$\begin{aligned}
&\mathbf{q} + \mathbf{P} = 0; \quad E_h = \sqrt{\mathbf{p}^2 + \mu^2} = \sqrt{p_\parallel^2 + p_\perp^2 + \mu^2}\,;\\
&s = (q + P)^2 = (q_0^2 + \sqrt{P^2 + M^2})^2\,;\\
&M_X^2 = (q + P - p)^2 = (q + P)^2 + p^2 - 2p(q + P)\,,
\end{aligned} \tag{3.176}$$

or

$$M_X^2 = s + \mu^2 - 2s^{1/2}(\mu^2 + p_\parallel^2 + p_\perp^2)^{1/2}\,.$$

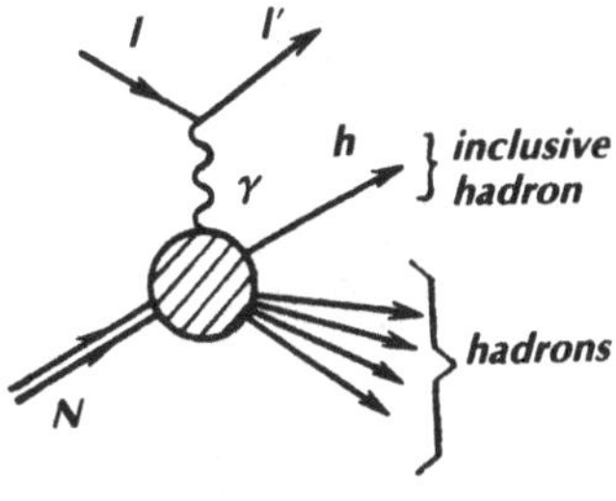

Figure 3.27. Diagram of an inclusive process in the one-photon approximation. (1) Inclusive hadron; (2) hadrons.

The hadron h can be described in this system by means of (for example) the following variables: $p_\perp^2$, the square of the transverse momentum component; the azimuthal angle φ, which is determined by the normals to the lepton scattering plane and the plane formed by the vectors $\mathbf{q}$ and $\mathbf{P}$; and the ratio

$$p_\parallel / p_{\max} = x_\parallel \,, \tag{3.177}$$

where $p_{\max}$ is the maximum of hadron h in reaction (3.174) at given values of q^2 and s. This maximum momentum can easily be found from (3.176) under the condition $M_x^2 = M^2$:

$$E_{\max} = (s + \mu^2 - M^2)/2s^{1/2}\,, \tag{3.178}$$

or

$$E_{\max} = \sqrt{p_{\max}^2 + \mu^2}\,. \tag{3.179}$$

The quantity $x_\parallel$ is approximately the same as the Feynman variable $x_\Phi = 2p_\parallel/\sqrt{s}$. In the limits $s \to \infty$, $E_{\max} \to p_{\max} \to s^{1/2}/2$, the value of $p_{\max}$ is close to $p_\parallel$.

In connection with the discussion of the kinematics, we recall the contents of the limiting fragmentation hypothesis which was advanced by Yang *et al.*[114] Fragmentation can be illustrated by the diagram in Fig. 3.28. According to this hypothesis, it is assumed that at very high energies of the impinging virtual photon the secondary particles arise as fragments of the target (in our case, a nucleon) and a photon. If we analyze the process illustrated in Fig. 3.28 in the c.m. frame (γ + nucleon), we would relate values $p_\parallel > 0$ ($x_\Phi > 0$, $x_\parallel > 0$) with the fragmentation of the impinging photon, while we would relate values $p_\parallel < 0$ ($x_\Phi < 0$, $x_\parallel < 0$) with the fragmentation of the target. In the region $x_\parallel < 0$ the value of $p_\parallel$ calculated in the laboratory frame does not depend on s, so that there is evidently justification for calling the process "limiting fragmentation." The region $p_\parallel \approx 0$ ($x_\parallel \approx 0$, $x_\Phi \approx 0$) is the "central region."

We turn now to the differential cross section for the virtual photoproduction of a hadron. At given values of q^2, s, and ε, the differential cross section for the production of a hadron in the intervals $dp_\perp^2$, $d\varphi$, and $dx_\parallel$ is given by [cf. (3.175)]

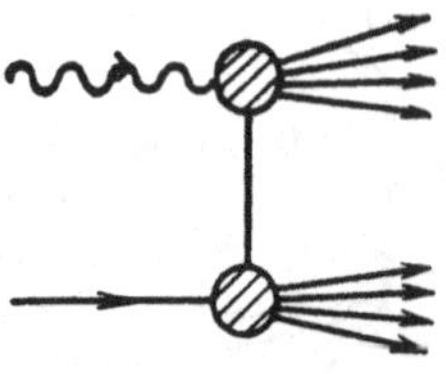

Figure 3.28. Diagram of a fragmentation process.

$$\frac{d^3\sigma(q^2,s)}{dp_\perp^2 d\varphi\, dx_\parallel} = \frac{1}{\Gamma}\frac{d^5\sigma}{dp_\perp^2 d\varphi\, dx_\parallel\, dq^2 ds}. \tag{3.180}$$

It is more convenient to use the ratio of the cross sections in (3.180) and (3.175), so that we can eliminate the arbitrary quantity Γ:

$$\frac{1}{\sigma_{\text{total}}(q^2,s)}\frac{d^3\sigma(q^2,s)}{dp_\perp^2 d\varphi\, dx_\parallel} = \frac{d^5\sigma}{dp_\perp^2 d\varphi\, dx_\parallel\, dq^2 ds}\Big/\frac{d^2\sigma}{dq^2 ds}. \tag{3.181}$$

It is a fairly simple matter to interpret (3.181). The left-hand side is the ratio of the cross sections for reactions (3.171) and $l + N \to l' +$ anything, i.e., the ratio of the number of scattered leptons with values q^2 and s which are measured in coincidence with the hadron in the intervals $dp_\perp^2$, $d\varphi$, and $dx_\parallel$, to the number of scattered leptons with the same values of q^2 and s for deep inelastic scattering. The left-hand side of expression (3.181) is usually written in a Lorentz-invariant form:

$$\frac{1}{\sigma_{\text{total}}} E\frac{d\sigma}{dp^3} = \frac{2E_{\text{c.m.}}}{p_{\max}^{\text{c.m.}}}\frac{d^3\sigma(q^2,s)}{dp_\perp^2 d\varphi\, dx_\parallel}\frac{1}{\sigma_{\text{total}}(q^2,s)}. \tag{3.182}$$

A function of a single variable is often used. This function, $F(x_\parallel)$, is found by integrating expression (3.182) over $p_\perp^2$ and taking an average over φ; it is called a "structure function" (which should not be confused with the structure functions $2MW_1$ and νW_2):

$$F(x_\parallel) = \frac{1}{\pi}\int_0^{2\pi} d\varphi \int_0^\infty dp_\perp^2 \frac{E_{\text{c.m.}}}{p_{\max}^{\text{c.m.}}}\frac{q^3\sigma(q^2,s)}{dp_\perp^2 d\varphi\, dx_\parallel}\frac{1}{\sigma_{\text{total}}(q^2,s)}. \tag{3.183}$$

From (3.182) we find yet another differential quantity:

$$\Delta N^\pm(x_\parallel^{(1)},x_\parallel^{(2)}) = \int_{x_\parallel^{(1)}}^{x_\parallel^{(2)}} dx_\parallel \int_0^{2\pi} d\varphi \int_0^\infty dp_\perp^2 \frac{d^3\sigma(q^2,s)}{dp_\perp^2 d\varphi\, dx_\parallel}\frac{1}{\sigma_{\text{total}}(q^2,s)}. \tag{3.184}$$

This is the number of charged ($\pm$) hadrons per interval $\Delta x_\parallel = x_\parallel^{(2)} - x_\parallel^{(1)}$. The cross section $d^3\sigma(q^2,s)/dp_\perp^2 d\varphi\, dx_\parallel$, which can be measured experimentally, and which appears in the integrals in (3.183) and (3.184), is usually approximated by an expression of the type

$$\frac{d^3\sigma(q^2,s)}{dp_\perp^2 d\varphi\, dx_\parallel} \sim \exp(-bp_\perp^2)(A + B\cos\varphi + C\cos 2\varphi). \tag{3.185}$$

In (3.185) it is assumed that the quantities b, A, B, and C depend on $x_{\parallel}$. The term with $\cos\varphi$ describes an interference between the amplitudes for the scattering of longitudinal and transverse photons, while the term with $\cos 2\varphi$ describes the presence of any polarization dependence for the transverse part of a virtual photon.

If approximation (3.185), with a single exponential function, is unsatisfactory, one turns to an approximation of the same form but with two exponential functions. The experimental data indicate that an approximation by two exponential functions is preferable. In this quark-parton model of the nucleon, these exponential functions can be interpreted in the following way: One of them—the dominant one—arises from the production of inclusive hadrons at valence quarks, while the other, which is manifested only at small values of $x_{\parallel}$, arises from the production of hadrons at sea quarks. We can make the replacement $x_{\parallel} \to x_{\Phi}$ everywhere in (3.180)–(3.184). The notation is not standard in the literature, and one sees expressions for the structure functions in terms of both $x_{\parallel}$ and x_{Feyn}.

The notation for the variable z is more standard (the meaning of this variable is similar to that of the variables $x_{\parallel}$ and x_{Feyn}),

$$z = E_h / \nu \tag{3.186}$$

(E_h is the energy of the inclusive hadron in the laboratory frame of reference), and for the rapidity y,

$$y = \tfrac{1}{2} \ln \frac{E_{\text{c.m.}} + p_{\parallel}}{E_{\text{c.m.}} - p_{\parallel}}, \quad \text{or} \quad y = \tfrac{1}{2} \ln \frac{E_{\text{lab}} + p_{\parallel}^{\text{lab}}}{E_{\text{lab}} - p_{\parallel}^{\text{lab}}}, \tag{3.187}$$

where $E_{\text{c.m.}}$ is the energy of the inclusive particle in the c.m. frame of the virtual photon and the nucleon. The difference between the rapidities of the two particles is an invariant under a Lorentz transformation with respect to an axis directed along the momentum of the virtual photon or along $p_{\parallel}$. The difference between the rapidity of the hadron in which we are interested here and that of any other hadron produced in the given event gives us a "relatively" comprehensive picture of the overall event. We say "relatively" here because although the transverse momenta of the hadrons are small they nevertheless distort the results of an analysis of this case. For each hadron one usually takes the difference $\Delta y = y_{\max} - y$, where $y_{\max}$ is the maximum possible rapidity for the given hadron. Since the spectrum of virtual photons covers a wide energy interval, this difference is a more objective characteristic of the inclusive hadron than the rapidity itself.

Most of the experimental results available today are structure functions which depend on only a single variable ($p_{\perp}^2$, z, x_{Φ}, or y). This circumstance simplifies a comparison of these structure functions with the existing quark-parton models. The structure functions are written symbolically as integrals of the expressions $(E/\sigma)d^3\sigma/dp^3$ or $(1/\sigma)d^3\sigma/dp^3$ [see (3.181) and (3.182)],

Figure 3.29. Comparison of experimental data on π^+ mesons (circles) and π^- mesons (crosses) with theoretical curves.

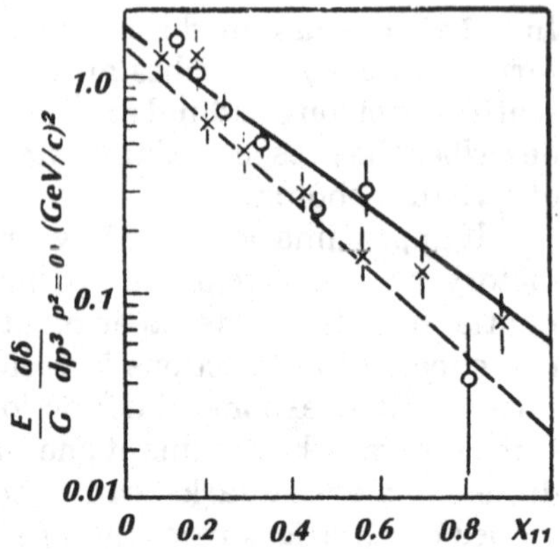

$$F(\Delta y) = \frac{1}{\pi} \int_0^{2\pi} d\varphi \int_0^{p^2_{\perp \max}} dp_\perp^2 \, \frac{E}{\sigma} \frac{d^3\sigma}{dp^3}; \quad F(z) = \frac{z}{\pi\sigma} \frac{d\sigma}{dz},$$

and also $\Delta N(x_\parallel^{(1)}, x_\parallel^{(2)})$ [see (3.184)] and $F(x_\parallel)$ [see (3.183)]. The physical meaning of the structure functions $F(z)$ and $F(\Delta y)$ is clear from their definitions.

Let us take a slightly more detailed look at one of the most important structure functions: $F(x_\parallel)$. This function has been studied, for example, in experiments on the scattering of electrons with an energy $E_e = 19.5$ GeV by protons and deuterons (SLAC), on the scattering of electrons by protons and deuterons at low energies (below 2 GeV, at Cornell), on the scattering of μ mesons by protons at $E = 14$ GeV (SLAC), and on the scattering of μ mesons by protons and deuterons at high energies ($E_\mu = 147$ GeV; Fermilab). Figures 3.29 and 3.30 illustrate the agreement of all the experimental data with regard to the dependence of $d\sigma$ and F on $x_\parallel$. It should nevertheless be noted that while the dependence $F(x_\parallel)$ is precisely the same for the inclusive particles of both signs (a sort of scaling) at high energies, at low energies the structure function for positive hadrons lies above that for negative hadrons. The structure function $F(x_\parallel)$ found in experiments on deep inelastic scattering of neutrinos agrees well with data on the scattering of μ mesons at high energies. There is of course much theoretical interest in the description of the structure functions within the quark-parton model.

Let us consider the "impulse approximation." Here it is assumed that (1) the energy of the virtual photon, which collides with one of the quarks of the nucleon, is completely absorbed by the quark and (2) the $q\bar{q}$ pair produced in the nucleon by the virtual photon produces a damped chain-reaction excitation within the nucleon.

In the damping process, one of the quarks (or antiquarks) produced by the photon combines with an antiquark (or quark) from the sea of the nucleon and forms a boson. The other antiquark (or quark) produced by the photon and a quark (or antiquark) from the sea of the nucleon continue to combine with other sea quarks, forming bosons, until the excitation of the nucleon is eliminated completely. The elimination of the excitation of the nucleon should ter-

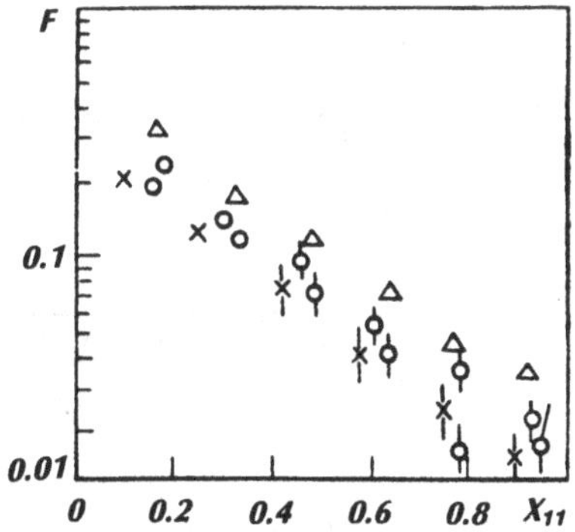

Figure 3.30. Comparison of the functions $F(x_\parallel)$ obtained in the SLAC and Fermilab experiments for hadrons with charges of both signs.

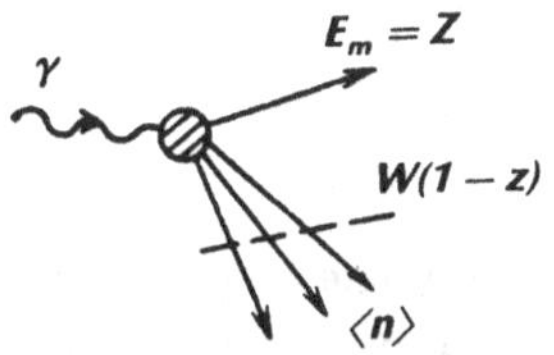

Figure 3.31. Diagram used to eliminate one particle from the data in Fig. 3.32.

minate in a recombination of an odd number (3) of valence quarks into a nucleon. Only those quarks and antiquarks with which the interaction occurs participate in the production of bosons. The others do not participate in this process. The production of π mesons, $K\bar{K}$ pairs, $N\bar{N}$ pairs, etc., can take this path in a nucleon.

It follows in particular that the multiple production of hadrons is a multistep process. This conclusion in turn has several consequences. For example, the law regarding the average multiplicity in the production of charged particles, $\langle n_{\text{ch}} \rangle$, should not change if one particle is excluded from a multistep process (Fig. 3.31). The quantity $\langle n_{\text{ch}} \rangle$ must obviously be smaller in absolute value, and the dependence of $\langle n_{\text{ch}} \rangle$ on the total energy W should not change in shape. These conclusions are confirmed well by experimental data[105] (Fig. 3.32), although, admittedly, the energy interval $3.2 < W < 3.6$ GeV is too narrow to support any claim of a profound test of this consequence. Let us go back to the function $F(x_\parallel)$ and examine its properties according to the quark-parton model. In the impulse approximation the fragmentation of a quark should evidently not depend on just how this quark was excited. The fragmentation of a quark depends on only the fraction of the energy which it transfers to a hadron. In this case the hadronic structure function $F(x_\parallel)$ would naturally be determined in the form

$$F^{\pm}(x,x_\parallel) = \sum_i e_i^2 x\, f_i(x) D_i^{\pm}(x_\parallel) / \sum_i e_i^2 x\, f_i(x)\,. \qquad (3.188)$$

Figure 3.32. Experimental data on $\langle n_{ch}\rangle$ in the interval $3.2 < W < 3.6$ GeV after the elimination of one particle in accordance with the diagram in Fig. 3.31. There is a good agreement with the theoretical curve.

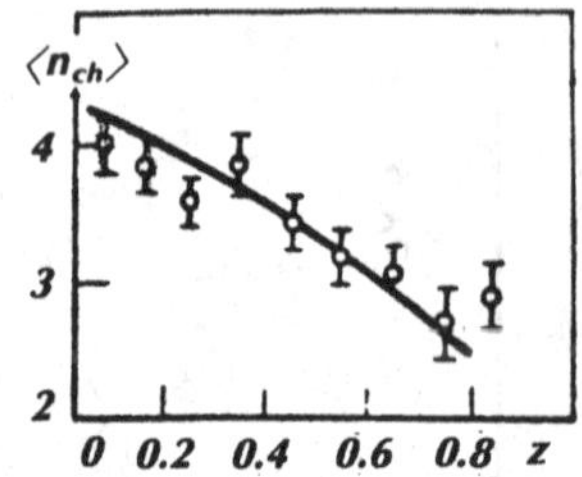

The function $D^{\pm}(x_{\|})$ here describes the properties of the fragmentation of quarks into a positive or negative hadron with a momentum fraction $x_{\|}$.

The division by

$$\nu W_2(x) = \sum_i e_i^2 x\, f_i(x)$$

means that we are averaging the function $D^{\pm}(x_{\|})$ over all of the constituent quarks of the hadron; the product $f_i(x) D^{\pm}(x_{\|})$ in the numerator in (3.188) expresses the independence of the quark fragmentation from the energy (ν) acquired by the quark from the lepton.

The structure functions $F^{\pm}(x, x_{\|})$ can be made independent of the scaling variable $x = Q^2/2M\nu$ by taking the following approach. In a trivalent quark model of a nucleon, the charge-conjugation operation for quarks leads to the relation

$$D_u^+(x_{\|}) = D_{\bar{u}}^-(x_{\|}) .$$

Here the indices u and $\bar{u}$ denote quarks and antiquarks, respectively. On the basis of isospin invariance (under the assumption that the quarks fragment exclusively into π mesons), we can write

$$D_u^+(x_{\|}) = D_{\bar{d}}^-(x_{\|}) .$$

Relations of this sort can of course be written for all the quark species; when we do this, we easily see that the number of independent D functions for the π mesons is 3:

$$\begin{aligned} &D_u^+ = D_{\bar{d}}^- = D_{\bar{u}}^- = D_{\bar{d}}^+ \;;\quad D_d^+ = D_u^- = D_{\bar{u}}^+ = D_{\bar{d}}^- \;; \\ &D_s^+ = D_s^- = D_{\bar{s}}^+ = D_{\bar{s}}^- \;. \end{aligned} \tag{3.189}$$

We assume that the fraction of strange quarks in the fragmentation of quarks into π mesons is small. We substitute relations (3.189) into (3.188):

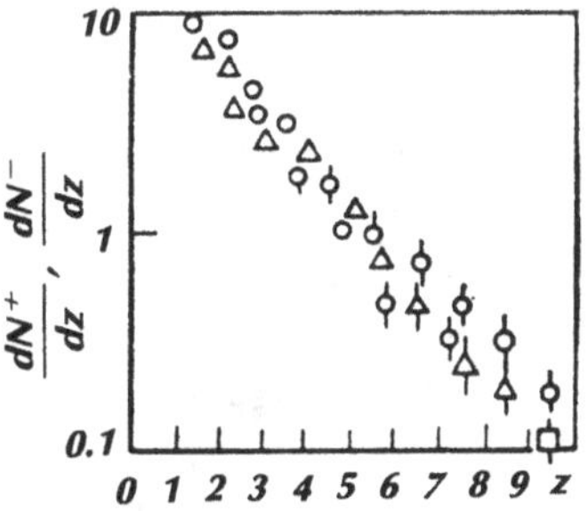

Figure 3.33. Plot of dN/dz vs z for charged hadrons according to various experiments (SLAC). The discrepancy at $z \geqslant 0.6$ can be attributed to the production of ρ^0 mesons.

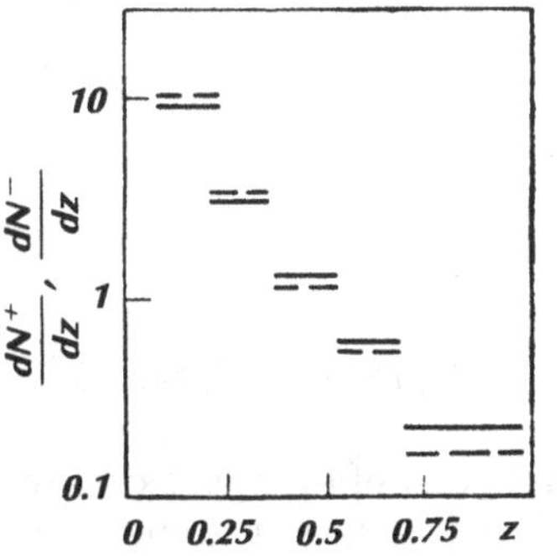

Figure 3.34. Plot of dN/dz vs z for charged hadrons according to a study of the deep inelastic scattering of μ mesons by hydrogen (solid lines) and deuterium (dashed lines). The contribution from the ρ^0 meson has been eliminated ($W^2 > 150$ GeV^2, $\omega > 100$).

$$F^{+}(x,x_{\|}) = \frac{(4/9)[f_u D_u^{+}(x_{\|}) + f_{\bar{u}}(x) D_{\bar{u}}^{+}(x_{\|})]}{(4/9)[f_u(x) + f_{\bar{u}}(x)]}$$

$$\rightarrow \frac{+(1/9)[f_d(x) D_d^{+}(x_{\|}) + f_{\bar{d}}(x) D_{\bar{d}}^{+}(x_{\|})]}{+(1/9)[f_d(x) + f_{\bar{d}}(x)]};$$

$$F^{-}(x,x_{\|})$$

$$= \frac{(4/9)[f_u(x) D_u^{-}(x_{\|}) + f_{\bar{u}}(x) D_{\bar{u}}^{-}(x_{\|})] + (1/9)[f_d(x) D_d^{-}(x_{\|}) + f_{\bar{d}}(x) D_{\bar{d}}^{-}(x_{\|})]}{(4/9)[f_u(x) + f_{\bar{u}}(x)] + (1/9)[f_d(x) + f_{\bar{d}}(x)]}$$

We form the sum of the functions $F^{+}(x, x_{\|})$ and $F^{-}(x, x_{\|})$. We then see that we have

$$F^{+}(x,x_{\|}) + F^{-}(x,x_{\|}) = D_u^{+}(x_{\|}) + D_u^{-}(x_{\|}); \tag{3.190}$$

i.e., the sum of the structure functions for the production of positively and negatively charged π mesons does not depend on the Bjorken variable x. It

Figure 3.35. Comparison of the inclusive spectrum of π^0 mesons (open circles; DESY) with mean values of the $\pi^+ + \pi^-$ spectrum (filled circles and plus signs; SLAC). There is a slight discrepancy at $z \gtrsim 0.8$.

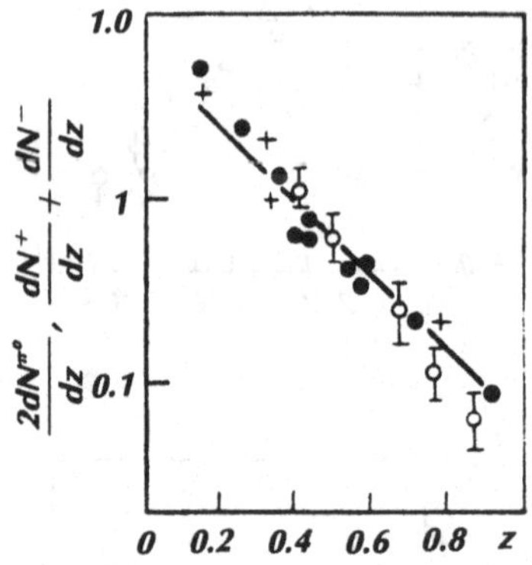

depends on only $x_\parallel$. It should be noted that at $x_\parallel > 0.15$ the variable $x_\parallel$ is nearly the same as the variables z and x_{Feyn}.

On the basis of expression (3.190) we can thus speak of an approximate scaling of the sum of structure functions F^+ and F^- with respect to the variables $x_\parallel$, x_Φ, or z. We give without proof the corresponding expressions for the structure functions F^+ and F^- which appear in neutrino reactions in deep inelastic scattering by a proton:

$$\begin{aligned} F_\nu^+ &= D_u^+(x_\parallel); \quad F_\nu^- = D_u^-(x_\parallel); \\ F_\nu^+ + F_\nu^- &= D_u^+(x_\parallel) + D_u^-(x_\parallel). \end{aligned} \tag{3.191}$$

Comparing (3.190) and (3.191), we see that the sums of the positively and negatively charged inclusive π mesons which are produced in the deep inelastic scattering of electrons (or muons) by protons and of neutrinos by protons are equal. The experimental data agree well with these conclusions (Figs. 3.33–3.36). Because of the hypothesis that the fragmentation of quarks is independent of the method by which they are excited, the behavior of the functions $D^+(x_\parallel)$ (which describe the inclusive spectra of $\pi^\pm$ mesons produced in e^+e^- collisions) as functions of $x_\parallel$ must be the same as in interactions of leptons with nucleons. We can apparently say that there is a universal fragmentation of quarks over a broad energy range ($3 < W < 150$ GeV), which does not depend on the method by which the quark is excited. As Fig. 3.36 shows, the region $z, x_\parallel > 0.8$ is an exception to this rule.

In this subsection we have touched on some characteristic features of inclusive hadron spectra which are attracting progressively more interest from researchers. The quark-parton model and its impulse approximation

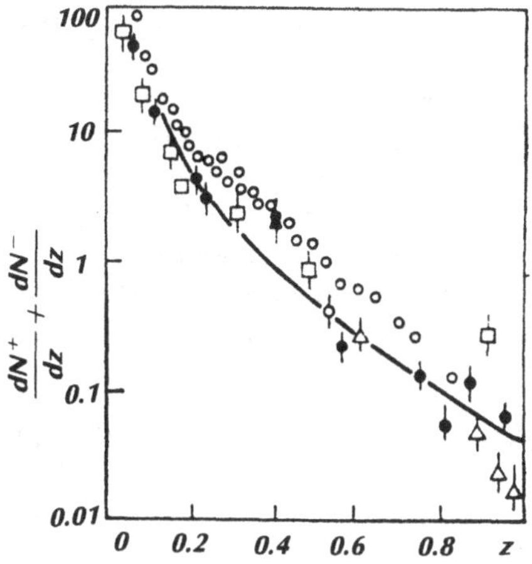

Figure 3.36. Plot of dN/dz for charged hadrons found in a study of e^+e^- annihilation and deep inelastic scattering of neutrinos by protons. (Squares) Fermilab data on the scattering of neutrinos by protons at energies $30 < E_\nu < 200$ GeV; (filled circles, triangles) DESY data ($W = 3.6$ GeV); (open circles) SPEAR data ($W = 7$–7.8 GeV); (line) theoretical fit by the Feynman–Field model.

constitute a first attempt to reach an understanding of the physical phenomena which occur in inclusive processes. The problem of the fragmentation of quarks and gluons into hadrons is also being studied in connection with research on the spectra of inclusive particles in e^+e^- interactions. The Feynman-Field model and certain questions related to the study of inclusive spectra are discussed in Chap. 4.

11. Study of the structure of hadrons in deep inelastic scattering of polarized leptons by polarized nucleons

Recent years have seen an awakening of interest in processes with polarized beams and targets in connection with the development of "frozen" polarized targets, intense sources of polarized electrons, and channels of high-energy beams of μ mesons.

For example, an intense source of polarized electrons has been developed at SLAC. The intensity in the pulse is $10^9/(1.5\,\mu s)$ (120 pulses per second), the polarization is 0.85 ± 0.08, and the stability is high.

The μ beams are automatically polarized along the direction in which the μ mesons are moving (parallel or antiparallel to this direction).

Table 3.4 shows some data on frozen targets and other polarized targets which are presently being used or which are being developed. A high intensity of a polarized beam is a source of noise in experiments with polarized targets, since the beam depolarizes the target as a result of radiative damping. In experiments with an unpolarized target there are no restrictions on the intensity of the polarized beam.

The availability of polarized targets and high-energy beams of leptons presents excellent opportunities for studying the structure of nucleons in electromagnetic interactions of particles and for studying such questions as the jet nature of the behavior of particles produced in deep inelastic scattering, tests of the conservation of *CP* parity, tests of sum rules, and so on.

We are interested here in research on the structure of nucleons in the deep inelastic scattering of polarized leptons by polarized nucleons.

The differential cross section for this process, summed over all the final hadronic states and also over the polarizations of the final lepton in the laboratory frame of reference, is

$$d^2\sigma/dE'd\Omega = [\alpha^2 E'/q^4(kp)]L_{\mu\nu}W_{\mu\nu}\,, \tag{3.192}$$

where

$$\begin{aligned} L_{\mu\nu} &= 2[k_\mu k'_\nu + k'_\mu k_\nu + (1/2)g_{\mu\nu}k^2] + 2i\xi\varepsilon_{\mu\nu\lambda\sigma}q^\lambda k^\sigma\,; \\ W_{\mu\nu} &= \left(-g_{\mu\nu} + \frac{q_\mu q_\nu}{q^2}\right)W_1(\nu,q^2) \\ &+ \left[p_\mu - \frac{q_\mu(pq)}{q^2}\right]\left[p_\nu - \frac{q_\nu(pq)}{q^2}\right]\frac{1}{M^2}W_2(\nu,q^2) \\ &+ i\varepsilon_{\mu\nu\lambda\sigma}q^\lambda\{Ms^\sigma G_1(\nu,q^2) + (1/M^2)[s^\sigma(qp) - p^\sigma(sq)]G_2(\nu,q^2)\}\,. \end{aligned} \tag{3.193}$$

Here k and k' are the 4-momenta in the initial and final states of the leptons; ξ is the helicity of the lepton in the initial state (if the mass of the lepton is ignored, only the longitudinal polarization contributes to the cross section); p and q are the 4-momenta of the initial nucleon and of the virtual photon, respectively; and s is the spin of the initial nucleon.

The hadron tensor $W_{\mu\nu}$ evidently contains comprehensive information on the excitation of a polarized nucleon by a virtual photon. The functions $G_1(\nu, q^2)$ and $G_2(\nu, q^2)$, like $W_1(\nu, q^2)$ and $W_2(\nu, q^2)$, are real and are called "structure functions." They satisfy the symmetry conditions

$$\begin{aligned} &W_{1,2}(-\nu,q^2) = -W_{1,2}(\nu,q^2)\,; \\ &G_1(-\nu,q^2) = -G_1(\nu,q^2); \quad G_2(-\nu,q_2) = G_2(\nu,q^2)\,. \end{aligned} \tag{3.194}$$

The functions $G_{1,2}$ are sometimes replaced by the functions X_1 and X_2:

$$X_1 = MG_1 + \nu G_2; \quad X_2 = -G_2/M\,. \tag{3.195}$$

If the nucleon is polarized in the direction perpendicular to the lepton scattering plane, the asymmetry of the scattered leptons will be zero; this conclusion follows from the invariance under time reversal. A nonvanishing asymmetry arises when the nucleon is polarized in the lepton scattering plane. In this case, one introduces the following asymmetry coefficient of the differential cross section for nucleons which are polarized antiparallel and parallel to the momentum of the impinging lepton, **k**:

$$A_\parallel = \frac{d^2\sigma(\uparrow\downarrow) - d^2\sigma(\uparrow\uparrow)}{d^2\sigma(\uparrow\downarrow) + d^2\sigma(\uparrow\uparrow)}\,. \tag{3.196}$$

For nucleons which are polarized in the lepton scattering plane in the direction perpendicular to the momentum of the impinging lepton, the corresponding asymmetry coefficient is

$$A_{\perp} = \frac{d^2\sigma(\uparrow\rightarrow) - d^2\sigma(\uparrow\leftarrow)}{d^2\sigma(\uparrow\rightarrow) + d^2\sigma(\uparrow\leftarrow)}. \tag{3.197}$$

The cross sections $d^2\sigma(\uparrow\downarrow)$ and $d^2\sigma(\uparrow\uparrow)$ are given by

$$\frac{d^2\sigma}{dE'd\Omega} = \frac{\alpha^2\cos^2(\theta/2)}{4E^2\sin^4(\theta/2)}\left[W_2 + 2\tan^2\frac{\theta}{2}W_1 \pm 2\tan^2\frac{\theta}{2}(E+E'\cos\theta)MG_1 \pm 8EE'\tan^2\frac{\theta}{2}\sin^2\frac{\theta}{2}G_2\right], \tag{3.198}$$

where the plus sign is used for the $(\uparrow\downarrow)$ combination of the electron and nucleon spins, and the minus sign is used for the $(\uparrow\uparrow)$ combination. Here θ is the angle through which the lepton is scattered.

From relations (3.196) and (3.198) we can easily find an expression for the asymmetry coefficient $A_{\parallel}$ in terms of the structure functions W_1, W_2, G_1, and G_2:

$$A_{\parallel} = \frac{(E+E'\cos\theta)MG_1 + Q^2G_2}{W_1 + (1/2)\cot^2(\theta/2)W_2}.$$

The expression for $A_{\perp}$ is

$$A_{\perp} = \frac{E'\sin\theta(MG_1 + 2EG_2)}{W_1 + (1/2)\cot^2(\theta/2)W_2}.$$

(Relations for $A_{\perp}$ and $A_{\parallel}$ are derived in Ref. 43.) In the general case in which the nucleon is polarized in the electron scattering plane at an angle φ from the direction of the momentum of the impinging electron, the asymmetry of the differential cross section is

$$\frac{d^2\sigma(\varphi)/dE'd\Omega - d^2\sigma(\varphi+\pi)/dE'd\Omega}{d^2\sigma(\varphi)/dE'd\Omega + d^2\sigma(\varphi+\pi)/dE'd\Omega} = \frac{[E\cos\varphi + E'\cos(\varphi-\theta)]MG_1(\nu,q^2) - 2EE'[\cos\varphi - \cos(\varphi-\theta)]G_2(\nu,q^2)}{W_1 + (1/2)\cot^2(\theta/2)W_2}.$$

Using the asymmetry coefficients $A_{\parallel}$ and $A_{\perp}$, we can also write the differential cross section as

$$d^2\sigma/dE'd\Omega = (\overline{d^2\sigma}/dE'd\Omega)(1 + \xi s_{\parallel}A_{\parallel} + \xi s_{\perp}A_{\perp}),$$

where the overbar means an average over spin states, ξ is the degree of longitudinal polarization of the lepton, $s_{\parallel}$ is the longitudinal component of the nucleon polarization along the momentum of the impinging lepton, and $s_{\perp}$ is the transverse component of the nucleon polarization. This component is directed perpendicular to the momentum of the impinging electron in the scattering plane of the lepton.

We will study only the coefficient $A_{\parallel}$ below, so we will omit the subscript on it.

The cross section for the scattering of polarized leptons by polarized nucleons is related to the cross section for virtual forward Compton scattering. To describe the virtual Compton scattering, we introduce the special amplitudes $T_{\lambda'_\gamma\lambda'_N,\lambda_\gamma\lambda_N}$, where λ_γ and λ_N are the helicity states of the photon and the nucleon.

We choose a coordinate system in which the z axis is directed along the momentum of the virtual photon, while the x axis lies in the electron scattering plane. The virtual photon has three helicity states: $\lambda_\gamma = 0, 1, -1$. We will use the respective symbols $\cdot$, $\rightarrow$, $\leftarrow$. The nucleon has two helicity states: $+1/2$, $-1/2$, which we represent by $\leftarrow$, $\rightarrow$, respectively. Since the helicity of a state should be conserved in a reaction, we find the four following independent amplitudes:

$$
\begin{aligned}
T_{1/2} &= T(1,1/2\rightarrow 1,1/2){:}(-\rightarrow\leftarrow)\Rightarrow(-\rightarrow\leftarrow);\\
T_{3/2} &= T(1,-1/2\rightarrow 1,-1/2){:}(-\rightarrow\rightarrow)\Rightarrow(-\rightarrow\rightarrow);\\
T_L &= T(0,1/2\rightarrow 0,1/2){:}(\cdot\leftarrow)\Rightarrow(\cdot\leftarrow);\\
T_{TL} &= T(0,-1/2\rightarrow 1,1/2){:}(\cdot\rightarrow)\Rightarrow(-\rightarrow\leftarrow).
\end{aligned}
$$

The imaginary parts of these four amplitudes can be expressed, within coefficients, in terms of the structure functions $W_{1/2}$ and $G_{1/2}$ by means of the relationship between the imaginary part of the amplitude for the Compton scattering of a virtual photon, Im $T_{\mu\nu}$, and the hadron tensor $W_{\mu\nu}$:

$$\text{Im}\, T_{\mu\nu} = 4\pi^2\alpha W_{\mu\nu}\,.$$

For the cross sections for the scattering of virtual photons by nucleons with transition amplitudes $1/2\rightarrow 1/2$ ($T_{1/2}$) and $3/2\rightarrow 3/2$ ($T_{3/2}$), we introduce the notation $\sigma_{1/2}\sim\text{Im}\, T_{1\,1/2,1\,1/2}$ and $\sigma_{3/2}\sim\text{Im}\, T_{1\,-1/2,1\,-1/2}$. We also introduce $\sigma_{Lt}\sim\text{Im}\,\text{T}_{1\,1/2,0\,-1/2}$. It can be shown that the cross sections $\sigma_{1/2}$ and $\sigma_{3/2}$ are related to σ_t [see (3.14)]:

$$\sigma_t = (1/2)(\sigma_{1/2}+\sigma_{3/2})\,. \tag{3.199}$$

The quantities $\sigma_{1/2}$, $\sigma_{3/2}$, and σ_t are always positive, since they are squares of corresponding amplitudes for the Compton scattering of a virtual photon by a nucleon. The quantity σ_{Lt} contains an interference of the scattering amplitudes for the transverse and longitudinal components of the virtual photon and can have either sign.

The total cross sections for the absorption of a virtual photon with different polarizations are expressed in terms of the structure functions W_1, W_2, G_1, and G_2 as follows:

$$
\begin{aligned}
\sigma_t &= (4\pi^2\alpha/K)W_1\,;\\
\sigma_L &= (4\pi^2\alpha/K)[(1+\nu^2/Q^2)W_2 - W_1]
\end{aligned}
\tag{3.200a}
$$

[these expressions are found from (3.14); the quantity K is defined in the same place];

$$\begin{aligned}\sigma_{Lt} &= (4\pi^2\alpha/K)\sqrt{2Q^2}(MG_1 + \nu G_2)\,;\\ \sigma_{1/2} - \sigma_{3/2} &= (4\pi^2\alpha/K)(M\nu G_1 - Q^2G_2)\,.\end{aligned} \tag{3.200b}$$

We also introduce the following notation, which was used in Ref. 20:

$$\begin{aligned}&A = D(A_1 + \eta A_2);\quad A_1 = \frac{\sigma_{1/2} - \sigma_{3/2}}{\sigma_{1/2} + \sigma_{3/2}};\quad A_2 = \frac{2\sigma_{Lt}}{\sigma_{1/2} + \sigma_{3/2}}\,;\\ &D = (E - E'\varepsilon)/E(1 + \varepsilon R) = (1 - \varepsilon^2)^{1/2}\cos\psi/(1 + \varepsilon R)\,;\\ &\eta = \frac{\varepsilon\sqrt{Q^2}}{E - E'\varepsilon} = \left(\frac{2\varepsilon}{1+\varepsilon}\right)^{1/2}\tan\psi \approx \tan\psi;\quad R = \frac{\sigma_L}{\sigma_t}\,,\end{aligned} \tag{3.201}$$

where ψ is the angle between the spin direction of the proton and the momentum of the virtual photon. In terms of the new notation we have

$$\begin{aligned}\frac{d^2\sigma}{dE'd\Omega} &= \left(\frac{d\sigma}{d\Omega}\right)_M\left(\frac{1}{\varepsilon(1 + \nu^2/Q^2)}\right)W_1[1 + \varepsilon R\\ &\pm (1 - \varepsilon^2)^{1/2}\cos\psi A_1 \pm \sqrt{2\varepsilon(1 - \varepsilon)}\,\sin\psi A_2]\,.\end{aligned} \tag{3.202}$$

The plus and minus signs in (3.202) correspond to the configurations ↓↑ and ↑↑ of the spins of the electron and the proton. Expressions (3.196)–(3.202) are used frequently in the recent literature.

The difficulties in obtaining experimental data with polarized beams and targets are still severe, and the number of experiments which have been carried out on deep inelastic scattering of leptons by nucleons is small. On the other hand, there have been many theoretical studies,[115–122] based on several models, which predict different types of behavior for the asymmetry coefficient. In a series of experiments[20] carried out at SLAC, measurements revealed a value $A/D \approx A_1$ [see (3.201)]. Since η is small, the contribution of the term ηA_2 to the asymmetry is small. Figure 3.37 shows the behavior $A_1(x, Q^2)$ for various values of x (the scaling independence of the function A_1 from Q^2 is demonstrated here). Figure 3.38 compares some theoretical curves with experimental data. Although the experimental data were obtained under the assumption of an undisrupted scaling, a comparison with theoretical predictions can still be made, because of the large errors in the experimental values of the coefficient A_1. The experimental data agree with the model which predicts $A_1 \approx 1$ and $x = 1$.

Let us take a brief look at the theoretical models. In the model of Ref. 112 (curve 1 in Fig. 3.38) we have the following relationships:

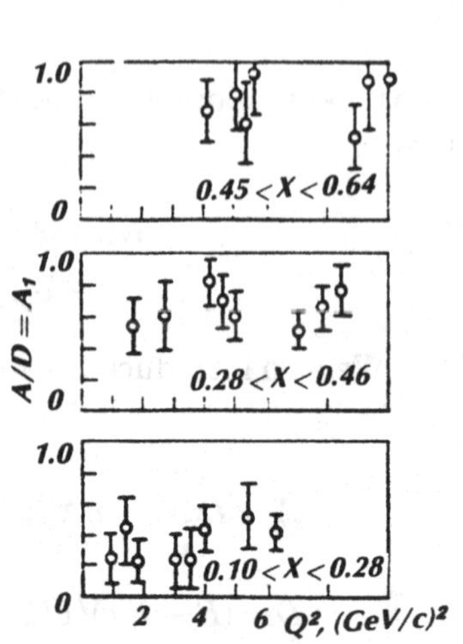

Figure 3.37. Asymmetry coefficient A/D as a function of Q^2 for various values of x.

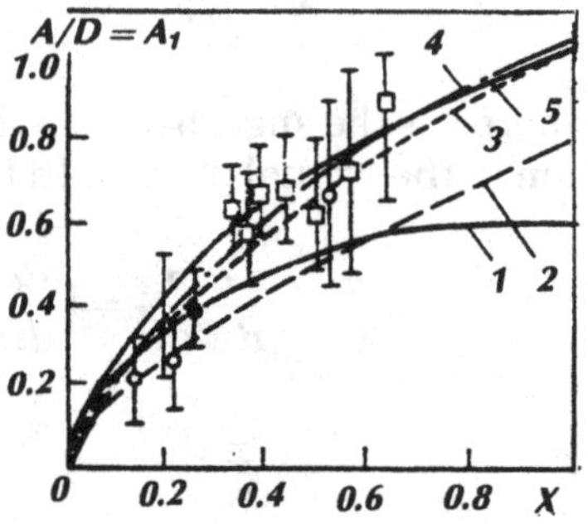

Figure 3.38. Comparison of experimental data with the predictions of various models.[112,117–119,121] Curves 1–5 are explained in the text proper.

$$\Delta W_2^{p,n}(\nu,q^2) = x \sum_{i=1}^{3} e_i^2 G_i^{p,n}(x) \equiv F_2^{p,n}(x)\,;$$

$$2MW_1^{p,n}(\nu,q^2) = \sum_{i=1}^{3} e_i^2 G_i^{p,n}(x) \equiv F_1^{p,n}(x)\,;$$

$$M^2\nu G_1(\nu q^2) = g_1(x); \quad M\nu^2 G_2(\nu,q^2) = g_2(x)\,;$$

$$g_1(x) = \tfrac{1}{4} \sum_i e_i^2 [G_i^{\uparrow}(x) - G_i^{\downarrow}(x)]; \quad g_2(x) = 0\,,$$

where $G_i^{\uparrow}$ and $G_i^{\downarrow}$ are distribution functions which are similar to the functions G_i but which correspond to helicities of the partons which are the same as or the opposite of the helicity of the nucleon.

If we assume that only valence quarks contribute to the structure functions for polarized scattering, and if we use the relations[120]

$$(1/2)[G^{\uparrow}_{1v}(x) - G^{\downarrow}_{1v}(x)] = (2/3)G_{1v}(x)\,;$$
$$(1/2)[G^{\uparrow}_{2v}(x) - G^{\downarrow}_{2v}(x)] = -(1/3)G_{2v}(x)\,,$$

we find

$$g_1^p(x) = (5/18)G^p_{2v}(x); \quad g_1^n(x) = 0\,.$$

We can use these relations to calculate the asymmetry coefficients $A_{\parallel}$ and $A_{\perp}$. In this approach, both the differential cross section for the scattering of unpolarized particles by an unpolarized target and the differential cross section for the scattering of polarized leptons by polarized nucleons are expressed in terms of distribution functions (3.156). The latitude in the choice of these functions is thus narrowed.

In the model of Ref. 112 there is no difference between the momentum distributions of the u and d quarks within a nucleon. In principle, such a distinction can be introduced. If we do this, we find an excellent agreement with experimental data (curve 5 in Fig. 3.38).

It might be suggested that not only the Regge trajectory of an A_2 meson but also the daughter trajectories might contribute to the asymptotic behavior of the cross section for the deep inelastic scattering of leptons by nucleons. These assumptions underlie the model of Ref. 121 (curves 2 in Fig. 3.38). Several models[117,118] introduce a rotation in spin space (a Melosh transformation) of such a nature that the spin-dependent matrix elements turn out to be sensitive to this rotation, while quantities averaged over the spins are out. The introduction of a transformation of this sort leans on the hypothesis that the valence quarks are different from the sea quarks. The valence quarks are surrounded by the sea of quarks and gluons, between which interactions can occur; displacements of the spins of the quarks through certain angles are then possible. This is a sort of relativistic generalization of the model of a nucleon which can be made by various approaches. One is to introduce a Melosh transformation which incorporates the shift of the spins of the valence quarks. This transformation establishes the relationship between the valence quarks and the sea quarks and leads to the greatest deviation from the predictions of the Kuti-Weisskopf model at small momenta of the sea partons.

In the course of the Melosh transformation an additional adjustable parameter is introduced into the model: the angle which the spin of the constituent quark makes with the spin of the nucleon itself. This parameter can be utilized to improve the agreement of the model with experimental data in a description of polarization phenomena in deep inelastic scattering of leptons by nucleons. It turns out that this model gives a good description of the experimental data (curve 3 in Fig. 3.38).

Schwinger's model of sources is based on a completely different physical interpretation of the observable phenomena.[119] In his theory of sources, Schwinger explains Adler's and certain other sum rules for the scattering of neutrinos by unpolarized nucleons and Bjorken's sum rules for the scattering of polarized electrons by polarized nucleons. Schwinger introduces an explicit expression for the asymmetry coefficient A/D. For the ideal case of deep inelastic scattering ($\sigma_L/\sigma_t \to 0, q^2/\nu^2 \to 0$), we have

$$A/D \approx XX^{-1/2}(0.9\omega + 1.1)/(\omega + 1).$$

This expression agrees well with experimental data (curve 4 in Fig. 3.38). In the limit $\omega \to 1$, we have $A/D \to 1$.

It can be concluded that certain quark-parton models apparently give a faithful description of the x dependence of the coefficient A_1, but at the present level of experimental accuracy it is difficult to say which of these models gives the more reliable description of the data.

The study of polarization phenomena in deep inelastic scattering is essentially still in its infancy.

12. Parity breaking in deep inelastic scattering of polarized leptons

Up to this point we have been discussing deep inelastic scattering in the one-photon approximation (Fig. 3.39a). The corrections for the scattering amplitude in the approximation of the exchange of a Z^0 boson (Fig. 3.39b) lead to parity breaking in the deep inelastic scattering of polarized leptons by nucleons or to a P-odd asymmetry. This effect arises because of the presence of the γ_5 matrix in the weak-interaction Hamiltonian [see (3.203) below].

One of the earliest experiments was carried out at the Institute of High-Energy Physics (Serpukhov; Ref. 123) in beams of muons with a momentum of 20 GeV/c. The purpose of that experiment was to observe an interference between the amplitude with one-photon exchange,

$$\sqrt{4\pi\alpha}\langle \mu'|\gamma_A|\mu\rangle(1/q^2)J_A^{\text{had}},$$

and the amplitude in which the lepton and nucleon exchange a Z^0 boson (the neutral weak-interaction current),

$$(G_0^{(\mu)}/\sqrt{2})\langle \mu'|\gamma_\lambda(1+\gamma_5)\mu\rangle J_A^{\text{had}}, \tag{3.203}$$

where $G_0^{(\mu)}$ is the constant of the parity-breaking interaction of leptons with hadrons, which must be determined experimentally. The effective cross section for the process is

$$d^2\sigma/d\nu\, dQ^2 = (d^2\sigma/dq^2 d\nu)_{em}\,[1 + s_\mu(a)5\times 10^{-4}Q^2(G_0^{(\mu)}/G_F)],$$

where the subscript em specifies the cross section calculated in the approximation of one-photon exchange, s_μ is the longitudinal polarization of the μ meson, $3<a<4$, and $G_F \sim 10^{-5}/M_p$ is the Fermi coupling constant. The quantity which is measured experimentally is defined by

$$R = [\sigma(+)-\sigma(-)]/[\sigma(+)+\sigma(-)],$$

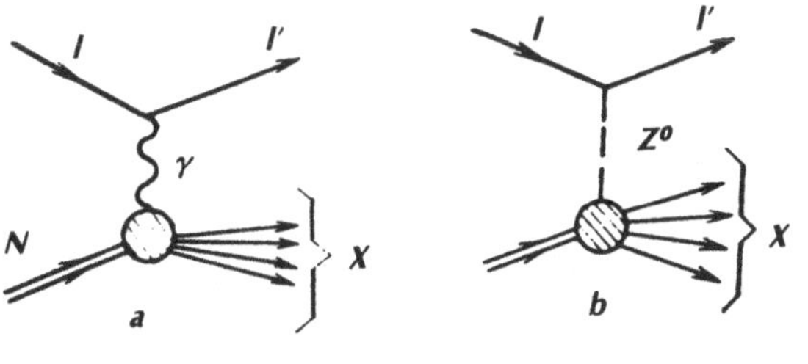

Figure 3.39. Diagrams whose incorporation leads to parity breaking.

where $\sigma(+)$ is the effective cross section measured for μ mesons with positive helicity, and $\sigma(-)$ is the same for μ mesons with negative helicity. The experimental data yielded

$$G_0^{(\mu)} = (+6 \pm 10) G_F$$

and a value of zero for R, within the experimental error. A parity-breaking effect with a magnitude agreeing with the prediction of the Weinberg–Salam theory was observed in an experiment carried out at SLAC on the deep inelastic scattering of longitudinally polarized electrons by unpolarized hydrogen and deuterium targets.[124] These measurements were carried out in the region $1 \leqslant Q^2 \leqslant 1.9\ (\text{GeV}/c)^2$. The asymmetry was defined as the difference and sum of the number of electrons with a positive helicity, $\langle Y(+) \rangle$ (the direction of the electron spin is the same as that in which the electron is moving), and the number with a negative helicity, $\langle Y(-) \rangle$ (the spin of the electron is antiparallel to the direction in which the electron is moving), which are scattered (at an angle of 4°):

$$A_{\text{expt}} = [\langle Y(+) \rangle - \langle Y(-) \rangle] / [\langle Y(+) \rangle + \langle Y(-) \rangle] .$$

The measured asymmetry A is related to A_{expt} by

$$A_{\text{expt}} = |P_e| A \cos 2\Phi_p ,$$

where $|P_e|$ is the degree of polarization of the electron beam, and the angle Φ_p depends on the experimental conditions. Since we have $A \sim Q^2$, it is pertinent to measure the value of A/Q^2. The experimental value of this quantity, measured at a deuterium target with $Q^2 = 1.6\ (\text{GeV}/c)^2$, is $(-9.5 \pm 1.6) \times 10^{-5}$ $(\text{GeV}/c)^2$. This value corresponds to the point $y = 1 - E'/E = 0.21$, where E' is the energy of the scattered electron and E is the energy of the impinging beam. For a hydrogen target we have $A/Q^2 = (9.7 \pm 2.7) \times 10^{-5}\ (\text{GeV}/c)^2$.

In some earlier experiments with muons and electrons, parity breaking had not been found, because the error of those experiments was too large to allow such a small effect to be detected.

Parity breaking in deep inelastic scattering has been analyzed in several theoretical papers (see, e.g., Ref. 125).

Chapter 4 One-photon e^+e^- annihilation

1. Introductory comments

The first accelerator with colliding e^+e^- beams was constructed in Italy in 1960 (Ref. 126). It was supposed that this accelerator would allow a test of QED at high energies (high at the time). It soon became clear, however, that colliding e^+e^- beams can be used successfully to study the hadronic states which arise in the course of the one-photon annihilation of an e^+e^- pair. Colliding-beam accelerators ("colliders") with an energy of 2×0.5 GeV ($E_{\text{c.m.}} \approx 1$ GeV) were soon constructed in the USSR, the USA, and France (Table 1.1). Since then, the experimental physics of colliding beams has occupied a leading position in elementary particle physics, and it has made huge contributions to research on the new J/ψ and Υ families of particles and the τ, D, F, and B mesons. It has also laid the foundation for several fundamental positions of QCD.

What are the advantages and disadvantages of e^+e^- colliders in comparison with ordinary accelerators?

One advantage is the dramatic increase in the energy in the laboratory frame of reference (Table 3.1); another is the possibility of studying unique phenomena. Among the disadvantages are the large radiative loss of the electron beam and the low luminosity of such accelerators. In experiments with a beam of accelerated particles at an ordinary accelerator, the target is ordinary matter in which the density of nuclei is determined by Avogadro's number. In an accelerator with colliding e^+e^- (or $p\bar{p}$) beams the target is instead a beam of accelerated electrons (or protons), with a density several orders of magnitude lower than the density of nuclei in ordinary matter. As a result, the number of collisions (or events) observed in a colliding-beam experiment is several orders of magnitude lower than in an ordinary accelerator. For an ordinary accelerator with a beam of intensity n, which is incident on a fixed target with ρ nuclei per cubic centimeter and with a length l, the number of events, N, can be found from the formula

$$N = \sigma n l \rho = L\sigma \,, \tag{4.1}$$

where L is the luminosity, and σ is the cross section of the process under study. With $n \sim 10^{12}\ \text{s}^{-1}$, $l\rho \sim 10^{23}\ \text{cm}^{-2}$, and a typical value of the cross section for an electromagnetic interaction ($\sigma \sim 10^{-30}$–10^{-31} cm^2), the number of events which can be detected in 1 s is $N = 10^4$–$10^5\ \text{s}^{-1}$. A typical cross section for a strong interaction is $\sigma \sim 10^{-27}$ cm^2; in this case the number of events

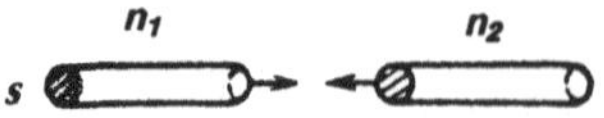

Figure 4.1. Diagram of a collision of colliding e^+e^- beams.

becomes 10^8 cm^{-1}. Where necessary, it can be raised by two or three orders of magnitude. To calculate the number of events for a collider, let us consider a collision of two bunches of particles which collide at a frequency determined by the frequency (f) at which the bunch revolves along its orbit. We denote by n_1 and n_2 the numbers of particles in the bunches, by s the cross-sectional area of a bunch (Fig. 4.1), and by k the number of bunches in the beam. The number of events is then given by

$$N = (n_1 n_2 f/s)k\sigma = L\sigma .$$

The luminosity L can be expressed in terms of the currents (i) of the charged e^+e^- beams. Since we have $i = nef\,k$, we find

$$N = i_1 i_2 \sigma / e^2 s\, f\, k .$$

Typical parameter values for e^+e^- colliders are $k = 1$, $f \sim 10^6$ s^{-1}, $i \approx 3\times10^{17}$ s^{-1}, and $s \approx 0.01$ cm^2. Using these values, we find $L = 10^{35}/0.01\times10^6 \sim 10^{31}$ cm$^{-2}\times$s^{-1}. For electromagnetic cross sections, $\sigma \sim 10^{-30}$–10^{-31} cm^2, we thus find $N \sim 1$–10 s^{-1}. This result is four or five orders of magnitude smaller than the number of events which can be achieved at ordinary accelerators.

Major difficulties arise in the use of polarized beams of electrons and positrons if they are polarized antiparallel (in the case of electrons) or parallel (positrons) with respect to the accelerating magnetic field. Major difficulties also result from synchrotron radiation. However, since the radiation depends on the energy of the e^+e^- beams, the degree of polarization of the beam is determined substantially by the beam energy. The polarization is disrupted if the frequency at which the beam orbits in the accelerating ring is close to a resonant frequency of the accelerator. Polarized e^+e^- beams are accordingly possible only under certain operating conditions of the accelerator.

Despite the low luminosity and the serious technical difficulties which arise in the construction and alignment of colliders, the latter offer exceptional opportunities for studying unique phenomena in the elementary-particle physics, as we mentioned earlier. For example, the discovery and study of the new J/ψ and Υ families of particles; the discovery of the τ, D, F, and B mesons; a search for heavy leptons and new quarks; a proof of the existence of the gluon; a study of the $\gamma\gamma$ interaction and of the properties of the hadronic jets in e^+e^- annihilation—all this and more have proved possible in experiments in colliding e^+e^- beams.

In this chapter of the book we will consider only those final states which arise in the course of the one-photon annihilation of e^+e^- beams: $e^+e^- \to \gamma^* \to X$ (Fig. 1.4; diagrams of class IV). Results pertaining to research on the form factors of elementary particles in the time-like region of the square of the 4-momentum transfer, $q^2 > 0$ (one-photon annihilation of e^+e^- particles into $\pi\pi$, $\mathrm{K\bar{K}}$, and $\mathrm{N\bar{N}}$ pairs), were discussed in Chap. 2.

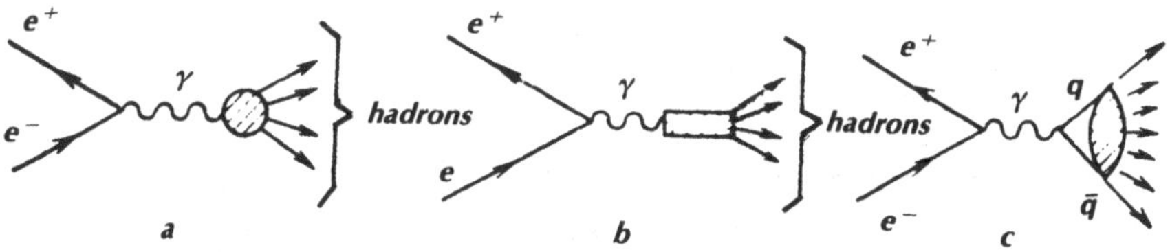

Figure 4.2. (a) One-photon annihilation into hadrons, which can occur either (b) through the production of vector mesons ($J^P = 1^-$), followed by their decay, or (c) through the conversion of a photon into a $q\bar{q}$ pair, followed by a fragmentation of quarks into hadrons.

2. The *R* ratio; tests of quark models; predictions of quantum chromodynamics

The one-photon annihilation of e^+e^- beams into hadrons,

$$e^+e^- \to \gamma^* \to \text{hadrons}\,, \tag{4.3}$$

should be acknowledged as an exceedingly important subject of research. The ratio

$$R = \sigma(e^+e^- \to \text{hadrons})/\sigma(e^+e^- \to \mu^+\mu^-) \tag{4.4}$$

allows a classic test of quark models and QCD predictions at high energies.

Process (4.3) can go by various paths (Fig. 4.2, a–c). We would naturally expect that at energies of e^+e^- beams close to the masses of the vector mesons ρ, ω, φ, etc., the R ratio would be described by a resonance curve. Figure 4.3 shows experimental data obtained at the accelerators at Orsay, Frascati, Novosibirsk, Stanford, and Hamburg at energies $\sqrt{s} = W = E_{\text{c.m.}} < 2.5$ GeV. The cross section for the production of vector mesons (ρ, ω, φ) with a mass M_V, followed by a decay into some final state f, can be described by the Breit–Wigner formula:

$$\sigma(e^+e^- \to f) = (3\pi/s)\Gamma_{ee}\Gamma_f/\left[(\sqrt{s} - M_V)^2 + \Gamma^2/4\right]\,. \tag{4.5}$$

Near resonances, R is tens of times greater than the cross section for the production of a $\mu^+\mu^-$ pair. The experimental data on the behavior of the total cross section for the process $e^+e^- \to$ hadrons in the resonance region can be used to calculate the partial width for the vector-meson decay $V \to ee$ and the coupling constant e/γ_r for the coupling of the photon with the vector meson. Specifically, using (4.5), the formal equation

$$\pi\delta(x) \to \lim_{\sigma \to 0}\,[\sigma/(x^2 + \sigma^2)]\,, \tag{4.6}$$

and an approximate expression for the cross section $\sigma_{ee} \to V$ at the peak, $\sigma_{ee \to v} \approx \sigma_{\text{peak}}\,\Gamma/2$, we find

$$\Gamma_{ee} = \frac{M_V^2\Gamma}{12\pi}\,\sigma_{\text{peak}} = \frac{\alpha^2(\Sigma e_i^2)}{9}\,M_V^2 \int \frac{dM}{M^2}\,, \tag{4.7}$$

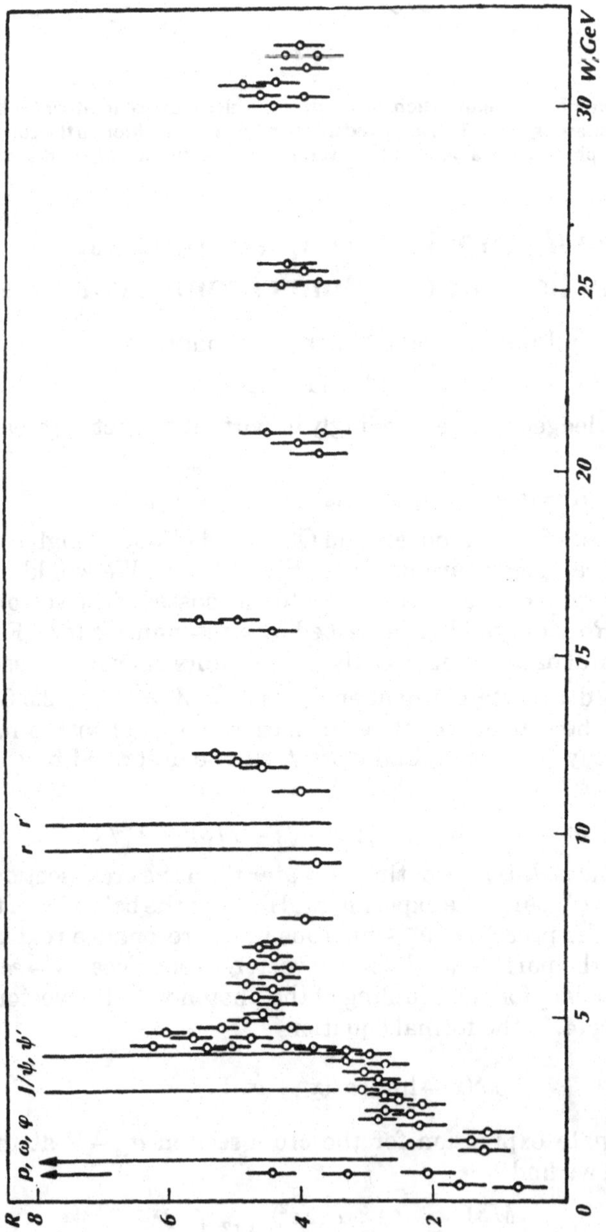

Figure 4.3. Orsay, Frascati, Novosibirsk, Stanford, and Hamburg data on the *R* function.

where σ_{peak} is the cross section for the production of the resonance at the peak. The latter equality can be proved under the assumption that a vector meson is produced as a quark-antiquark pair:

$$\sigma_{\text{peak}}\Gamma = \int \sigma_V\, dM = \int \sigma_{q\bar{q}}\, dM = \sigma_{q\bar{q}}(M_V)\int \frac{M_V^2}{M^2}\, dM$$

or

$$\sigma_{\text{peak}} = \sigma_{q\bar{q}}(M_V)\frac{M_V^2}{\Gamma}\int \frac{dM}{M^2}.$$

Knowing the cross section for the process $e^+e^- \to \mu^+\mu^-$ [see (1.53a)] we can immediately write the cross section for the production of a $q\bar{q}$ pair: $e^+e^- \to q\bar{q}$. All that we have to do is replace the square of the electric charge, e^2, in (1.53a) by the sum of the squares of the charges of the quarks of all species: $\alpha^2 \to \alpha^2\Sigma e_i^2$. We thus write

$$\sigma_{q\bar{q}} = (4/3)(\pi\alpha^2/s)\Sigma e_i^2\,, \tag{4.8}$$

from which we find

$$\sigma_{\text{peak}} = \frac{4\pi\alpha^2}{3}\Sigma e_i^2\frac{1}{\Gamma}\int \frac{dM}{M^2}.$$

The partial width Γ_{ee} is related to the constant e/γ_v by

$$\Gamma_{ee} = \frac{\alpha^2}{\gamma_v^2/4\pi}\frac{M_V}{1.2},$$

so we have

$$\gamma_v^2/4\pi = \left((4/3)(\Sigma e_i^2)M_V \int dM/M^2\right)^{-1}.$$

The coupling constant $\gamma_v^2/4\pi$ is found experimentally both in the region $Q^2 > 0$ (in the process $e^+e^- \to \gamma^* \to V$) and in the region $Q^2 < 0$ (in the electroproduction of vector mesons). Analysis of the experimental data on the production of ρ, ω, and φ mesons leads to the conclusion that the constant $\gamma_v^2/4\pi$ does not depend on Q^2 (or on the mass of the virtual photon) (Table 4.1). The independence of the coupling constant γ_v from Q^2 is one of the basic assumptions of the vector-dominance models.

At energies above the resonances, the cross section for process (4.3) is described by the diagram in Fig. 4.2c; i.e., it is assumed that the photon initially converts into a quark-antiquark pair, which then fragments into a hadronic state. It is assumed here that the quarks are Dirac particles of spin 1/2 with a fractional charge e_i, where i is the flavor of the quark, and that the subsequent annihilation of the $q\bar{q}$ pair into hadrons goes with a unit probability.

We thus have $\sigma(e^+e^- \to \text{hadrons}) \equiv \sigma(e^+e^- \to q\bar{q})$. Hence

$$R = \sigma(e^+e^- \to \text{hadrons})/\sigma(e^+e^- \to \mu^+\mu^-) = \Sigma e_i^2\,.$$

There is a variety of quark models.

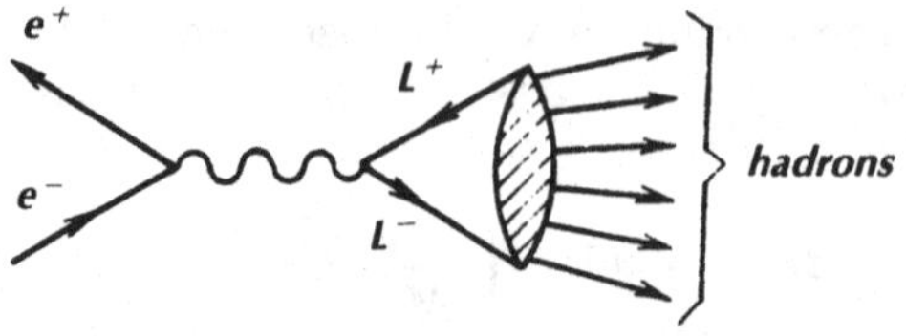

Figure 4.4. Diagram describing the additional contribution of heavy leptons to R.

- A model with three colorless quarks (u, d, s) which have fractional electric charges and another model in which these quarks have three additional color degrees of freedom.
- A model with four colorless quarks (u, d, s, c) which have fractional electric charges and another model in which these quarks have three additional color degrees of freedom.
- The Han–Nambu model with three colored quarks (u, d, s) with integer charges.
- The Han–Nambu model with four colored quarks (u, d, s, c) which have integer charges.

In the Han–Nambu model the electric charges of the colored quarks take on the following values:

Color	u	d	s	c
1st	0	-1	-1	0
2nd	1	0	0	1
3rd	1	0	0	1

In this model we thus have $\Sigma e_i^2 = I^2 + I^2 + (-1)^2 + (-1)^2 = 4$ in the case of three quarks or $\Sigma e_i^2 = 6$ in the case of four.

The different quark models predict different values of R (Table 4.2).

Comparing lines 2 and 4 (or 1 and 3; or 5 and 6) in Table 4.2, we see that an abrupt change in R as a function of the energy of the e^+e^- beams would indicate a threshold for the production of the quark of a certain species—the introduction of a new quark flavor, so to speak. A discontinuous behavior of R might be caused by the threshold for the production of a pair of heavy leptons. To the diagrams in Fig. 4.2 we might add a diagram of the type in Fig. 4.4. The cross section for the production of a pair of heavy leptons (L^+L^-) is the same as that for the production of a $\mu^+\mu^-$ pair when the mass of the μ meson is replaced by that of the L lepton. It should be noted, however, that not all L^+L^- pairs annihilate into hadrons; some of them undergo leptonic decays. In this case we have

$$R = \frac{\sigma(e^+e^- \to \text{hadrons}) + \sum_i^N \sigma(e^+e^- \to L_i^+ L_i^- \to \text{hadrons})}{\sigma(e^+e^- \to \mu^+\mu^-)} = \Sigma e_i^2 + \widetilde{N}, \qquad (4.9)$$

where $\widetilde{N}$ is a correction to the number of species of heavy leptons L_i.

An additional contribution to R (not discontinuous) comes from QCD corrections. Taking these corrections into account in the lowest order in α_s

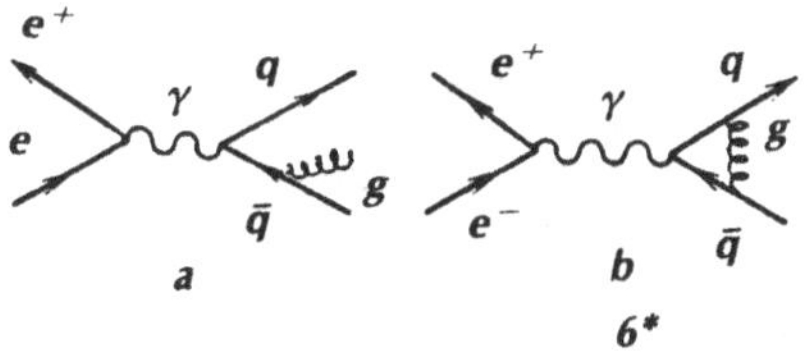

Figure 4.5. First-order radiative corlections in QCD.

reduces to incorporating gluon radiative corrections of the types shown in Fig. 4.5, a and b. The correction enters the expression for R as an additional term

$$R_{\text{QCD}} = R(1 + \alpha_s/\pi), \tag{4.10}$$

where $\alpha_s = 12\pi/(33 - 2f)\ln(s/\Lambda^2)$ (f is the number of quark flavors and Λ is a fundamental constant in QCD). The corrections of first and second orders in α_s are small.[127] For $f = 3$, $\Lambda = 0.5$ GeV, and $\sqrt{s} = 3.6$ GeV, we have $R_{\text{QCD}} \approx 2.3$ (instead of 2; Table 4.2); for $f = 4$, $\Lambda = 0.5$ GeV, and $\sqrt{s} = 0.5$ GeV, we have $R_{\text{QCD}} = 3.9$ (instead of 10/3; Table 4.2). In the limit $Q^2 \rightarrow \infty$, we have a correction $\alpha_s/\pi \rightarrow 0$, and R_{QCD} tends toward the values given in Table 4.2. The correction of second order in the coupling constant $(\alpha_s/\pi)^2$ depends on the renormalization scheme, but it is always smaller than the correction proportional to α_s/π.

Turning to the experimental results, we will attempt to explain them with the help of the QCD rules for calculating R which we have just established. Figure 4.3 shows data obtained by various laboratories around the world. Figure 4.6 shows data for the energy interval $3.5 < E_{\text{c.m.}} < 5$ GeV, which has been studied in extreme detail at the SPEAR and DORIS accelerators.

Comparing the predictions of the quark models (Table 4.2) with the results in Figs. 4.3 and 4.6, we reach the following conclusions: At $E_{\text{c.m.}} \leqslant 2.5$ GeV the data can be interpreted unambiguously on the basis of the model with three colored quarks (u, d, s). In this region we have $R \approx 2$ (except for the positions of the resonances ρ, ω, and φ). The Han-Nambu model predicts $R = 4$, which is twice the observed value. The energy interval $2.5 < E_{\text{c.m.}} < 5$ GeV is saturated with resonance states (Fig. 4.6); near $E_{\text{c.m.}} \approx 4$ GeV, R is discontinuous, jumping to a value of 4.5. This jump is explained as resulting from the attainment of the threshold for the production of a new quark. The three-color model of quarks with fractional charges yields the value $R = 10/3$ under the condition that a fourth (charmed) quark is included. Since this value is still below the experimental value, it might be suggested that near $E_{\text{c.m.}} \approx 4$ GeV the threshold for the production of some unknown heavy lepton is reached, along with the threshold for the production of a fourth quark. In this manner we could raise the theoretical prediction of R to 13/3. By incorporating radiative corrections we could increase this value yet a bit further and achieve a completely satisfactory agreement with experiment.

A candidate for the role of this heavy lepton might by the τ lepton, but its mass is $m_\tau = 1.78$ Mev, so that the threshold for the production of a $\tau\bar{\tau}$ pair would lie at an energy $E_{\text{c.m.}} \approx 3.6$ GeV, which is well below the discontinuity in

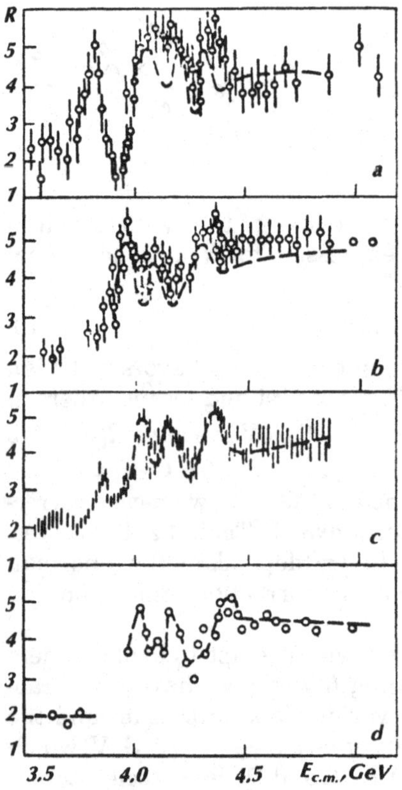

Figure 4.6. Experimental data provided on R by various laboratory groups. (a) SLAC–LBL; (b) DASP; (c) DELCO; (d) PLUTO.

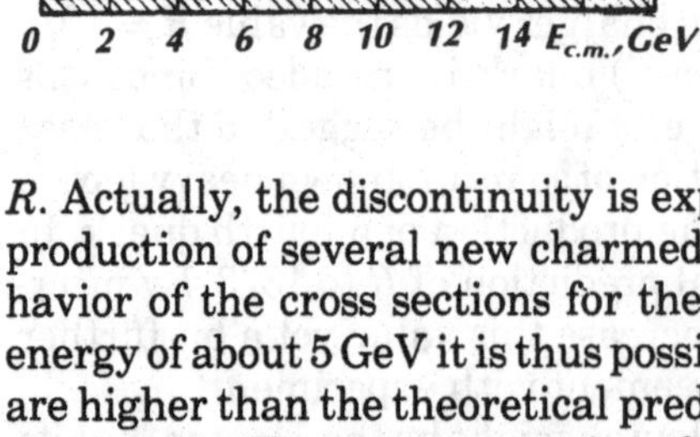

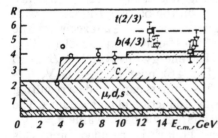

Figure 4.7. Experimental data of the MARK J group (triangles), the PLUTO group (circles), and the TASSO group (squares) on R at energies $E_{c.m.} = 13$ and 17 GeV. Data from the PLUTO group at $E_{c.m.} < 10$ GeV are shown for comparison; the area under the wavy line represents QCD corrections.

R. Actually, the discontinuity is explained in terms of the threshold for the production of several new charmed mesons and in terms of a resonance behavior of the cross sections for the production of these particles. Up to an energy of about 5 GeV it is thus possible to explain why the experimental data are higher than the theoretical predictions. The data obtained at the PETRA accelerator (Fig. 4.7) at energies of 13 and 17 GeV yield a value of R higher than the theoretical value $R_{\text{theo}} \approx 3.9$ (the QCD corrections and the four tricolor quarks are taken into account) and may be interpreted as the appearance

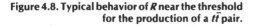

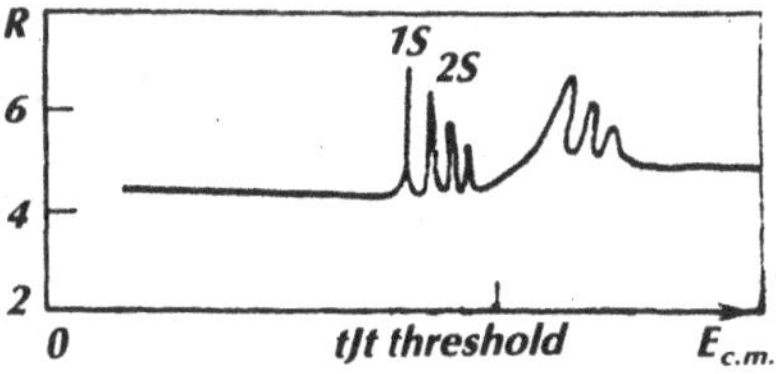

Figure 4.8. Typical behavior of R near the threshold for the production of a $t\bar{t}$ pair.

of a fifth flavor: a b quark with a charge of 1/3. Experimentally, R remains essentially constant ($R \approx 4$) over the interval $17 \leqslant E_{\text{c.m.}} \leqslant 32$ GeV and can be explained in terms of a quark model with five tricolor quarks (u, d, s, c, b) with QCD corrections. The existence of only five quarks, however, leaves the theoreticians unsatisfied. There is a symmetry between leptons and quarks, and this symmetry points to the need for a sixth quark, t:

$$\begin{pmatrix} \nu_e & \nu_\mu & \nu_\tau \\ e & \mu & \tau \end{pmatrix} \rightarrow \begin{pmatrix} u & c & t \\ d & s & b \end{pmatrix}.$$

The quartet of quarks $\binom{u}{d}$ and $\binom{c}{s}$ is required for describing many processes which occur in strong, electromagnetic, and weak interactions. The fifth quark, b, was introduced on a fairly reliable basis after the discovery of the Υ family of particles. The b quark, with a charge of $-1/3$, should be "teamed up" with another quark (a t quark) with a charge[10] of $+2/3$. If a t quark with a charge of $+2/3$ does exist, then near the threshold for the production of a $t\bar{t}$ pair we would expect to find the standard picture of the behavior of R, as seen near the thresholds for the production of $c\bar{c}$ and $b\bar{b}$ quark pairs (Fig. 4.8). The jump in R must be equal to $\Delta R = 3(2/3)^2 +$ QCD corrections $= 4/3 + 15$–20% of $4/3 \approx 5/3$. Above the threshold we might therefore expect R to reach values of 5.7–6. Experimental data[45,128] (Fig. 4.3) show that up to energies $E_{\text{c.m.}} = 36.6$ GeV there is no discontinuity in the behavior of R and that a t quark with a charge of $+2/3$ and a mass $m_t \leqslant 19$ GeV is ruled out entirely (within 12 standard deviations). However, the data available do not equally ambiguously rule out the existence of a t quark with a charge of 1/3. The search for a sixth quark is accordingly one of the most important problems in the experimental physics of elementary particles. Searches for a t quark feature prominently in the programs of research for the new generation of accelerators with colliding e^+e^- beams.

3. Some general characteristics of the final state in the reaction $e^+e^- \rightarrow$ hadrons

In this subsection, as in Sec. 10 of Chap. 3, we will examine some characteristics of the angular and energy distributions of the hadrons in the final state. However, while in Chap. 3 we were seeking information on the structure of the nucleon from various types of hadron distributions, the study of the final state in the present subsection makes it possible to test several fundamental questions of QCD. For example: Does a gluon exist? What are its quantum numbers? What are the properties of quarks? Which quark species participate in reactions of a given type? Is the interaction of quarks with gluons a point-like interaction? What is the nature of this interaction?

A QCD approach has already been developed for analyzing the hadronic states which arise in the one-photon annihilation of e^+e^- beams. Along this approach there are many more model-based arguments of a "pictorial" nature (i.e., expressed in terms of perturbation-theory diagrams); as of yet, there are still few rigorous quantitative theoretical results. To some extent, the situation can be blamed on the difficulty in calculating higher-order corrections in QCD.[129] The incorporation of these corrections often leads to fundamental changes in the conclusions and estimates generated in a lower-order approximation. The reader should not be surprised by the large number of diagrams presented below. These diagrams replace accurate calculations in several cases. They also create the illusion of an understanding of the phenomena which are occurring and inspire a confidence that all of these diagrams can be calculated and that the calculations will lead to the expected agreement with experimental data.

Information on the final hadronic state in the annihilation of e^+e^- beams has been obtained in two steps. First (up to 1978), while the energy of the beams in the c.m. frame did not exceed 10 GeV, a picture of the phenomena was drawn, but there remained a group of questions whose answers had to await an increase in the energies of the e^+e^- beams. In the next step (1978–1981), when c.m. energies up to 30–35 GeV were achieved at the PETRA and PEP accelerators, a significantly deeper understanding of the phenomena was achieved, new facts arose, and new questions arose. We hope to find answers to these questions at the accelerators of the next generation. We will cover the various topics here in the order indicated.

For one-photon annihilation of e^+e^- beams into hadrons, $e^+e^- \to \gamma^* \to$ hadrons, we can generate an estimate of the total cross section and of the angular distribution of an individual hadron. For example, the final state must have the quantum numbers of the photon: $J^{PC} = 1^-$. It follows that the orbital angular momentum L of the colliding electrons and positrons is restricted to the values of 0 and 2. Since the orbital angular momentum L and the interaction radius r are related by $L \sim rE_{\text{c.m.}}$, where $E_{\text{c.m.}}$ is the energy of the colliding beams in the c.m. frame, we find $r \sim 1/E_{\text{c.m.}}$, so that the total cross section for hadron production is

$$\sigma_{\text{total}} \approx \alpha^2 \pi r^2 = \alpha^2 \pi / E_{\text{c.m.}}^2 \approx 60/E_{\text{c.m.}}^2 ,$$

where $E_{\text{c.m.}}$ is expressed in giga-electron-volts, and σ is in units of 10^{-33} c.m.^{2}. The cross section thus falls off rapidly with increasing energy. The expected angular distribution of an individual (inclusive) hadron h ($e^+e^- \to \gamma^* \to h + X$) would be (we are allowing for the spin and parity of the virtual photon γ^* in the intermediate state)

$$d\sigma/d\Omega = a + b\cos^2\theta .$$

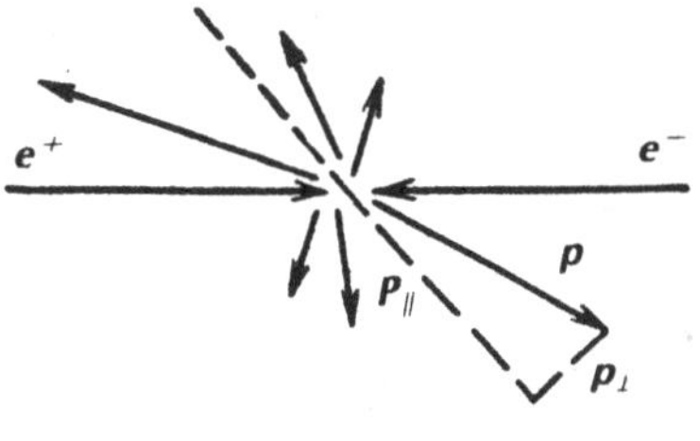

Figure 4.9. Diagram of a two-jet event. The dashed line (the jet axis) is the direction of the emission of the quark and the antiquark; the solid lines are the tracks of charged hadrons. The neutral component is not shown.

In other words, there would be a minimum at the point $\theta = \pi/2$; this distribution would be substantially different from the angular distribution of the hadron in *pp* collisions, which has a sharp forward-scattering peak.

The quark-parton model makes excellent predictions regarding the final-state distribution of hadrons. It was shown in Sec. 2 of Chap. 4 that at energies $E_{\text{c.m.}}$ above the resonance states this process is described by the diagram in Fig. 4.2b. Let us assume that a quark-antiquark pair in an intermediate state fragments into hadrons in a special way. Specifically, for each individual hadron there is a limitation on the momentum component transverse with respect to the emission direction of the "related" quark; here "related" means that quark (or antiquark) whose fragmentation gave rise to this particular hadron (the reason for the limitation is unknown at this point). Under such conditions the hadrons will then form in two jets in the final state (Fig. 4.9). Just what actually occurs in each e^+e^- annihilation event remains unclear, because the neutral hadronic component is not observable.

The effect of decays of resonance hadronic states on the analysis of the jet behavior of hadrons can be avoided by selecting cases in which the number of charged particles in the final state is more than three. Even at energies $E_{\text{c.m.}} \approx 9$ GeV, where the jet nature should be manifested quite clearly because of obvious considerations, the number of events having a clearly defined two-jet picture is exceedingly small. For this reason, a special procedure is introduced for analyzing each separate case of e^+e^- annihilation into hadrons. The first step is to find the axes of the jets. The axis of a jet can be found mathematically for each case by reducing the tensor $T_{\mu\nu}$, constructed with the help of the momenta of all the visible charged hadrons, to its principal axes:

$$T_{\mu\nu} = \sum_i (\delta_{\mu\nu} p_i^2 - p_\mu^i p_\nu^i),$$

where the summation is over all the charged particles which can be detected. The diagonalization of this tensor leads to a spatial determination of three eigenvectors and three eigenvalues $(\lambda_1, \lambda_2, \lambda_3)$, each of which is the sum of the squares of the transverse components of the momenta with respect to the direction of the eigenvector. Among the eigenvalues λ_i there is a smallest one (e.g., $\lambda_1 \geqslant \lambda_2 \geqslant \lambda_3$). The associated eigenvector is called the "axis" of the jet. We then find a natural definition of the concept of "sphericity" S, as the extent of the deviation of the hadrons from an isotropic spatial distribution for each event:

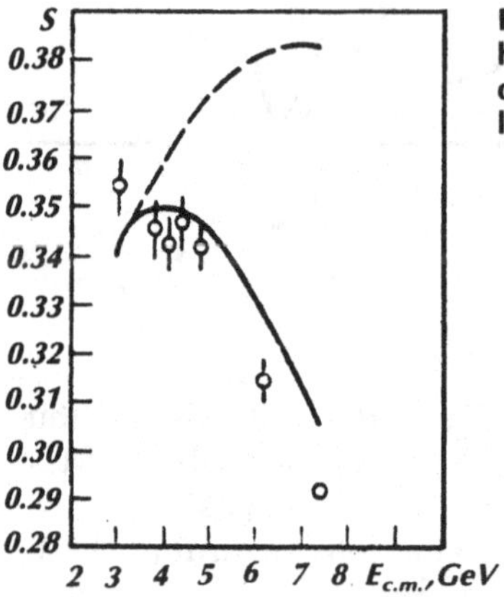

Figure 4.10. First piece of evidence for jet behavior of hadrons in a reaction $e^+e^- \to$hadrons. (Points) experimental data; (solid line) Monte Carlo calculation of jets; (dashed line) Monte Carlo calculation of the phase space.

$$S \equiv \frac{3\lambda_3}{\lambda_1 + \lambda_2 + \lambda_3} = \frac{3 \min \sum_i p_{\perp i}^2}{2 \sum_i p_i^2}. \tag{4.11}$$

Since the momenta appear quadratically in the expression for S, the particles with large momenta have a large statistical weight, and the sphericity characteristic therefore depends on the particular features of the fragmentation. Accordingly, other characteristics are also used, in addition to the sphericity, to analyze two-jet events: the "spherocity"

$$S' = \left(\frac{4}{\pi}\right)^2 \min \left(\frac{\sum_i p_{\perp i}}{\sum_i |p_i|} \right)^2 \tag{4.12}$$

and the "thrust"

$$T = \frac{\max \sum_i |p_{\parallel i}|}{\sum_i |p_i|}. \tag{4.13}$$

The direction of the axis of an event can easily be discerned from the definition of the characteristics S' and T. For T, for example, the axis is determined by a maximization of the sum of the quantities $p_{\parallel}$ with respect to an arbitrary direction in space. In this subsection we will use the characteristics S and T exclusively. Some new characteristics of jets will be presented further on. If we require that $p_{\perp}$ remain constant, then S and $1-T$ will decrease with increasing energy. There are limits on the ranges of these quantities: $S \to 1$, $T \to 1/2$ for an isotropic distribution of hadrons and $S \to 0$, $T \to 1$ for an ideal jet event.

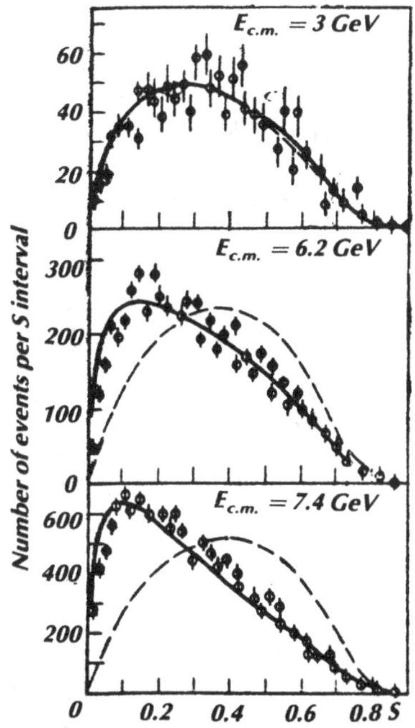

Figure 4.11. Number of cases per sphericity interval of 0.02 for various energies. (Points) experimental data from the SLAC-LBL group; (solid lines) Monte Carlo calculations of the jet behavior for limited values of $\langle p_\perp \rangle$; (dashed line) Monte Carlo calculations of the phase space.

The first experimental data obtained at the SLAC accelerator, in 1975, were analyzed with the goal of proving the existence of jets. Figure 4.10 shows that the idea of an ordinary invariant phase distribution of the hadrons in the final state (the dashed line) is sharply at odds with the experimental data. Figure 4.11 shows that the jet nature of the events becomes progressively more apparent as the energy of the e^+e^- beams is raised. These results were subsequently verified in experiments at the DORIS accelerator, in which the jet nature of events was tested at energies from 3.6 to 9.4 GeV. The value of S falls off sharply with increasing $E_{\text{c.m.}}$, reaching 0.25 at $E_{\text{c.m.}} = 9.4$ GeV.

Measurements of the angular distribution of the jet axes as a function of the polar angle θ are of interest here. It can be shown that if the spin of the quarks is 1/2, and if the jets form by the scheme of Fig. 4.2c, then the angular distribution of the jet axes will be described by $1 + \cos^2\theta$. Figure 4.12 compares experimental data on the angular distribution of the jet axes found from an analysis with the help of (4.13) with a curve of $1 + \alpha\cos^2\theta$, where α is an adjustable parameter. For $E_{\text{c.m.}} = 7.7$ GeV we find the value $\alpha = 0.76 \pm 0.3$, while for $E_{\text{c.m.}} = 9.4$ GeV we find $\alpha = 1.63 \pm 0.6$. It follows that the spin of the quarks is 1/2 (Ref. 130).

Considerable importance is placed on a study of the inclusive spectra of hadrons which appear in e^+e^- annihilation. In the reaction

$$e(\mu) + p(N) \to e(\mu) + X$$

the momentum spectrum of the inclusive hadrons is examined with respect to the direction of the momentum of the virtual photon (Sec. 10 of Chap. 3).

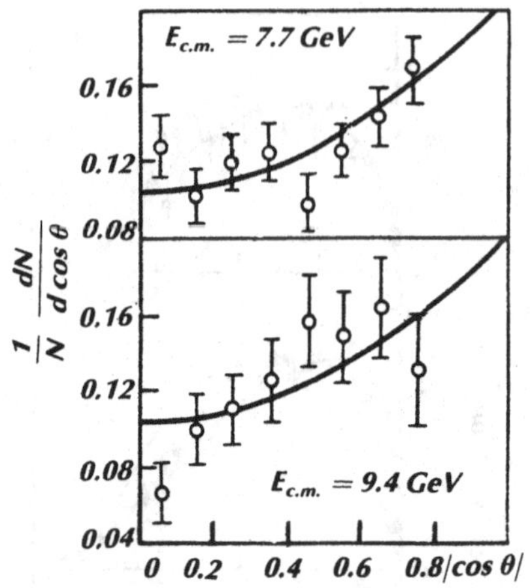

Figure 4.12. Comparison of data from the PLUTO group on the jet characteristic T with a curve of $1 + \alpha \cos^2 \theta$.

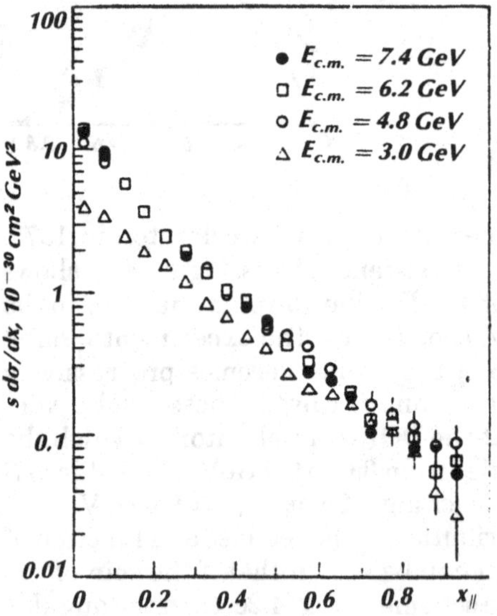

Figure 4.13. Scaling of the function $s\,d\sigma/dx_{\parallel}$ as a function of the variable $x_{\parallel}$.

Similarly, the momentum spectrum of the inclusive hadrons in the jets which appear in one-photon e^+e^- annihilation is examined with respect to the axes of jets identified by means of expression (4.11) or (4.13). Experimentally, one studies various distributions $s\,d\sigma/dx$ and $s\,d\sigma/dx_{\parallel}$ as functions of x and $x_{\parallel}$, respectively, where $s = 4E^2_{\text{c.m.}}$, $x = 2p/E_{\text{c.m.}}$, and $x_{\parallel} = 2p_{\parallel}/E_{\text{c.m.}}$; and various distributions $(1/\sigma)\,d\sigma/dp_{\perp}$ and $(1/\sigma)\,d\sigma/dy$ as functions of $p_{\perp}$ and y, respectively, where $y = (1/2)\ln[(E + p_{\parallel})/(E - p_{\parallel})]$. It follows from the experimental data in Fig. 4.13 that the distribution of hadrons in the longitudinal component of the momentum of the inclusive particles exhibits a scaling, which sets

Figure 4.14. Function $(1/\sigma)d\sigma/dy$ as a function of the variable y. The notation is the same as in Fig. 4.13. Cases with $x_{max} > 0.5$ are shown, and it is assumed that all the hadrons in jets are pions.

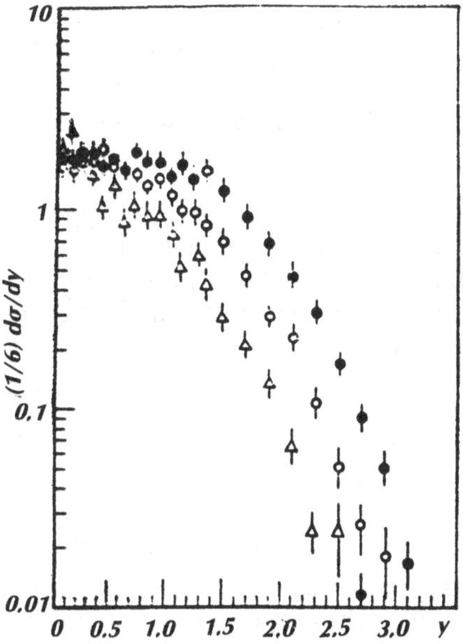

Figure 4.15. Kinematic variables used in describing jets.

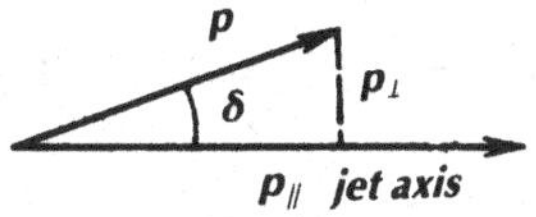

in at about $E_{c.m.} \geqslant 4$ GeV. The distribution in the rapidity y (Fig. 4.14) is reminiscent in shape of the corresponding distribution in pp collisions.

All of the experimental data at energies above the production of resonances but below 10 GeV thus demonstrate a jet distribution of the hadrons produced in the one-photon annihilation of e^+e^- beams, the existence of a spin of 1/2 for quarks, and a scaling of the momentum distribution of the inclusive hadrons.

At higher energies of the e^+e^- beams we can expect qualitative changes in the jet distribution of hadrons because of the emission of virtual gluons. While the transverse components of the hadron momenta found with respect to the jet axis are approximately constant quantities, as we raise the energy of the e^+e^- beams the jets will contract progressively closer to the jet axis. An average value of half the cone angle of the jet can be estimated on the basis of elementary considerations (Fig. 4.15): $\langle\delta\rangle \approx \langle p_\perp\rangle / \langle p_\parallel\rangle$.

The emission of a gluon by a quark with a momentum p should lead to increases in the angle δ and in the transverse component of the hadron momentum, $p_\perp$. If the increase in δ with increasing beam energy is slower than the increase in the energy, we would expect a "flattening" of the spatial distribution of hadrons. Experiments provide a qualitative confirmation of this

Fermi's statistical model, which was proposed in 1950 for describing multiple production of particles, predicts an increase $\langle n_{\rm ch}\rangle \sim s^{1/4}$ and gives a good description of the data in Fig. 4.16; on the other hand, it incorrectly predicts the following measure of the dispersion of the charged particles:

$$D_{\rm ch} = (\langle n_{\rm ch}^2\rangle^2)^{1/2} .$$

We now consider the inclusive hadron spectrum in the variable $x = E_h/E_{\rm c.m.}$, i.e., a reaction $e^+e^- \to h + X$, where h is any hadron, at energies above 10 GeV. The differential cross section for the production of a hadron with a momentum **p** and an energy E_h at an angle θ from the beam collision axis is

$$d^2\sigma/dx\,d\Omega = (\alpha^2/s)\beta x\,[m\overline{W}_1(x,s) + (1/4)\beta^2 x\nu\,\overline{W}_2(x,s)\sin^2\theta\,] , \qquad (4.15)$$

where $x = E_h/E_{\rm c.m.} = 2E_h\,\sqrt{s}$, $\beta = |p|/E$, m is the mass of the hadron, and $\nu = E_h\sqrt{s}/m$ is the energy of the virtual photon in the rest frame of the hadron.

The structure functions $\overline{W}_1$ and $\overline{W}_2$ are intimately related to the structure functions for the deep inelastic scattering of leptons by nucleons. This relationship follows from an analysis of the tensors $\overline{W}_{\mu\nu}$ and $\overline{W}_{\mu\nu}$, which describe the structure of the hadron vertex in the processes $e^+e^- \to$ hadrons and $e + N \to e + X$, respectively. For the process $e^+e^- \to$ hadrons we have

$$\overline{W}_{\mu\nu} \equiv 4\pi^2 E \sum_n \langle 0|J_\mu(0)|p,n\rangle\langle n,p|J_\mu(0)|0\rangle(2\pi)^4\delta(q-p-p_n)$$

$$= -M\left(g_{\mu\nu} - \frac{q_\mu q_\nu}{s}\right)\overline{W}_1(s,\nu) + \frac{1}{M}\left(p_\mu - \frac{pq}{s}\,q_\mu\right)\left(p_\nu - \frac{pq}{s}\,q_\nu\right)\overline{W}_2(s,\nu) ,$$

while for the process $e + N \to e + X$ the tensor $W_{\mu\nu}$ is written

$$W_{\mu\nu} = 4\pi^2 E \sum_n \langle\, p|J_\mu(0)|n\rangle\langle n|J_\nu(0)|p\rangle(2\pi)^4\delta(q+p-p_n)$$

$$= -M\left(g_{\mu\nu} - \frac{q_\mu q_\nu}{q^2}\right)W_1(q^2,\nu) + \frac{1}{M}\left(p_\mu - \frac{pq}{q^2}\,q_\mu\right)\left(p_\nu - \frac{pq}{q^2}\,q_\nu\right)W_2(q^2,\nu)$$

[see (3.11)].

In the field theory the tensors $\overline{W}_{\mu\nu}$ and $W_{\mu\nu}$ are related by crossing-symmetry relations:

$$\overline{W}_{\mu\nu}(q,p) = -\;W_{\mu\nu}(q,-p) \text{ if } h \text{ is a fermion} ,$$

$$\overline{W}_{\mu\nu}(q,p) = W_{\mu\nu}(q,-p) \text{ if } h \text{ is a boson} .$$

It follows that for a fermion (for example) the structure functions $\overline{W}_1$ and $\overline{W}_2$ from (4.15) are related to the structure functions W_1 and W_2 for deep inelastic scattering of leptons by nucleons [see (2.12)] by

$$\overline{W}_1(q^2,\nu) = -\;W_1(q^2,-\nu); \quad \nu\overline{W}_2(q^2,\nu) = \nu W_2(q^2,-\nu) .$$

Let us go back to expression (4.15). If scaling holds, the functions $\overline{W}_i$ should depend on x alone, and the cross section $s\,d^2\sigma/dx\,d\Omega$ will not depend on the energy. If the energy of the hadron is sufficiently large in comparison with its

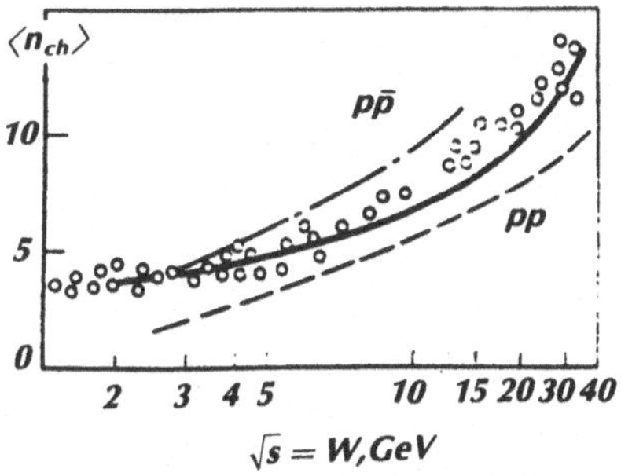

Figure 4.16. Average charged multiplicity as a function of the energy in the c.m. frame (data from Frascati, Stanford, and Hamburg).

conclusion and the quantitative changes have been explained satisfactorily on the basis of QCD.

Before we take up these fine characteristics of jets associated with proofs of the existence of gluons and a determination of their spin (Sec. 4 of Chap. 4), let us consider some other general properties of the final state which is formed in one-photon annihilation of e^+e^- beams. The JADE group at the PETRA accelerator[45] measured the fraction of the energy which was carried off by photons and other neutral particles in the reaction $e^+e^- \to X$. The results are shown in Table 4.2A.

We clearly see that the fraction of the energy which corresponds to other neutral particles increases with increasing energy of the beams, while the fraction of the energy which corresponds to photons remains essentially constant (within the error), although there is an unmistakable tendency toward an increase.

Yet another characteristic is the average charged-particle multiplicity $\langle n_{ch} \rangle$. At low energies we see a logarithmic increase in $\langle n_{ch} \rangle$, but at a higher energy the picture changes substantially. Figure 4.16 compares experimental data with a theoretical curve derived in the leading-log approximation:

$$n = n_0 + a \exp(b\sqrt{\ln s/\Lambda^2}) . \tag{4.14}$$

In (4.14) we have $b = 1.77$; the values of a and n_0 are found experimentally. The experimental data are described well over the entire energy range by expression (4.14) with $n_0 = 2.0 \pm 0.2$, $a = 0.27 \pm 0.01$; $b = 1.9 \pm 0.2$; and $\Lambda = 0.3$ GeV. The dashed and dot-dashed lines above and below the solid line in Fig. 4.16 show the energy dependence of $\langle n_{ch} \rangle$ observed in $p\bar{p}$ annihilation and pp collisions. The fact that the average multiplicity is observed to increase more rapidly than logarithmically at $E_{c.m.} \geqslant 10$ GeV can be explained qualitatively in terms of either an additional fragmentation of a gluon emitted by one of the quarks of the $q\bar{q}$ pair into hadrons or a phase-space effect.

It can also be seen from Fig. 4.16 that over the entire energy range the values found for $\langle n_{ch} \rangle$ in pp interactions lie below the values found for $\langle n_{ch} \rangle$ in e^+e^- annihilation; at $2E_{c.m.} = \sqrt{s} \lesssim 5$ GeV the data from $p\bar{p}$ annihilation agree with the data from e^+e^- annihilation and are described by $\langle n_{ch} \rangle \approx (2.1 \pm 0.85) \ln s$.

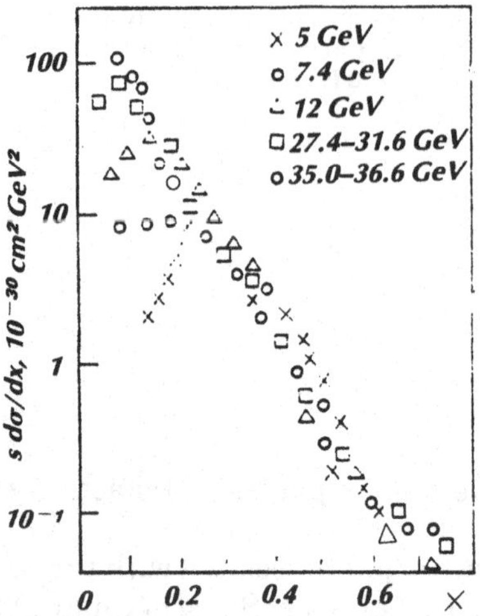

Figure 4.17. Scaling of the function $s\,d\sigma/dx(x)$ for the inclusive production of charged particles.

mass, its mass can be ignored ($\beta \approx 1$), and expression (4.15) can be written as follows,[131] after an integration over $d\Omega$:

$$s\,d\sigma/dx \approx 4\pi\alpha^2 x[m\overline{W}_1 + (1/6)x\nu\overline{W}_2]\,. \tag{4.16}$$

The cross section $d\sigma/dx$ can be expressed in terms of a function which we will call the "quark fragmentation function" D_q^h. Since we are assuming that the one-photon e^+e^- annihilation into hadrons occurs in two steps, $e^+e^- \to \gamma^* \to q\bar{q} \to$ hadrons, i.e., since we are assuming that the photon first converts into a $q\bar{q}$ pair, and then the quarks fragment into hadrons, we write $d\sigma/dx$ in terms of the function D_q^h as follows:

$$d\sigma/dx = \sigma_{q\bar{q}} 2D_q^h(x)\,.$$

Using (4.8) for σ_{qq}, we find

$$\frac{d\sigma}{dx} = \frac{8\pi\alpha^2}{3s}\,\Sigma e_q^2 D_q^h(k)\,. \tag{4.17}$$

Comparing (4.16) and (4.17), we find an expression for $D_q^h(x)$ in terms of the structure functions $\overline{W}_i$. We will not attempt to determine the function D_q^h either from a comparison with experimental data or from QCD (we will return to this point in Sec. 4 of Chap. 4). Let us look at the experimental data presently available. It can be seen clearly from Fig. 4.17 that at values $x = p/p_b > 0.2$ the scaling holds very accurately over a broad energy range ($5 \leqslant E_{\text{c.m.}} \leqslant 36.6$ GeV). At $x < 0.2$, on the other hand, the number of hadrons increases sharply with increasing energy. This effect can be explained qualitatively as follows: With increasing energy of the e^+e^- beams, the gluons emitted by the quarks acquire a fairly large energy and themselves fragment into low-energy hadrons. Furthermore, in carrying off part of the energy of a quark the gluons set the stage for a fragmentation of quarks into hadrons with low energies.

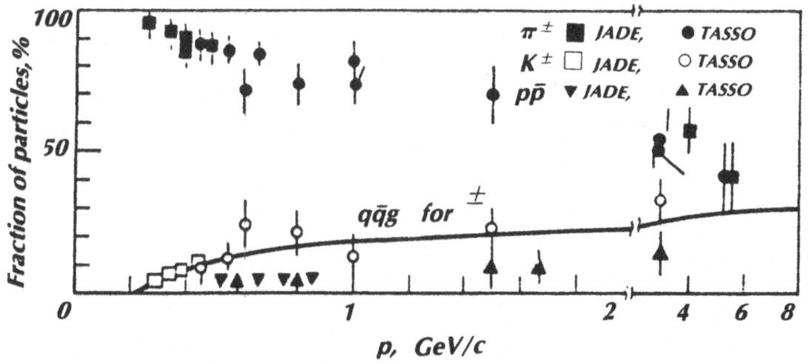

Figure 4.18. Relative numbers of charged π and K mesons and protons at $E_{c.m.} = 30$ GeV.

Finally, yet another global characteristic of an inclusive hadron spectrum is the contribution of charged K mesons, pp pairs, and other particles to the cross section $d\sigma/dx$. Figure 4.18 shows data obtained at the PETRA accelerator at an energy $E_{c.m.} = 30$ GeV. These results indicate an extremely interesting dependence of the species of the product particles on their momentum: Essentially only π mesons are produced with small momenta; the relative number of π mesons decreases with increasing momentum, while the relative numbers of $K\bar{K}$ and $p\bar{p}$ pairs increase; and at $p \gtrsim 3$ GeV/c the ratios of the relative numbers of particles become approximately $\pi^\pm : K^\pm : p^\pm \sim 55:35:10$. We turn now to one of the fundamental problems concerning the substantiation of QCD: the search for evidence of the existence of gluons.

4. Evidence for the existence of gluons; the spin of gluons

With increasing energy of the colliding e^+e^- beams (above 10 GeV), gluons play a progressively more significant role in the qualitative changes in the behavior of jets, as was mentioned in Sec. 3 of Chap. 4. In QCD, as in QED, there are corrections of various orders in the constant α_s for the emission of virtual gluons. In the lowest order in α_s, these are corrections for the emission of a single virtual gluon followed by its fragmentation into hadrons (the process $e^+e^- \to q\bar{q}g \to$ hadrons) or corrections of a vertex type to the reaction ($e^+e^- \to q\bar{q} \to$ hadrons). If the coupling constant α_s is known, the correction of lowest order in α_s can be calculated accurately by the QCD rules. The function describing the fragmentation of quarks and gluons into hadrons remains unknown. Table 4.3 shows corrections to various physical quantities which are measured in e^+e^- annihilation. It can be seen from this table that a test of the QCD predictions in the lowest order in α_s which are associated with the emission of a single virtual gluon is a simple matter: It reduces to a study of an additional energy dependence of the cross section for the corresponding process.

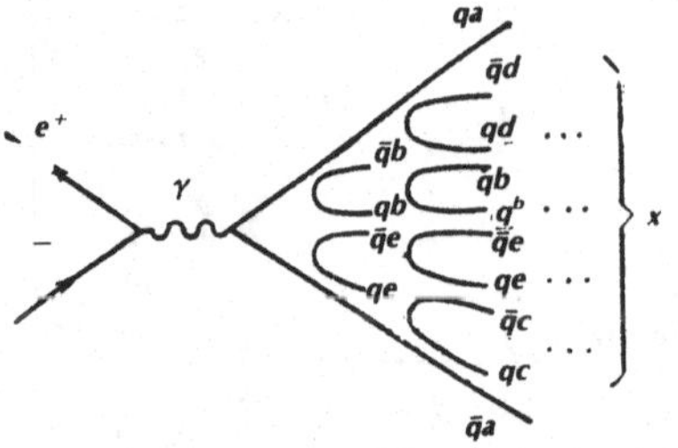

Figure 4.19. Formation of a final state X, which can occur in several steps, according to the Feynman–Field model.

The function describing the fragmentation of quarks into hadrons is in accordance with the Feynman–Field model,[132] which gives a relatively simple parametrization of the jet formation process. Let us examine this function in more detail.

It is assumed that the hadrons in the final state form by the scheme illustrated in in Fig. 4.19. The chain of the formation of particles from quarks may be distinct for each individual particle: Some particles form immediately while others may be formed in the third, fourth, etc., stage of the combining of $q\bar{q}$ pairs of various types, with the result that π, ρ, and K mesons and other particles appear in the final state. The number of particles which appears in the final state is of course a number which is allowed by the energy and momentum conservation laws. It is assumed that on each occasion the appearance of $q\bar{q}$ quark pairs from vacuum occurs in the proportions

$$\gamma_u{:}\gamma_d{:}\gamma_s = u\bar{u}{:}d\bar{d}{:}s\bar{s} = 2{:}2{:}1$$
$$(\gamma_u = \gamma_d, \quad \gamma_u + \gamma_d + \gamma_s = 1)\,. \tag{4.18}$$

A trial function for the fragmentation of a quark into hadrons is introduced in the model:

$$f(\eta) = 1 - a + 3a\eta^2\,. \tag{4.19}$$

This function is chosen in such a way that the fragmentation function of interest, $D_q^h(z)$, gives a good description of data on the deep inelastic scattering of leptons by nucleons and e^+e^- annihilation. The quantity η in (4.19) is the fraction of the momentum which the emitted hadron leaves the primary quark. The function $f(\eta)$ is normalized by

$$\int_0^1 f(\eta)d\eta = 1.$$

For an initial quark with a momentum q_0, the probability that the first emitted meson will have a momentum ξ_1 in the interval $d\xi_1$ is $f(1-\xi_1/q_0)d\xi_1/q_0$; the probability that the second meson will have a momentum ξ_2 in the interval $d\xi_2$ is $f(1-\xi_2/q_1)\,d\xi_2/q_1$, where $q_1 = q_0 - \xi_1$; etc. Consequently, the probability for the appearance of a sequence of n mesons, $P_n(\xi_1,...,\xi_n)$, where the nth meson has a momentum ξ_n (a momentum fraction $z_n = \xi_n/q_0$) in the interval $d\xi_n$, is

$$P_n(\xi_1,\xi_2,\ldots,\xi_n)d\xi_1 d\xi_2 \cdots d\xi_n = \prod_{i=1}^{n} f(\eta_i)d\eta_i\,, \tag{4.20}$$

where

$$\eta_i = q_i/q_{i-1} = 1-\xi_i/q_{i-1}\,; \quad \xi_i = q_{i-1}-q_i\,; \quad d\eta_i = d\xi_i/q_i\,;$$

$$q_i = q_0 - \sum_{k=1}^{i} \xi_k\,.$$

Once we have proposed a method of this sort for the production of the particles in the jet, the rest of the calculation follows rigorous mathematical rules. If we define the function $F(z)$ as the probability for finding any primary meson with a momentum fraction z in the interval dz in the jet (regardless of the number of steps of the appearance of $q\bar{q}$ pair; Fig. 4.19), this function must satisfy the integral equation ($q_0 = 1$)

$$F(z) = f(1-z) + \int_0^1 f(\eta)\,F\left(\frac{z}{\eta}\right)\frac{d\eta}{\eta}\,. \tag{4.21}$$

Equation (4.21), which is analogous to the Lipatov-Altarelli-Parisi equation, can be solved. Working from the given function $f(\eta)$, we find the function $F(z)$ explicitly. The trial function (4.19) and the value of the parameter a are chosen on the basis of a comparison with the experimental charged-particle distribution $D_q^{h+}(z) + D_q^{h-}(z)$ observed in the deep inelastic scattering (Sec. 10 of Chap. 3).

To finally find the relationship between the function $F(z)$ and $D_q^h(z)$, it is necessary to consider the diversity of quark flavors, the parity, the isospin, and the strangeness of the mesons. As a result, we find

$$D_q^h(z) = A_q^h f(1-z) + B^h \overline{F}(z)\,, \tag{4.22}$$

where

$$\overline{F}(z) = F(z) - f(1-z)\,; \tag{4.23}$$

and A_q^h and $B^h = \sum_q \gamma_q A_q^h$ are numbers whose values for π, ρ, and K mesons are listed in Table 4.4.

Now that we know how to calculate the function D_q^h, we can carry out numerical calculations of the corrections, qualitative aspects of which can be seen in Table 4.3. Figure 4.20 compares the values of $\langle p_\parallel \rangle$, $\langle p_\perp \rangle$, and $\langle p_\perp^2 \rangle$ with theoretical curves which have and have not been corrected for the emission of a gluon (the solid and dashed lines, respectively). The data in Fig. 4.20 indicate an increase in $\langle p_\perp \rangle$ with increasing energy. For example, while at $13 \lesssim E_{\rm c.m.} \lesssim 17$ GeV we have $\langle p_\perp \rangle \approx 0.3$ GeV/c, at $E_{\rm c.m.} \gtrsim 30$ GeV the value of $\langle p_\perp \rangle$ increases to 0.45 GeV/c (the Feynman–Field model postulates $\langle p_\perp \rangle \approx$ const ≈ 0.3 GeV/c). This increase is explained satisfactorily in QCD and is one possible piece of evidence for the existence of gluons.

Some extremely convincing pieces of evidence in favor of the existence of a gluon come from the so-called seagull effect, which can be outlined as follows: We assume that one of the quark pairs produced in the process

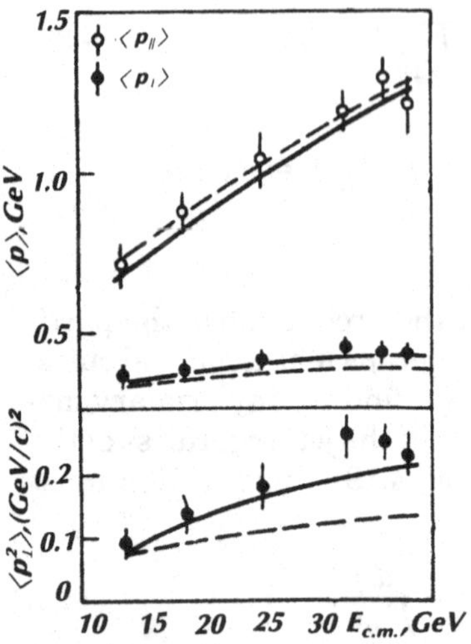

Figure 4.20. Energy dependence of the mean values $\langle p_\perp \rangle$, $\langle p_\parallel \rangle$, and $\langle p_\perp^2 \rangle$ in jets (PLUTO data).

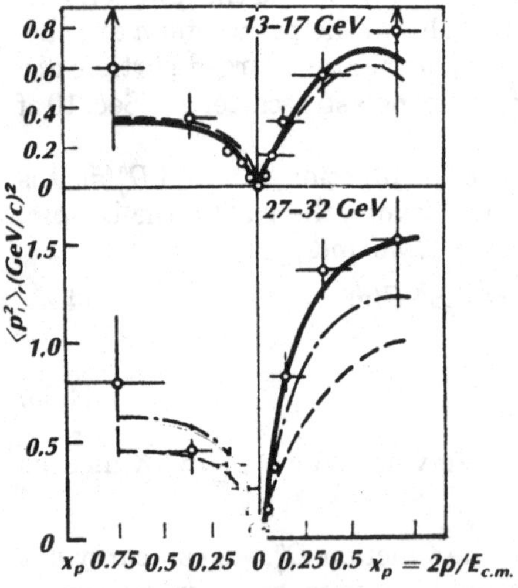

Figure 4.21. Comparison of the data found by the PLUTO group on the seagull effect in jets with theoretical models. (Dashed lines) $\sigma_q = 0.25$ GeV; (dot-dashed lines) $\sigma_q = \langle p_\perp \rangle = 0.35$ GeV; (solid lines) QCD calculations.

$e^+e^- \to \gamma^* \to q\bar{q}$ emits a fast gluon, which then fragments into hadrons. The probability for the emission of two gluons (one by each quark, for example) is small. In the case of the emission of one gluon we observe an asymmetry of the two-jet event. For example, we orient all the events along the thrust (T) axis in such a way that a narrow jet is always on the same side of this axis (the left side, say). A diagram for hadrons on which $\langle p_\perp^2 \rangle$ is plotted separately for narrow and broad jets will then exhibit a seagull effect (Fig. 4.21): Narrow jets will have small values of $\langle p_\perp^2 \rangle$, i.e., a large thrust (the left side of the figure), while broad jets will have large values of $\langle p_\perp^2 \rangle$ (the right side of the figure). The seagull effect is explained by the argument that a gluon emitted by one of the quarks moves in a direction close to that in which the related quark is moving and carries off part of the energy of the quark; during the subsequent fragmentation of the gluon and the corresponding quark there is an emission of low-energy hadrons in a nearly isotropic distribution. The effect is to substantially increase $\langle p_\perp^2 \rangle$.

The behavior of the cross section $d\sigma/dp_\perp^2$ is described in the Feynman–Field model by the exponential function

$$d\sigma/dp_\perp^2 \sim \exp(-p_\perp^2/2\sigma_q^2), \tag{4.24}$$

where $\sigma_q = 0.3$ GeV. In the seagull effect the narrow and broad jets should obviously be described by different parameters σ_q.

The increase in the transverse momentum both with increasing energy of the e^+e^- beams and in the seagull effect agrees well with a QCD prediction.

Additional pieces of evidence for the existence of gluons come from observations of a planar structure of events. The emission of a gluon should result in a "flattening" of the distribution of hadrons produced in jets. The characteristics S and T have previously been used to analyze jets [see (4.11) and (4.13)]. The MARK J group introduced the concept of "oblateness" for an analysis of jets. We will define T in a slightly different way:

$$T = \max\left[\sum_i |E^i \mathbf{e}_1| / \sum_i |E^i|\right], \tag{4.25}$$

where E^i is the energy flux measured for each fixed direction i, and $\mathbf{e}_1$ is the direction of the maximum energy flux. We choose the plane perpendicular to $\mathbf{e}_1$ and introduce the new quantity Major:

$$\text{Major} = \max \frac{\sum_i |E^i \mathbf{e}_2|}{\sum_i |E^i|}, \quad \mathbf{e}_1 \mathbf{e}_2 = 0. \tag{4.26}$$

To determine the plane of the event, we construct a third axis $\mathbf{e}_3$: $\mathbf{e}_3 \perp \mathbf{e}_2$ and $\mathbf{e}_3 \perp \mathbf{e}_1$. We call the energy flux along this axis "Minor." The oblateness O is then defined as follows:

$$O = \text{Major} - \text{Minor}. \tag{4.27}$$

Obviously, O should be greater in the emission of a gluon than in the process $\gamma^* \to q\bar{q}$. This is what is observed experimentally, and the value predicted for the oblateness by QCD agrees with experimental data.

A significant number of events with a planar structure have been observed. It was important to determine how the hadrons are distributed in the plane of the event: isotropically from the center of the event or in the form of jets. If they are distributed in jets, then two- and three-jet structures should be seen most frequently, according to QCD. In particular, if the hadrons are distributed as three jets, it would be difficult to devise any mechanism for their production other than their appearance as a result of the fragmentation of two quarks and one gluon into hadrons.

For an analysis of events under the assumption of a three-jet structure, the concept of "triplicity" is introduced. This quantity is defined by

$$T_3 = \frac{\max\left\{\left|\sum_{N_1} p_i\right| + \left|\sum_{N_2} p_i\right| + \left|\sum_{N_3} p_i\right|\right\}}{\sum_{N_1+N_2+N_3} |p_i|}. \quad (4.28)$$

The particles participating in an event are allotted to all possible versions of the three groups N_1, N_2, and N_3. The numerator in (4.28) is maximized in terms of this allotment. That allotment which leads to the maximum is defined as a three-jet event. The momenta of each of the three groups are used to reconstruct the axes of the three jets. From these axes one can determine other characteristics of interest. We might add that a three-jet nature is inferred exclusively from the charged particles: Neutral particles are not analyzed.

Analysis of events with a planar structure by means of expression (4.28) unambiguously points to the existence of three-jet events. Quantum chromodynamics, which predicts the existence of three-jet events, agrees well with the experimental data. Despite the evidence in favor of the existence of a gluon, it has yet to be proved that the gluon is a colored, neutral, vector, spin-1 particle with a gauge coupling constant. An attempt can be made to extract information on the color of the gluon from the decays of particles consisting of colored quarks. Information on the spin of the gluon and the coupling constant α_s can be found from an analysis of three-jet events. Just as an analysis of the angular distribution of the jet axes in two-jet events established that the spin of a quark is 1/2, analysis of the angular distribution of the axes of the two lower-energy jets in three-jet events (in their c.m. frame, under the assumption that one jet is formed as a result of fragmentation of a gluon emitted by the related quark) reveals the spin of the gluon and its nature (vector or scalar).[133] Differential cross sections are calculated for the following processes: $e^+e^- \to q\bar{q}$ + vector gluon, $e^+e^- \to q\bar{q}$ + scalar gluon, and $e^+e^- \to$ three vector gluons. The cross sections are written in a form convenient for experimental analysis (in terms of the angles between the jet axes). The variables $x_i = 2E_i/E_{\text{c.m.}} = E_i/E_b$ ($E_{\text{c.m.}} = 2E_b$) are introduced, where E_i is the energy of one of the quarks or the gluon. We obviously have a sum $x_1 + x_2 + x_3 = 2$. The quantities x_i are assigned the order $x_3 \leqslant x_2 \leqslant x_1$, which implies $2/3 \leqslant x_1 \leqslant 1$. The upper boundary $x_1 = 1$ is obvious, while the lower boundary on x_1 follows from the equations $x_1 = x_2 = x_3$, $3x_1 = 2$. If we assume that the hadron masses are small in comparison with $E_{\text{c.m.}}$, expressions for x_i follow automatically from the trigonometric formulas for triangles (Fig. 4.22a):

$$x_i = 2 \sin \theta_i / (\sin \theta_1 + \sin \theta_2 + \sin \theta_3). \quad (4.29)$$

The quantity x_1 is by definition the same as T [see (4.13)]. In analyzing experimental data it is convenient to use the angle $\tilde{\theta}$ (Fig. 4.22b):

$$\cos \tilde{\theta} = (x_2 - x_3)/x_1 = (\sin \theta_2 - \sin \theta_3)/\sin \theta_1. \quad (4.30)$$

Figure 4.22. Kinematics of a three-jet event. (a) In the c.m. frame of the colliding e^+e^- beams; (b) in the c.m. frame of the low-energy quark and the gluon.

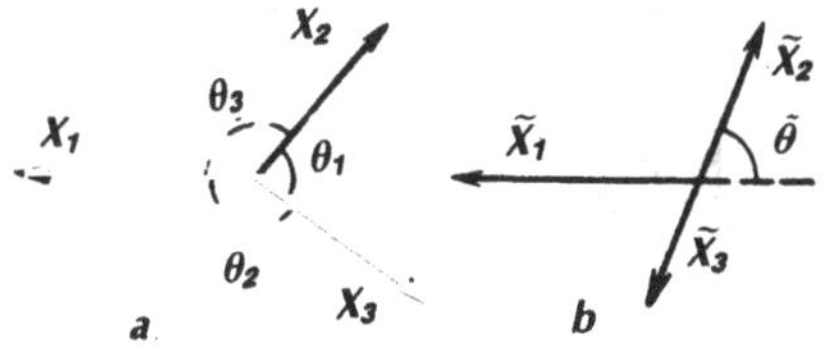

After an average is taken over the production angles with respect to the direction of the e^+e^- beams, the following distributions in x_i are derived[133]:

for vector gluons,

$$\frac{1}{\sigma_0}\left(\frac{d\sigma}{dx_1dx_2}\right)_v = \frac{2\alpha_s}{3\pi}\left[\frac{x_1^2+x_2^2}{(1-x_1)(1-x_2)} + \text{terms found through a cyclic permutation of the indices 1, 2, 3}\right]; \tag{4.31}$$

for scalar gluons,

$$\frac{1}{\sigma_0}\left(\frac{d\sigma}{dx_1dx_2}\right)_s = \frac{\tilde{\alpha}_s}{3\pi}\left[\frac{x_3^2}{(1-x_1)(1-x_2)} + \text{terms found through a cyclic permutation of the indices 1, 2, 3}\right].$$

Since the value $x_1 = 1$ leads to a divergence in (4.31), values $x_1 < 1$ are used in the analysis. The experimental data obtained by the TASSO group[134] have been analyzed in the region $1 - x_1 > 0.10$. This analysis is complicated by several circumstances: The expressions for the vector and scalar gluons differ significantly only at large values of x_1, and at these values it is difficult to distinguish two-jet events from three-jet events; the cross sections vary rapidly with x_1; the QCD corrections cannot be calculated satisfactorily; and effects unrelated to perturbation theory must be incorporated in the analysis.

The effect of the coupling constants α_s and $\bar{\alpha}_s$ on the results of the analysis was reduced by utilizing distributions normalized to the number of events lying in the kinematic region selected for the analysis. The difference between the vector and scalar cases thus reduces to simply a difference in the shape of the distributions of three-jet events in the angle $\tilde{\theta}$ (Fig. 4.23). The experimental data agree well with the distribution of vector gluons and disagree with the distribution of scalar gluons by more than three standard deviations. The confidence level for the vector gluons is 79%, and that for the scalar gluons 0.2%.

The PLUTO group analyzed three-jet events in a slightly different way for the same purpose: to resolve the question of the vector nature of the gluon.[135] The distribution of x_1 was studied by means of the formula (for vector gluons)

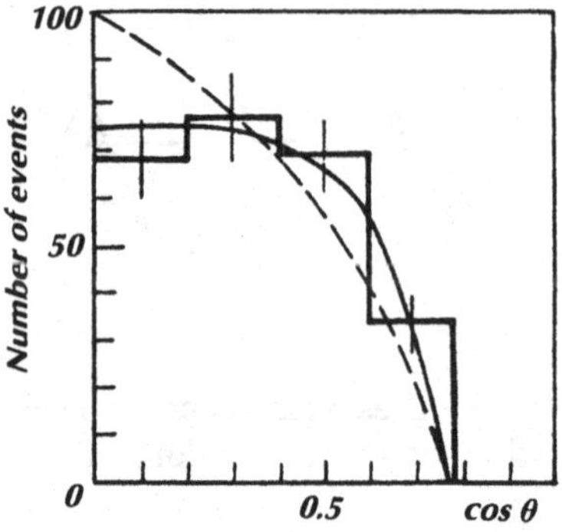

Figure 4.23. Data from the TASSO group in the region $1 - x_1 > 0.10$. (Dashed line) theoretical prediction for a scalar gluon; (solid curve) QCD prediction for a vector gluon; (broken line) constructed from experimental points.

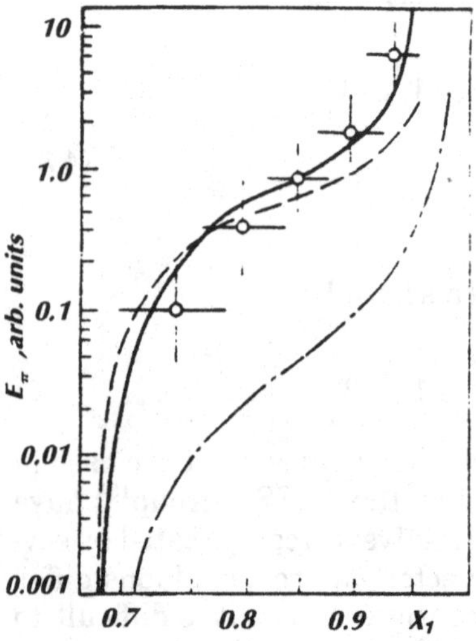

Figure 4.24. Energy distribution of the fastest π^- meson for (solid line) a vector gluon and (dashed line) a scalar gluon. (Dot-dashed line) constituent interchange model.

$$\frac{1}{\sigma}\frac{d\sigma}{dx_1} = \frac{\alpha_s}{3\pi}\frac{1}{1+\alpha_s/\pi}\frac{1}{1-x_1}\left[\frac{4}{x_1}(3x_1^2 - 3x_1 + 2)\ln\frac{2x_1 - 1}{1 - x_1}\right.$$
$$\left. - 6(3x_1 - 2)(2 - x_1)\right]. \tag{4.32}$$

Figure 4.24 compares the experimental data with theoretical curve (4.32) and with the theoretical predictions for scalar gluons. As in the preceding analysis, the pole at the point $x_1 = 1$ was eliminated by choosing the interval $2/3 < x_1 < 0.95$. The statistical base for the result derived is of course degraded. However, the conclusions that the gluon is a vector particle of spin 1 seems extremely convincing. The hypothesis of a scalar gluon is supported by neither an analysis of three-jet events in e^+e^- annihilation nor research on the decay of the $\Upsilon(9.46)$ particles (Sec. 8 of Chap. 4).

Figure 4.25. Choice of a 4-momentum transfer q^2 for determining the coupling constant $\alpha_s(q^2)$.

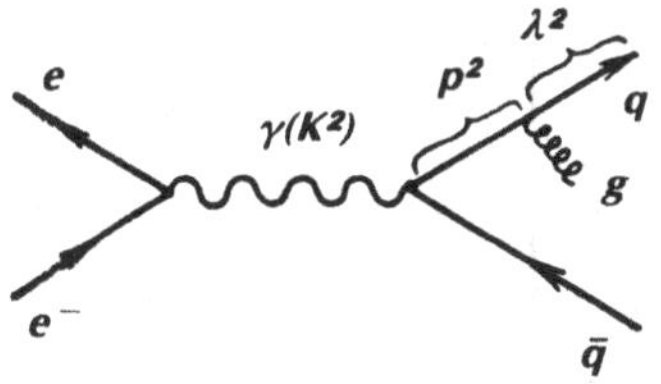

5. Determination of the strong-interaction constant α_s

The strong-interaction constant α_s can be found directly from the number of three-jet events and the ratio of this number to the number of two-jet events. Equation (4.31) contains a functional dependence on the first power of α_s, since it was derived under the assumption that three-jet events arise from the emission of one gluon by one of the quarks (the process $e^+e^- \to q\bar{q}g$). An integration can be carried out over one of the variables in (4.31). The integrated cross section, normalized to the total cross section for e^+e^- annihilation, will depend on α_s alone. It is of course assumed here that experimentalists are able to distinguish three-jet events quite well from two- and four-jet events. We will not take up the methodological problems of selecting events, but we should cite certain aspects of the process of selecting three-jet events: (a) It is frequently not possible to incorporate neutral particles in the final state; (b) it is necessary to identify jets which may arise in the decay of a b quark; and (c) the QCD and QED corrections, especially those for the emission of a high-energy photon in the initial state, must be taken into account as well.

Measurements of α_s have been carried out by all the experimental groups which have used the PETRA accelerator:

MARK J	TASSO	PLUTO	JADE
$0.19 \pm 0.02 \pm 0.04$	$0.17 \pm 0.02 \pm 0.03$	$0.15 \pm 0.03 \pm 0.02$	$0.18 \pm 0.03 \pm 0.03$

The first number listed is the value of the coupling constant, the second is the statistical error, and the third is the systematic error.

The mean value is $\alpha_s = 0.17 \pm 0.02$. Since the constant α_s depends on q^2 (or on the energy s) and on the QCD constant Λ,

$$\alpha_s(q^2) = g^2/4\pi = 12\pi/(33-2f)\ln(q^2/\Lambda^2)\,, \tag{4.33}$$

one can use experimental values of q^2 to find Λ. Unfortunately, it is not clear how we should choose the value of q^2 in the process $e^+e^- \to q\bar{q}g$. The variable p^2 has been selected (Fig. 4.25) [it would have been more accurate to determine $q^2 = (p-\lambda)^2$]. At energies $s \approx 30$ GeV^2 the mean value is $\langle p^2 \rangle = 140$ GeV^2. Substituting this value into (4.33), we find (for $\alpha_s = 0.17 \pm 0.2$) $\Lambda = 95^{+65}_{-35}$ MeV. Incorporating systematic errors results in an increase in the value: $\Lambda < 290$ MeV. Analysis of data on deep inelastic scattering (Chap. 3) yields $100 < \Lambda < 700$ MeV.

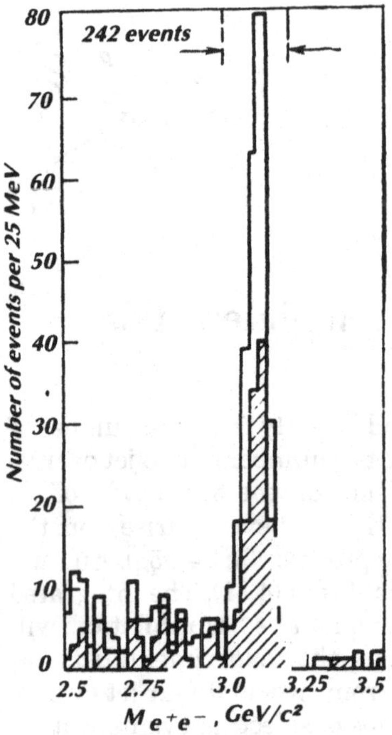

Figure 4.26. Mass spectrum of e^+e^- pairs in the reaction $p + \text{Be} \rightarrow e^+e^- + X$ at a normal current (the hatched region) and at a 10% current.

6. The J/ψ family of particles[128,136–140]

In some cases, the discovery of a new particle is a planned event. This was the case with the antiproton, which was discovered in 1955. However, not all the particles which have been predicted have been successfully observed. Theoreticians have predicted the existence of a pseudoscalar π^0 meson, but it has yet to be found. Searches for quarks in a free state with fractional charges have so far been unsuccessful. The discovery of the new particles which we will discuss in this subsection was not planned. These particles were found at experimental installations with a high energy (or momentum) resolution. Since their properties distinguished them from the usual hadronic and baryonic resonances with large decay widths, they were called "new" particles.

In 1974 a group from the Massachusetts Institute of Technology and the Brookhaven National Laboratory working at Brookhaven's 30-GeV proton accelerator studied the mass spectrum of the e^+e^- pairs produced in the reaction

$$p + \text{Be} \rightarrow e^+e^- + X, \tag{4.34}$$

finding a narrow peak near the mass value 3.1 GeV/c^2 (Fig. 4.26; Ref. 141). The group showed that this peak was due to the production of a new particle, which they called the "J particle." On the same day, experiments at the SPEAR accelerator also revealed a narrow resonance at a mass of 3.1 GeV/c^2 in an e^+e^- annihilation process in a study of the reactions

Figure 4.27. Cross sections for the reactions (a) $e^+e^- \to$hadrons, (b) $e^+e^- \to \mu^+\mu^-$ and (c) $e^+e^- \to e^+e^-$ vs the energy of the e^+e^- beam in the c.m. frame, W, in the mass region of the J/ψ particle. (b, c) Integration over angles in the interval $|\cos\theta| < 0.6$.

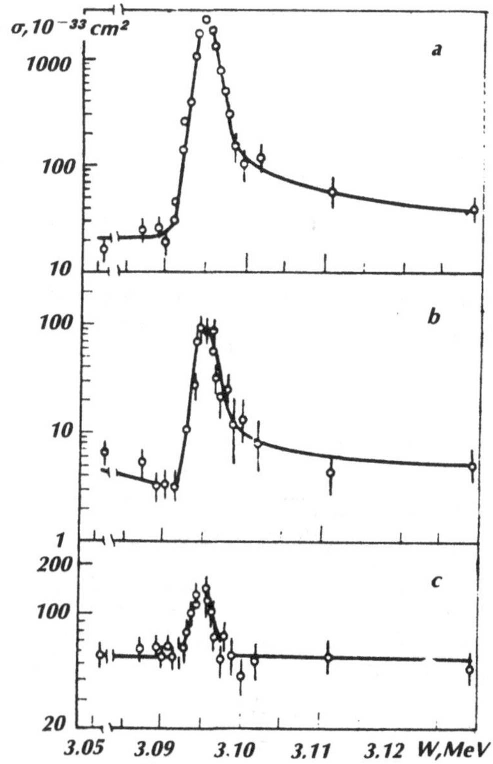

$$e^+e^- \to \text{hadrons}; \quad e^+e^- \to \mu^+\mu^-; \quad e^+e^- \to e^+e^- \tag{4.35}$$

(Fig. 4.27); that particle was called a "ψ resonance"[142] (this particle has now come to be called the "J/ψ particle"). A few days after the J/ψ particle was discovered, it was observed at accelerators at Frascati and Hamburg, and a succeeding resonance, ψ' (3685), was discovered at the SPEAR accelerator[143] (Fig. 4.28).

The most interesting property of the J/ψ and the ψ' particles is their small decay width. A decay of these particles into strongly interacting hadrons is not forbidden by the conservation laws, but the decay width has nevertheless turned out to be three or four orders of magnitude smaller than that of hadronic and baryonic resonances. Such a relation between decay widths can be explained by the Okubo–Zweig–Iizuka rule, which had been formulated earlier in order to explain the suppression of the decay $\varphi \to 3\pi$. The essence of this rule can be expressed in a graphic manner as follows: If the diagram describing the decay is connected (Fig. 4.29a), then the decay is not forbidden; if the diagram describing the decay is disconnected (Fig. 4.29b), then the decay is strongly suppressed (by a factor of two or three orders of magnitude or more). The scheme for the suppression of the decay $\psi \to \varphi\pi\pi$ is similar to that for the suppression of the decay $\varphi \to 3\pi$.

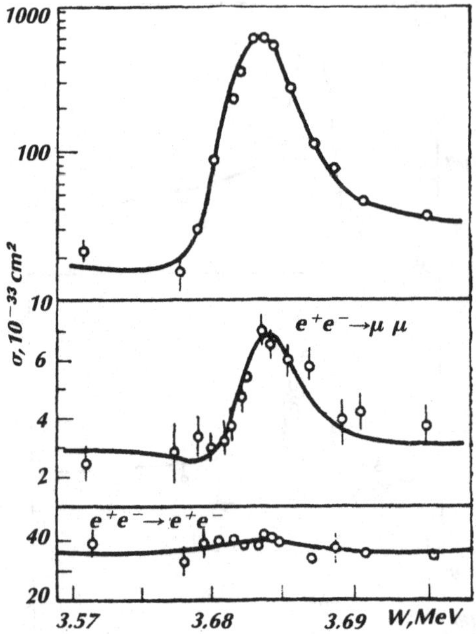

Figure 4.28. Same as in Fig. 4.27, but for the mass region of the ψ' particle.

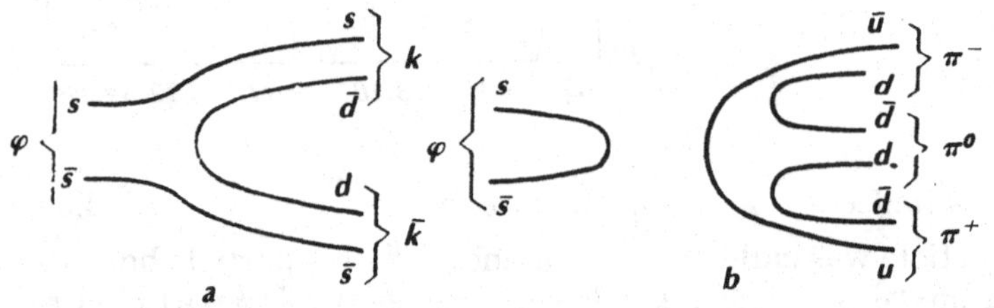

Figure 4.29. Graphic interpretation of the Okubo–Zweig–Iizuka rule.

The partial decay width of the J/ψ particles was found experimentally by the following approach. The resonance cross section for the process of e^+e^- annihilation through a narrow resonance with a mass M_0 into any final state f is given by [see (4.5)]

$$\sigma = \frac{\pi(2J+1)}{s} \frac{\Gamma_{ee}\Gamma_f}{(M_0 - \sqrt{s})^2 + \Gamma^2/4}, \tag{4.36}$$

where J and Γ_f are the spin and partial decay width of the resonance, and $\Gamma = \sum_f \Gamma_f$ is the total width. Let us assume that the spin of the J/ψ particle is 1 (below we will prove that this is the case). Armed with experimental values of the cross sections $\sigma_{\text{had}}^{\text{total}}$, $\sigma_{\mu\mu}$, and σ_{ee} (Fig. 4.27), we can then easily calculate all

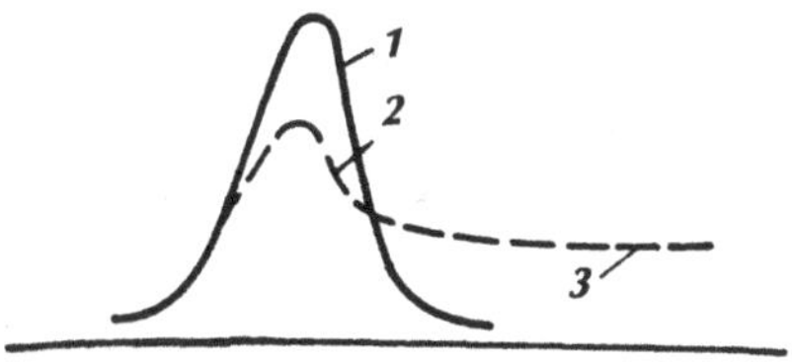

Figure 4.30. Change in the shape of a resonance curve caused by the formation of a radiation tail. 1, exact resonance curve; 2, with bremsstrahlung; 3, radiation tail.

the decay widths. Here we can replace (4.36) by the integral contributions of the cross sections [we are using (4.6) and (4.7)],

$$\int \sigma_{\text{had}}^{\text{total}}\, ds = \frac{6\pi^2}{M_0^2}\frac{\Gamma_{ee}\Gamma_h}{\Gamma};$$
$$\int \sigma_{\mu\mu}\, ds = \frac{6\pi^2}{M_0^2}\frac{\Gamma_{ee}\Gamma_{\mu\mu}}{\Gamma}, \quad \int \sigma_{ee}\, ds = \frac{6\pi^2}{M_0^2}\frac{\Gamma_{ee}^2}{\Gamma} \tag{4.37}$$

(the integration is carried out over the range of the resonance, $\Gamma \ll M_0$), and the equation $\Gamma = \Gamma_h + \Gamma_{\mu\mu} + \Gamma_{ee}$. In this manner we find the partial decay widths. Their values are listed in Table 4.5.

The data in Table 4.5 include the radiative corrections, which are substantial (about 50%). The effect of the radiative corrections can be explained in a qualitative way as follows: If the sum of the energies of the e^+e^- beams is precisely equal to the energy of a resonance, the emission of a photon by one of the particles in a bremsstrahlung process before the collision will cause the total energy of the e^+e^- pair to be too low for the production of the J/ψ particle. The cross section at the point of the resonance will therefore decrease. On the other hand, if the sum of the energies of the e^+e^- beams is above the energy of the resonance, the emission of a photon by one of the electrons may have the consequence that the remaining energy of the e^+e^- beams is precisely equal to the energy of the resonance; if so, there will be an additional contribution to the cross section for the production of the J/ψ particle. As a result, a "radiation tail" will appear and add to the production cross section at energies above the resonance (Fig. 4.30).

The spin and parity of the J/ψ particle were determined through a study of the interference of the contributions of the resonance amplitude and the ordinary single-photon amplitude to the differential cross sections for the process $e^+e^- \to \mu^+\mu^-$. The interference in this process is considerably greater than in the process $e^+e^- \to \mu^+\mu^-$. The experimental data were compared with the theoretical predictions of the differential cross section under the assumption that the spin and parity J^P of the new J/ψ particle could take on the values $0^\pm, 1^\pm, \ldots$. If the J/ψ particle is a spin-0 particle, the resonance amplitude (Fig. 4.31a) will not interfere with the one-photon amplitude (Fig. 4.31b):

$$J^P = 0^\pm: \quad \frac{d\sigma}{d\Omega} = \frac{2\alpha^2}{3s}\left\{\frac{3}{8}(1+\cos^2\theta) + \frac{g^4}{e^4}\frac{s^2}{(M_0^2 - s)^2 + M_0^2\Gamma^2}\right\}. \tag{4.38}$$

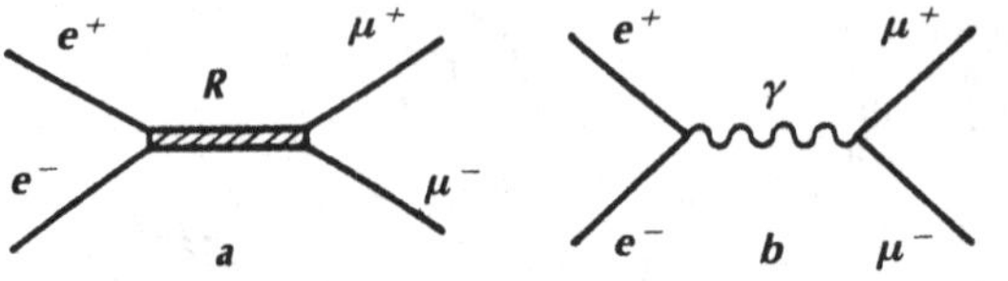

Figure 4.31. Diagrams representing (a) a resonance amplitude and (b) a QED amplitude.

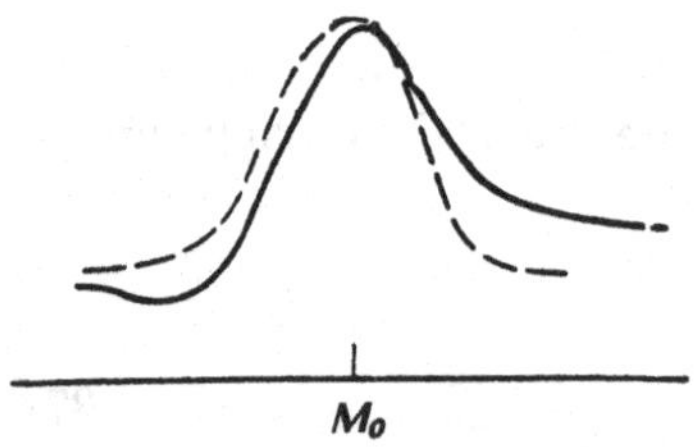

Figure 4.32. Effect of interference on the behavior of a cross section. (Solid line) with interference; (dashed) without.

If the J/ψ particle has spin and parity $J^P = 1^-$, the interference will change the shape of the cross section, as shown schematically in Fig. 4.32. Up to the energy of the resonance, the interference is destructive, while above it, it is constructive:

$$J^P = 1^-: \quad \frac{d\sigma}{d\Omega} = \frac{\alpha^2}{4s}(1+\cos^2\theta)\left\{1 - 2\frac{g_V^2}{e^2}\frac{(M_0^2 - s)s}{(M_0^2 - s)^2 + M_0^2\Gamma^2} + \frac{g_V^4}{e^4}\frac{s_2}{(M_0^2 - s)^2 + M_0^2\Gamma^2}\right\}. \tag{4.39}$$

The forward-backward angular distribution of the final products of the decay of the J/ψ particle is symmetric for the $J^P = 1^-$ state.

There is no interference for $J^P = 1^+$, but there is a forward-backward asymmetry in the angular distribution of μ mesons. There is no need to describe the shape of the interference or to discuss the forward-backward asymmetry of the angular distribution for other possible values of the spin and parity of J/ψ, since we now know that the spin and parity are $J^P = 1^-$. Comparison of the theoretical curves with experimental data made it possible to establish unique values for the spin and parity of J/ψ. In an effort to reduce the systematic errors, this comparison was carried out not for the differential cross section $\sigma(e^+e^- \to \mu^+\mu^-)$ itself but for the cross-section ratio $d\sigma(e^+e^- \to \mu^+\mu^-)/d\sigma(e^+e^- \to e^+e^-)$ (Fig. 4.33). It can be seen from this figure that the experimental data indicate that the interference is of a destructive nature below the resonance, and the experimental data also agree with the suggestion of a maximum interference. The data rule out the possibilities that the spin value is 0, 2, or higher. The forward-backward asymmetry in the angular distribution of μ mesons is zero near the resonance. The angular distributions of the $\mu^+\mu^-$ and e^+e^- pairs and all of the data presented above agree with the value $J^{PC} = 1^-$ for the J/ψ particle.

The G parity and the isospin of J/ψ were established in a study of its hadronic decays. We know that the G parity of a neutral nonstrange boson is

Figure 4.34. Diagrams with the appearance of π mesons in the final state. (a) Direct decay of J/ψ particles; (b) fragmentation of a photon through resonance states; (c) nonresonant fragmentation of a photon.

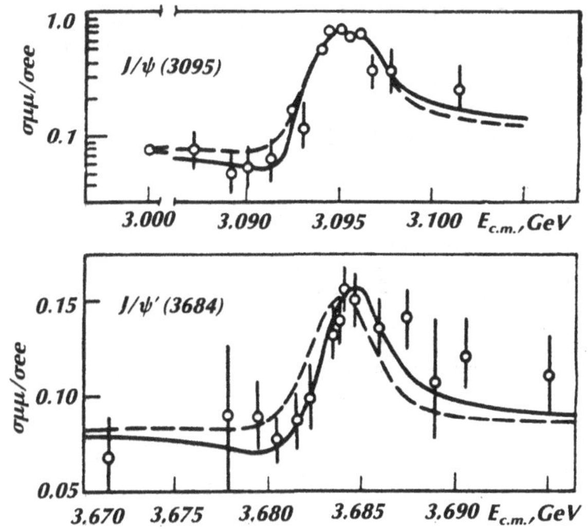

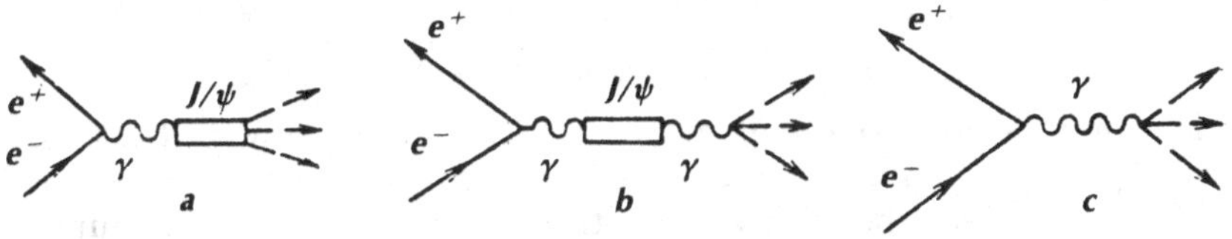

Figure 4.34. Diagrams with the appearance of π mesons in the final state. (a) Direct decay of J/ψ particles; (b) fragmentation of a photon through resonance states; (c) nonresonant fragmentation of a photon.

even (odd) if it decays into an even (odd) number of π mesons. Data obtained in 1974–1975 at the SPEAR and DORIS accelerators indicated that the decays into an odd number of π mesons clearly outnumbered the decays into an even number of π mesons. Since decays into an even number of π mesons also occurred, the G parity could not be accepted as a good quantum number. Analysis of the process $e^+e^- \to n$ of the π mesons showed that the π mesons in the final state may appear both as a result of the decay of J/ψ and as a result of a fragmentation of the photon (Fig. 4.34, b and c).

The diagram in Fig. 4.34c, however, makes only a small contribution to the amplitude for the process $e^+e^- \to$ hadrons, since the photon propagator does not exhibit a resonance behavior near the mass of the J/ψ particle. The resonance region is dominated by the diagram in Fig. 4.34a, while near the J/ψ resonance the dominant diagram is that in Fig. 4.34b, which contains the propagator of the J/ψ resonance. As usual, the cross section $\sigma_{tot}(e^+e^- \to \text{hadrons})$ away from the resonance, determined by the diagram in Fig. 4.34 b, has been normalized to the value of the cross section $\sigma(e^+e^- \to \mu^+\mu^-)$:

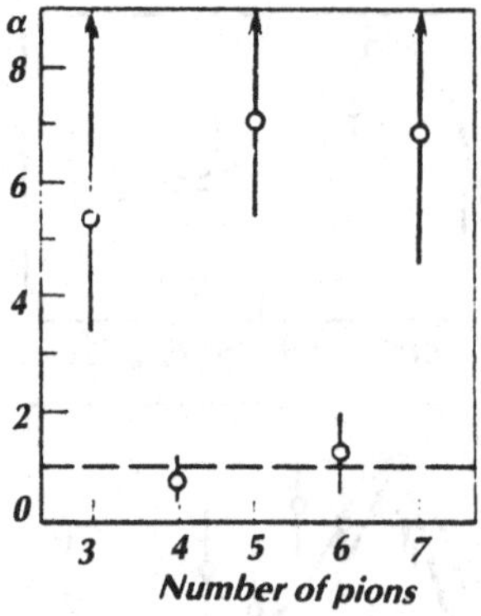

Figure 4.35. Plot of α vs the number of π mesons in the final state in the production of pions in the reaction $e^+e^-\to$hadrons.

$$\left|\frac{\sigma_{\text{total}}(e^+e^-\to\text{hadrons})}{\sigma(e^+e^-\to\mu^+\mu^-)}\right|_{\text{away}} = R_{\text{away}}\,.$$

It is assumed that R_{away} remains essentially constant in the vicinity of the resonance.

We introduce a corresponding ratio for the resonance region:

$$R_{\text{res}} = \sigma_{\text{total}}(e^+e^-\to\text{hadrons})/\sigma(e^+e^-\to\mu^+\mu^-)\,.$$

This ratio reflects all of the diagrams in Fig. 4.34. We now see that the difference $R_{\text{res}} - R_{\text{away}}$ must be equal to zero away from the resonance and different from zero near the resonance in the case of an even or odd number of π mesons for even or odd values of the G parity of the J/ψ particle.

Experimentally, the quantity $\alpha = R_{\text{res}}/R_{\text{away}}$ was measured (Fig. 4.35). We see that for an odd number of π mesons the value is $\alpha = 6$–7, while that for an even number of π mesons is $\alpha = 1$. We thus conclude $G = -1$.

The isospin I of the J/ψ particle can be found with the help of the formula for determining the charge parity C: $C = (-1)^I G$. Under the assumption that the charge parity of the J/ψ particle is equal to that of the photon ($C_\gamma = -1$), and knowing $G = -1$, we find $(-1)^I = 1$, i.e., $I = 0$ or 2 (higher even values of I are improbable). It follows from isotopic invariance that the decays of J/ψ into different charge combinations $\rho\pi$ obey the relations

$$\frac{\text{J}/\psi\to\rho^0\pi^0}{\text{J}/\psi(\rho^+\pi^- + \rho^-\pi^+)} = \begin{cases} 1/2 & \text{if } I = 0\,; \\ 2 & \text{if } I = 2\,. \end{cases}$$

The ratio found experimentally is 0.57 ± 0.17, which agrees with the value $I = 0$. Further evidence in favor of the value $I = 0$ follows from the observation of the decay J/$\psi\to p\bar{p}$. From an analysis of the experimental data we conclude that this decay does not involve intermediate states, that its branching fraction is high, and that in this case the isospin of the J/ψ particle must be even or zero.

The mass, the partial decay width, the spin, and the parity of the ψ' particle were determined in the same way as for the J/ψ. For ψ' the spin and parity

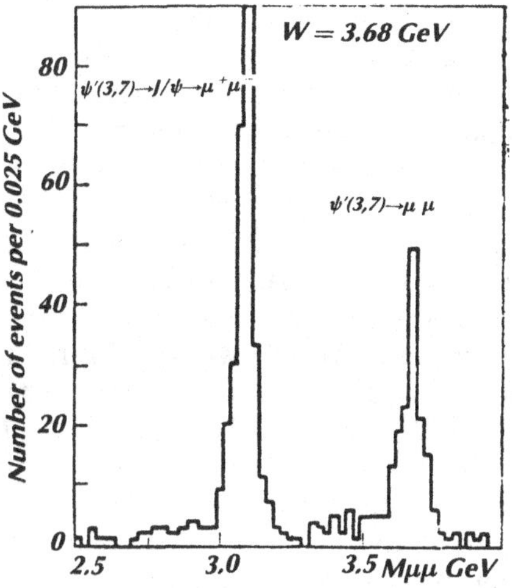

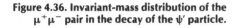

Figure 4.36. Invariant-mass distribution of the $\mu^+\mu^-$ pair in the decay of the ψ' particle.

are $J^P = 1^-$, and the charge parity $C = -1$. The G parity and the isospin were determined from the cascade decay

$$\psi' \to J/\psi + X, \tag{4.40}$$

which amounts to more than half of all of the decays of the ψ' particle. A cascade decay can take several paths. Figure 4.36 shows the results of a study of cascade decay (4.40) on the basis of the invariant mass of the $\mu^+\mu^-$ pair which arises in the decay of the ψ' and J/ψ particles. The expected number of events agrees with the theoretical prediction for $I^G = 0^-$. Comparison of the experimental ratios of the widths of the cascade decays of the other type with the theoretical predictions in Table 4.6 also points to the values $I^G = 0^-$.

To conclude this discussion of the determination of the quantum numbers of the J/ψ and ψ' particles, we consider the question of whether they belong to the $SU(3)$ group. If we assume that the decays of these particles into any two hadrons do not break $SU(3)$ symmetry, then we can find selection rules for decays into the pairs $\pi^+\pi^-$, K^+K^-, and $K^0 \sim K^0$ under the assumption that the J/ψ and ψ' particles belong to an $SU(3)$ singlet or an $SU(3)$ octet.

Table 4.7 shows experimental values of the ratio $\Gamma \to 2h/\Gamma \to X$ and the theoretical predictions of these ratios for both particles. It can be seen from Table 4.7 that the experimental ratios are very small and the corresponding decays are strongly suppressed, in agreement with the assumption that J/ψ and ψ' belong to an $SU(3)$ singlet.

The decay of J/ψ into $\pi\rho$ and $KK^*(892)$ pairs provides yet further confirmation that J/ψ belongs to the $SU(3)$ group. The amplitudes for these decays can be written in the form $A(\pi^\pm\rho) = A_1 - 2A_8$ and $A(K^\pm K^*) = A_1 + A_8$, where A_1 and A_8 are the singlet and octet parts of the amplitudes for the decays of J/ψ. The experimental data indicate that the ratios $|A_8|/|A_1|$ are small; this is evidence that J/ψ belongs to an $SU(3)$ singlet. The baryon decays of the J/ψ and ψ' particles ($J/\psi, \psi' \to p\bar{p}, \Lambda\bar{\Lambda}, \Xi^-\bar{\Xi}^-$) are strongly suppressed; this is further evidence that these particles belong to an $SU(3)$ singlet.

After the discovery of the J/ψ and ψ' particles, experimentalists turned their attention to a search for other vector mesons with a small decay width (narrow vector states). There was also interest in determining whether the J/ψ particle is the lowest (ground) state of a new family of vector particles. The energy region 1.9–7.7 GeV was studied thoroughly at e^+e^- colliders. It turned out that no new narrow vector mesons lie in this region. It was also shown that J/ψ is the lowest-lying state of a new family of vector particles. However, a broad resonance ψ'' was discovered near the mass value of 3770 MeV/c^2. Table 4.8 shows certain characteristics of ψ''. The basic decay schemes are $\psi'' \to D^0\bar{D}^0$ [$(49 \pm 21)\%$] and $\psi'' \to D^+D^-$ [$(50 \pm 38)\%$]. Since ψ'' decays almost exclusively into D mesons, it is frequently called a "D-meson factory."

The energy interval 4–4.5 GeV (Fig. 4.6) has turned out to be so complicated to analyze that physicists have not yet been able to reconcile the experimental results reported by the PLUTO, DASP, and DELCO groups (all from DESY) and the MARK J group (SLAC–LBL). Some groups believe that there are narrow peaks near 4.028 and 4.41 GeV. The DASP and PLUTO groups assert that there is a peak at 4.16 GeV.

Three particles have subsequently been observed: $\chi(3415)$, $\chi(3510)$, and $\chi(3550)$, which fall between the J/ψ and ψ' particles along the mass scale. The ψ' particle undergoes a radiative decay into one of these particles; in turn, $\chi(3550)$, $\chi(3510)$, and $\chi(3415)$ convert into J/ψ in a radiative decay. The radiative transitions have been utilized to establish several properties of these three "intermediate" particles of this family of new particles. We turn now to some of the details of these decays.

7. The new particles as bound states of heavy quarks

The properties of the new particle which were described in the preceding subsection can be explained in a fairly natural way by a model of bound states of heavy charmed c quarks.

Going to a model with four quarks (u, d, s, c), each of which has an additional degree of freedom in the form of color, means choosing $G = SU(4) \times SU(3)^c$, where $SU(3)^c$ is the color group, as the hadron symmetry group.

One can use a QCD perturbation theory and, by complete analogy with QED, examine the bound states of quarks, write Bethe–Salpeter equations for them, study the mass spectrum of the new particles, calculate the magnetic moments of quarks, etc.

We will not use a rigorous theory here. The problem of finding the mass spectrum of the new particles may be studied in the nonrelativistic approximation by solving a Schrödinger equation with a given potential. That approach is completely legitimate since the charmed quarks are heavy, and the

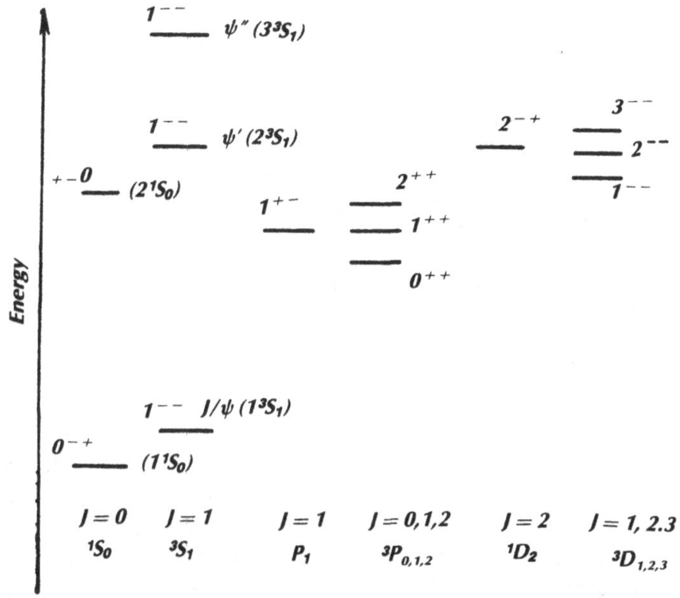

Figure 4.37. Level scheme predicted for the $c\bar{c}$ system with potential (4.41). The spectroscopic notation for the levels; $^{2S+1}L_J$, is used here.

binding energy in this problem is small. At short range, the interaction potential can be taken to be a Coulomb potential $\alpha_s(Q^2)/r$, where $\alpha_s(Q^2)$ is the constant of the strong interaction of gluons with quarks determined at a value of Q^2 equal to the mean value of the square mass of the $q\bar{q}$ system. Here r is the distance between quarks. At long range, we should choose a barrier potential, which will rule out the possibility of an emission of quarks from the system, since quarks in a free state have yet to be found. As this barrier potential we could choose a linear potential ar or a quadratic potential br^2. The potential is thus chosen in the form

$$V(r) = -(4/3)\alpha_s/r + V_0 + ar, \tag{4.41}$$

where V_0 is a constant.

With regard to potential (4.41), we note that (first) the spatial region responsible for the Coulomb interaction is small in comparison with the dimension of the $c\bar{c}$ system and (second) logarithmic corrections are added to the Coulomb dependence r^{-1}, as follows from the form of the coupling constant $\alpha_s(Q^2)$. Unfortunately, there is nothing we can say about the potential in the intermediate region between small and large values of r, although this intermediate region is particularly important for describing the properties of the J/ψ particles. It is chosen on the basis of the continuity of the potential and is found by fitting the parameters V_0 and a. A potential of the type in (4.41) leads to the level scheme for the $c\bar{c}$ system shown in Fig. 4.37.

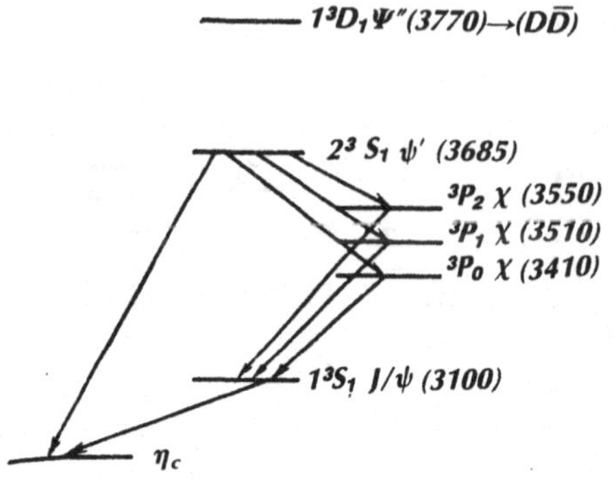

Figure 4.38. Experimental level scheme of the family of J/ψ particles.

The levels are classified on the basis of J^{PC} states, where $P=(-1)^{L+1}$, $C=(-1)^{L+S}$. In Fig. 4.37, there are two columns of states, with opposite eigenvalues of the charge-conjugation operator (depending on whether the value at $S=0$ or $S=1$ is chosen), for each value of L.

The experimental data indicate that there are no other vector particles with a small decay width below J/ψ on the energy scale. The J/ψ particle can thus be identified with the ground state of a bound $c\bar{c}$ quark pair, i.e., a $1\,^3S_1$ state. It is natural to identify ψ' with a $2\,^3S_1$ state. The particle ψ'', with quantum numbers 1^{--}, is identified with a $^3D_1$ level. Figure 4.38 is the level scheme observed experimentally for the family of J/ψ particles.

Figures 4.37 and 4.38 show three P-wave states $3P_{0,\,1,\,2}$ with positive parity and a positive charge-conjugation operator. These states were observed in an analysis of the decays

$$\psi' \to \gamma\chi\,; \tag{4.42}$$

It follows that for all of the P states we have $C=+1$.

The spin and parity of the P states can be determined from an analysis of their hadronic decay schemes, the angular distribution of the photons with respect to the beam direction in reaction (4.42), i.e., $W(\cos\theta)$, and the level scheme predicted by the $c\bar{c}$ model (Fig. 4.37). The angular distribution of photons is

$$W(\cos\theta)=1+a\cos^2\theta\,. \tag{4.43}$$

Analysis of the experimental data leads to the following values for a:

State:	$\chi(3.415)$	$\chi(3.510)$	$\chi(3.550)$
a:	1.4 ± 0.4	0.1 ± 0.4	0.3 ± 0.4

If the spin is 0, we have $a=1$. The value of a for the $\psi(3.41)$ particle thus agrees with a zero spin, while a zero spin is ruled out for the states $\chi(3.51)$ and $\chi(3.56)$. From experiments, we know that $\chi(3.41)$ decays into an even number of π mesons. Since the spin is 0, and we have $C=+1$, we conclude that $\chi(3.41)$ is an isoscalar particle with a positive parity ($J^{PC}=0^{++}$).

The absence of $\pi^+\pi^-$ and K^+K^- decays of $\chi(3510)$ indicates an abnormal series of values[11] $J^P=0^-,1^+\ldots$. From the coefficient a in (4.43) we conclude

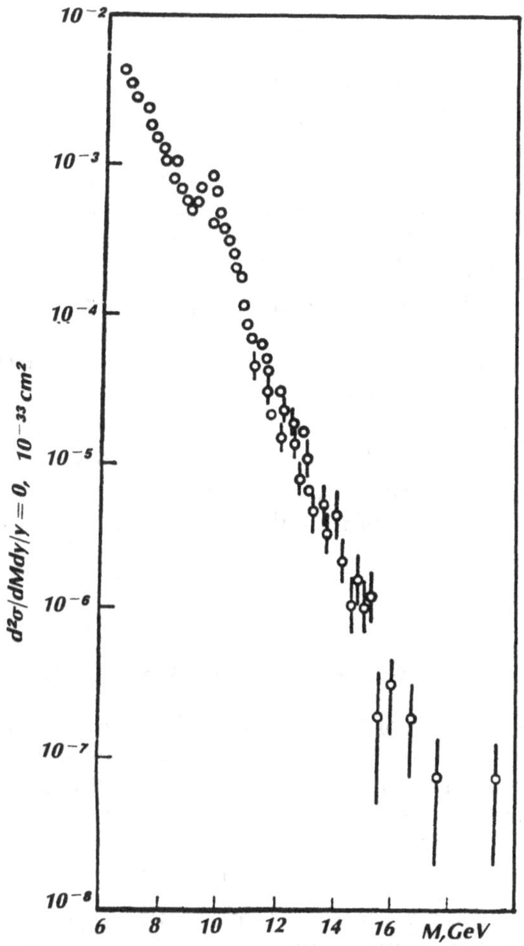

Figure 4.39. Cross section $d^2\sigma/dM\,dy|_{y=0}$ as a function of the invariant mass of the $\mu^+\mu^-$ pair as measured in the reaction $p^+\text{nucleus}\to\mu^+\mu^- X$.

$J\neq 0$. The isospin should be even, since $\chi(3510)$ decays into an even number of π mesons.

The particle $\chi(3.550)$ decays into an even number of π^- and K mesons; it should be an isoscalar particle and should belong to a normal series of J^P values. For it, however, we have $J\neq 0$, according to the values of the coefficients a in (4.43). The level scheme (Fig. 4.37) indicates the presence of a $J^{PC} = 1^{+-}$ state. Since $C=1$, this state cannot arise from γ decay (4.42). A decay of the type $\psi'\to\pi^0 + {}^1P_1$ is suppressed by isospin conservation. The decay $\psi'\to\pi\pi + {}^1P_1$ is forbidden kinematically, since the mass difference $\psi' - {}^1P_1$ is not large enough for a two-pion decay. So far, there is no experimental indication of the existence of a particle with $J^{PC} = 1^{+-}$.

We would like to say a few words about the level η_c, which may arise both as a result of decay (4.42) and in the decay (Fig. 4.39)

$$J/\psi \to \gamma\eta_c \,. \tag{4.44}$$

It follows from the form of potential (4.44) that there are several pseudoscalar states η_c which should have quantum numbers $1\,^1S_0, 2\,^1S_0, \ldots$ and should lie below the J/ψ and ψ' particles along the mass scale. The η_c states may decay into ordinary hadrons through a two-gluon intermediate state. The state $^1S_0$, with $C = +1$, may decay into two photons. Decay (4.44) can thus occur by the pathway $J/\psi \to \gamma\gamma\gamma$ or $J/\psi \to \gamma +$ hadrons. The invariant-mass distribution for any pair of photons in a decay of the first type may point to the existence of an η_c state.

Different experimental groups report different values for the mass of the η_c meson (e.g., 2.82 and 2.976 ± 0.020 GeV). In 1981 the Crystal Ball group (SLAC) observed and studied a resonance with a mass below that of J/ψ (Ref. 144). That resonance was observed in the decays of J/ψ and ψ' and was identified with the η_c meson. The following values were given for the mass and width: $M_{\eta_c} = (2984 \pm 4)$ MeV$/c^2$ and $\Gamma_{\eta_c} = (12.4 \pm 4.1)$ MeV. The same experimental group observed a new resonance in the inclusive photon spectrum of the decay of ψ'. This resonance is a good candidate for a $2\,^1S_0$ state (Fig. 4.37).

8. The Υ family of particles

The family of Υ particles was discovered[145] in 1977 by a collaboration of scientists (Columbia University and Fermilab) in experiments on the mass spectrum of the $\mu^+\mu^-$ pair produced in collisions of 400-GeV protons with Be, Cu, and Pt nuclei:

$$p + \text{nucleus} \to \mu^+\mu^- + X \,. \tag{4.45}$$

Figure 4.39 shows the results of one of these experiments. If we subtract the background from the $|d^2\sigma/dM\,dy|_{y=0}$ expression (y is the rapidity for the $\mu^+\mu^-$ pair in the c.m. frame, and M is the invariant mass of the $\mu^+\mu^-$ pair), we find that the spectrum of the invariant mass of the $\mu^+\mu^-$ pair has the form shown in Fig. 4.40. A peak clearly emerges near $M \approx 9.5$ GeV$/c^2$ here. Less obvious is a second peak near $M \approx 10$ GeV$/c^2$. After an improvement of the mass resolution and an increase in the statistical base,[146] it was shown that the experimental data indicate the existence of at least two Υ states. Indications—not very clear—of the existence of a third Υ state were also found. Table 4.9 shows the values found for the masses and the cross sections at the peak under the assumptions of the existence of two and three peaks.

We would like to know the decay widths of the new particles, which can be measured highly accurately in experiments with colliding e^+e^- beams. Unfortunately, in 1977 there were no e^+e^- colliders with sufficiently high energy. A successful effort was accordingly undertaken at DESY to raise the energy of the beams at the DORIS accelerator to 2×5 GeV. In the period May through July 1978, the decay widths of two Υ particles were measured [by

Figure 4.40. Cross sections $d^2\sigma/dM\,dy|_{y=0}$ versus the invariant mass of the $\mu^+\mu^-$ pair after the subtraction of the background.

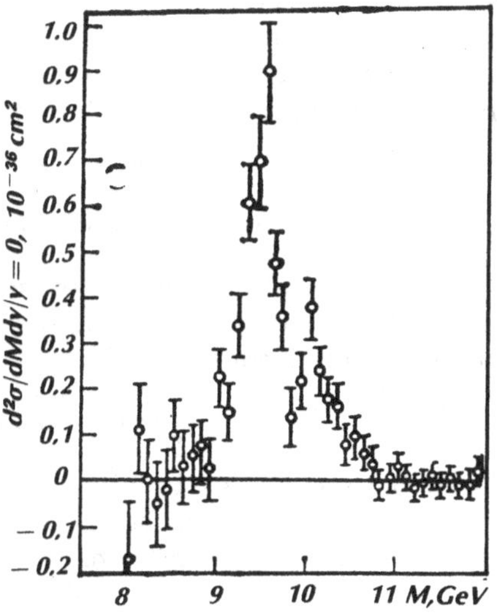

analogy with the measurement of the decay widths of the family of J/ψ particles; see (4.36) and (4.37)].

The integral contribution to the cross section couples the hadronic cross section with the total, hadronic, and leptonic decay widths:

$$\int \sigma_{\text{had}}\, dM = (6\pi^2/M^2)\Gamma_{ee}\Gamma_{\text{had}}/\Gamma\,. \tag{4.46}$$

Let us assume $\Gamma = \Gamma_{\text{had}} + \Gamma_{ee} + \Gamma_{\mu\mu} + \Gamma_{\tau\tau} = \Gamma_{\text{had}} + 3\Gamma_{ee}$. To distinguish Γ from Γ_{ee}, we need to know the cross section for one of the leptonic decay schemes, say, $\sigma_{\mu\mu}$. We can then use the relation

$$\int \sigma_{\mu\mu}\, dM = (6\pi^2/M^2)\Gamma_{ee}^2/\Gamma \tag{4.47}$$

to determine either Γ_{ee} or Γ; then using (4.46), we can determine the other unknown quantity. In this manner, the following values were found:

$$M_{\Upsilon} = (9462 \pm 10)\ \text{MeV}/c^2; \qquad \Gamma_{ee} = (1.29 \pm 0.09 \pm 0.13)\ \text{keV}\,;$$

$$M_{\Upsilon'} = (10\,015 \pm 10)\ \text{MeV}/c^2; \quad \Gamma_{ee} = (0.58 \pm 0.08 \pm 0.26)\ \text{keV}\,.$$

After the CESR accelerator was started up (Table 1.1), the CLEO and CUSB experimental groups discovered two more particles of this family. Table 4.10 lists the parameter values which have been found so far for four particles of the Υ family.

Since a c quark was used to explain the mass spectrum of the J/ψ family, the discovery of the Υ particles naturally led to the suggestion that there exists a new quark (b), bound states of which (i.e., the $b\bar{b}$ pair) might generate the new family of Υ particles. Under the assumption that the b quarks were spin-1/2 particles and that the interaction potential of the b quarks was of the same shape as potential (4.41), we find for the Υ family the same level scheme as has been discussed for the $c\bar{c}$ system (Fig. 4.37). The quantum numbers of the Υ family are measured by a method completely analogous to that used to measure the quantum numbers of the J/ψ particles. Here are the presently known quantum numbers of the Υ particles and the total decay widths:

$$\Upsilon(9460){:}(J^P)C = (1^-), \qquad \Gamma_\Upsilon = 0.060 \text{ MeV};$$
$$\Upsilon'(10\,020){:}(J^P)C = (1^-), \quad \Gamma_{\Upsilon'} < 12 \text{ MeV}.$$

In the nonrelativistic quark model, the leptonic decay width for vector states is proportional to the square of the charge of the quark from which the vector particle is constructed:

$$\Gamma_{\Upsilon_{ee}} = (16\pi^2\alpha^2/M^2)|^3S_1(0)|^2Q^2, \tag{4.48}$$

where $^3S_1(0)$ is the wave function of the $q\bar{q}$ system, and Q is the charge of quark q. The following fact has been established empirically: For the basic vector mesons (ρ, ω, φ, and J/ψ) we have a ratio

$$\Gamma_{V_{ee}}/\sum_i |C_iQ_i|^2 \approx 11,$$

where $\sum_i |C_iQ_i|^2$ is the average charge of the constituent quarks. If we apply this rule to the Υ particle, under the assumption that it is a bound state of a $b\bar{b}$ quark pair, we find $Q = 0.34 \pm 0.04$ or $Q = 1/3$ for the charge of the b quark.

A very recent attempt was made to study the jet structure which arises in the decay of a Υ particle into hadrons. The $b\bar{b}$ bound state is associated with three gluons in lowest-order QCD. In the decay of $b\bar{b}$ into gluons, the energy of the latter is sufficient for the observation of a jet structure of the hadrons which appear as a result of the fragmentation of the gluons. Proof of a three-jet structure in the decay of Υ particles is regarded as very important, since it provides yet another test of the existence of gluons and the legitimacy of QCD. Unfortunately, the energy distribution among the three gluons is not strictly symmetric. The decay of the Υ particles is more reminiscent of two-jet events; the fragmentation of gluons into hadrons has not been studied very well; and resonance events are difficult to distinguish from nonresonance events. All these circumstances have complicated the analysis. There is no need to dwell here on the details of the procedure for selecting and analyzing events. The general conclusions, which require further confirmation, are that the structure of events in which Υ particles decay into hadrons is quite different from a two-jet structure and that all the topological features of the structure of the events agree wth a three-jet picture and with the suggestion that the direct decay of the Υ particles goes through a three-gluon intermediate state. The model of a simple phase space does not explain many details of the decay of the Υ particles. Only QCD gives a satisfactory explanation of many observed facts. If it is assumed that the three-jet picture of the decays of the Υ particles is correct, then we must rule out a scalar nature of the gluons.

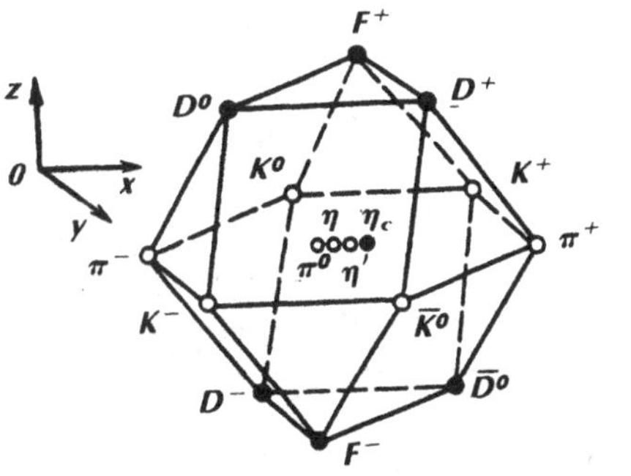

Figure 4.41. 16-plet of pseudoscalar mesons. The quantum number charm is plotted along the *z* axis; *Y* and I_3 are plotted along the *y* and *x* axes, respectively.

9. Charmed particles

The families of J/ψ and Υ particles are families with hidden values of the quantum numbers charm (c) and beauty (b), since they are determined as bound systems of $c\bar{c}$ and $b\bar{b}$ states. Mesons consisting of two quarks, one of which is a c quark, are called "charmed" or "mesons with an open charm." By analogy, mesons consisting of two quarks, one of which is a b quark, might be called "beautiful mesons." In this subsection we will consider only charmed mesons. We have already mentioned (Sec. 7 of Chap. 4) the need to switch from $SU(3)$ symmetry to $SU(4)\times SU(3)^c$ in a classification of particles in terms of the multiplets of a group. In this case, an $8+1$ nonet of mesons should be replaced by a $15+1$ 16-plet of mesons, as illustrated in Fig. 4.41. In addition to the ordinary mesons, which consist of u, d, and s quarks, the 16-plet contains one $c\bar{c}$ state and six charmed mesons ($C=\pm 1$). The charmed mesons are classified as follows:

$$
\begin{array}{llll}
C=0: & \eta_c = c\bar{c}; & - & - \\
C=+1: & D^+ = c\bar{d}; & D^0 = c\bar{u}; & F^+ = c\bar{s}\,; \\
C=-1: & D^- = \bar{c}d; & \bar{D}^0 = \bar{c}u; & F^- = \bar{c}s\,.
\end{array}
$$

There exist excited states of the D and F mesons, which are marked with an asterisk (e.g., $\bar{D}^{*0}$ or F^{*-}). These states convert into low-lying D and F states through a strong or electromagnetic decay.

Charmed baryons should have a charmed quark c as at least one of their constituent quarks. A total of ten combinations of the four quarks u, d, s, and c can be constructed in such a way that each contains at least one charmed quark:

	$S=0$	$S=1$	$S=2$
$C=1$:	$(cuu)^{++}, (cud)^+, (cdd)^0$	$(csu)^+, (csd)^0$	$(css)^0$
$C=2$:	$(ccu)^{++}, (ccd)^+$	$(ccs)^+$	–
$C=3$:	$(ccc)^{++}$	–	–

By analogy with the strange particles Λ, Σ, Ξ, and Ω, one introduces strange charmed baryons Λ_c, Σ_c, Ξ_c, and Ω_c. In the Weinberg-Salam weak-interac-

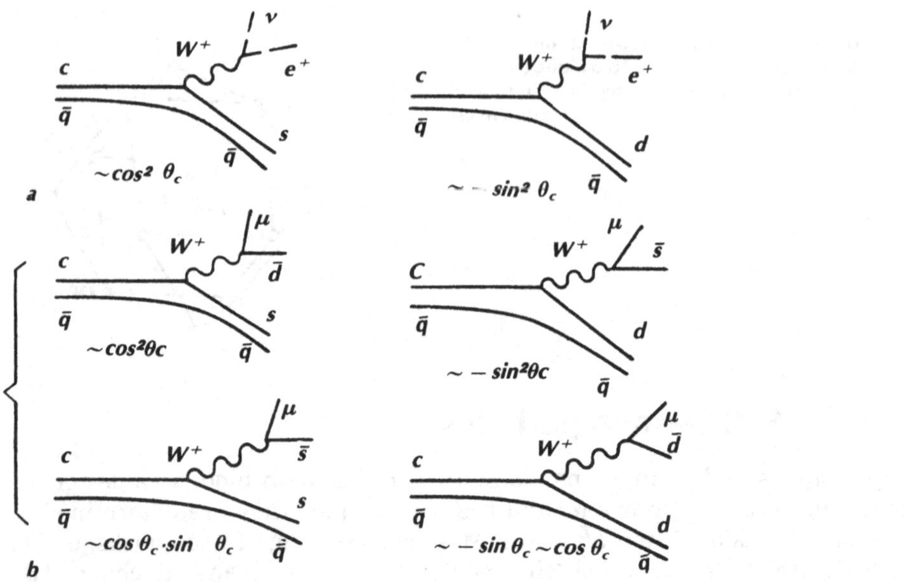

Figure 4.42. Decay schemes of charmed particles.

tion theory, the hadronic current is written as follows, in accordance with the mechanism of the Glashow-Illiopoulos-Maiani model[43,147]:

$$j_\alpha = \bar{u}\gamma_\alpha(1+\gamma_5)(d\cos\theta_c + s\sin\theta_c) + \bar{c}\gamma_\alpha(1+\gamma_5)(-d\sin\theta_c + s\cos\theta_c),$$

where u, d, s, and c are operators (spinors) representing the fields of the corresponding quarks; γ_α is the Dirac matrix; and θ_c is the Cabibbo angle. It follows from the form of this current that the transition of a c quark into an s or d quark is accompanied by the emission of a W^+ boson; the $c\bar{s}$ transition is characterized by a factor $\cos\theta_c$, and the cd transition by a factor $\sin\theta_c$. Since we have $\sin^2\theta_c \approx 0.05$, the contribution of processes with $|\Delta S| = 1$ is small.

Schemes for a semileptonic decay (D, $F \to$ leptons + hadrons) and for a hadronic decay of D and F mesons are shown in Fig. 4.42, a and b, respectively. The leptonic decays of these mesons are strongly suppressed because of nonconservation of the third component of the angular momentum. Estimates show that the width of a semileptonic decay of charmed mesons is 100–1000 times the width of a purely leptonic decay. As can be seen from Fig. 4.42, the schemes for the semileptonic decays of mesons consist for the most part of a single charged lepton and a strange meson (the factor $\cos\theta_c$). This correlation of leptons and K mesons is one of the distinctive features of the decays of D and F mesons. Another distinctive feature is the appearance of exotic decays of the type $D^+ \to K^-\pi^+\pi^-$. If it is assumed that the final state arises by virtue of isospin conservation (this is a natural assumption, since the final state contains only hadrons), the initial state can have an isospin of 3/2 or 5/2 (the isospins of the K and π mesons are 1/2 and 1, respectively). Such an isospin cannot be obtained from the sum of the isospins of a $q\bar{q}$ pair. We call such initial states "exotic." Decays of F mesons (4.43) should give rise to a significant fraction of $s\bar{s}$ states (like the η, η', and φ' mesons).

Tables 4.11 and 4.12 (Ref. 148) list the branching fractions of decays into two PS or two V mesons. These tables are presented to illustrate the situation; they are far short of constituting a comprehensive list of the possible decays.

The production of pairs of new particles with a large fraction of leptonic or semileptonic decays leads to the circumstance that a mixture of particles appears in the final state: electrons and muons or leptons and hadrons. The observation of such final states at a level above the background from electromagnetic interactions can—along with the higher orders of a QED perturbation theory—serve as evidence for the production of new particles.

The first pieces of evidence for the existence of D^0 and D^+ mesons were obtained in 1976 by the SLAC-LBL group. In the same year, the DASP and PLUTO groups observed semileptonic decays of charmed mesons and correlations between electrons and kaons at the DORIS accelerator.

F mesons have been observed in e^+e^- annihilation in only a single experiment so far (by the DASP group[149]). Some information on the D mesons is listed in Table 4.13.

The MARK II group (SPEAR) recently measured the branching ratio of the decay mode $D^0 \to \pi^-\pi^+$ with respect to the decay mode $D^0 \to K^+\pi^-$, finding 0.033 ± 0.015. For the corresponding branching ratio of the mode $D^0 \to K^+K^-$ with respect to the mode $D^0 \to K^+\pi^-$, the value 0.113 ± 0.030 was found. The Glashow–Illiopoulos–Maiani model predicts a value of 0.05 for these ratios; this value agrees within two standard deviations with the experimental data.

The DELCO and MARK II groups measured the ratio of the lifetimes of the D^0 and D^+ mesons:

$$\tau(D^+)/\tau(D^0) > 4(\text{DELCO})^1\,;$$

$$\tau(D^+)/\tau(D^0) = 3.08\,^{+4.1}_{-1.33}(\text{MARK II})\,.$$

The DELCO group also reports the lifetime of D mesons:

$$\tau(D^0) < 3.5\times10^{-13}s; \quad \tau(D^+) = (8 \pm 5)\times10^{-13}s\,.$$

The fact that the lifetimes of the D^0 and D^+ mesons are significantly different means that the quark which is the complement of the c quark plays an important role in the structure of D mesons. The data on the D^0 and D^+ lifetimes need to be confirmed.

The F meson should be recognized as an extremely interesting subject for research since it consists of a single charmed quark c and a single strange quark s. As was mentioned above, the decay schemes of the F meson which are shown in Fig. 4.43 are the most probable.

The DASP group[149] studied the reaction

$$\begin{array}{llll} e^+e^- \to FF^* \to & FF\gamma & & \\ & |\!\to & \pi\eta & \\ & & |\!\to & \gamma\gamma\,, \end{array} \tag{4.49}$$

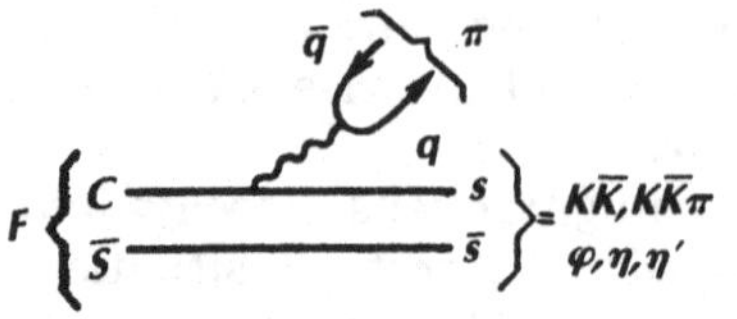

Figure 4.43. Most probable decay schemes of the *F* meson.

which is difficult to distinguish from the reaction

$$
\begin{aligned}
e^+e^- \rightarrow\ & F^*F^* \\
& |\!\rightarrow\ F\gamma \\
& \qquad |\!\rightarrow\ \pi\eta \\
& \qquad\qquad |\!\rightarrow\ \gamma\gamma\,.
\end{aligned}
\tag{4.50}
$$

A careful analysis yielded the following values for the masses of the F mesons:

$$
\begin{aligned}
M_{F^+} &= (2.03 \pm 0.06)\ \mathrm{GeV}/c^2\,; \\
M_{F^{*+}} &= (2.14 \pm 0.06)\ \mathrm{GeV}/c^2\,; \\
M_{F^*} - M_F &= (0.11 \pm 0.046)\ \mathrm{GeV}/c^2\,.
\end{aligned}
$$

The SLAC-LBL group observed a peak in the mass distribution of the $K^+K^-\pi^+$ system at a mass of $2.039 \pm 0.001\ \mathrm{GeV}/c^2$, which is identified as an F^+ meson.

Experiments on photoproduction with the OMEGA spectrometer at CERN yielded evidence for the existence of an F meson in the $\eta 3\pi$ and $\eta' 3\pi$ systems.

A search for charmed baryons in e^+e^- collisions was carried out by the SLAC–LBL group in the inclusive reactions

$$
e^+e^- \rightarrow (p + \bar{p})X; \quad e^+e^- \rightarrow (\Lambda + \bar{\Lambda})X\,. \tag{4.51}
$$

The behavior of the quantities $R(p+\bar{p})$ and $R(\Lambda + \bar{\Lambda})$ (Fig. 4.44),

$$
R(p+\bar{p}) = \frac{\sigma(e^+e^- \rightarrow (p+\bar{p})X)}{\sigma(e^+e^- \rightarrow \mu^+\mu^-)}; \quad R(\Lambda+\bar{\Lambda}) = \frac{\sigma(e^+e^- \rightarrow (\Lambda+\bar{\Lambda})X)}{\sigma(e^+e^- \rightarrow \mu^+\mu^-)}\,,
$$

indicates that they increase over the energy interval from 4.5 to 5.5 GeV. The point at which this growth begins coincides with the threshold for the production of the low-lying charmed baryon Λ_c, which had been observed previously in νN and γN interactions. In reactions (4.51), the charmed baryon Λ_c was sought on the basis of its presumed decay $\Lambda_c \rightarrow pK^-\pi^+$. A peak in the distribution of the invariant mass of the $pK^-\pi^+$ system was observed at a mass value of $2.285 \pm 0.002\ \mathrm{GeV}/c^2$. A mean value for the production of the Λ_c particle was extracted from data on $R(p+\bar{p})$ and $R(\Lambda+\bar{\Lambda})$ (Fig. 4.44): $\sigma(e^+e^- \rightarrow \Lambda_c^+ X) = (0.8 \pm 0.2) \times 10^{-33}\ \mathrm{cm}^2$.

The family of Υ particles is a family of particles with a hidden quantum number b. It may be that the decay of the Υ particles gives rise to hadrons with an open quantum number $b\,(\bar{b})$. For example, in the reaction

$$
\Upsilon(b\bar{b}) \rightarrow (b\bar{q}) + (\bar{b}q) \tag{4.52}
$$

Figure 4.44. Ratio of the cross sections for the production of $p\bar{p}$ and $\Lambda\bar{\Lambda}$ pairs to the cross section $\sigma(e^+e^- \rightarrow \mu^+\mu^-)$ (data from a SLAC–LBL group).

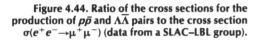

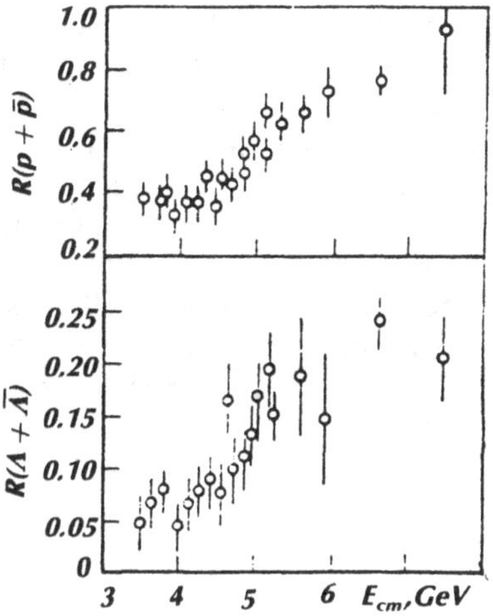

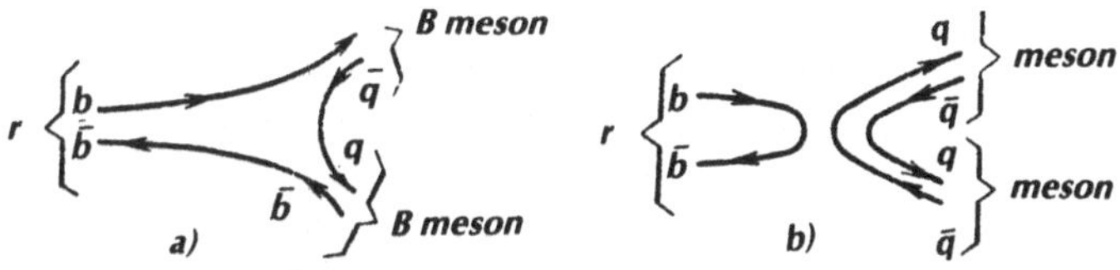

Figure 4.45. Coupled (a) and uncoupled (b) decay schemes of Υ particles.

two B bosons with an open quantum number b may appear. However, the decay of the $\Upsilon(b\bar{b})$ particles can also take another path, e.g.,

$$\Upsilon(b\bar{b}) \rightarrow (q\bar{q}) + (\bar{q}q)\,, \tag{4.53}$$

where q ($\bar{q}$) is not the b ($\bar{b}$) quark.

The sharp distinction between these two types of decays of Υ particles follows from the phenomenological Okubo–Zweig–Iizuka rule (Fig. 4.30). Following this rule, we say that if the family of Υ particles contains particles at which the decay width becomes large, this fact is an indication of the appearance of bosons with an open quantum number b (Fig. 4.45). The experimental data indicate that the decay width of the Υ'' particles is small and that of the Υ''' is large.

It may be that the sum of the masses of two B bosons (Fig. 4.45a)—the mass of $(b + \bar{b})$ plus the mass of $(q + \bar{q})$—is greater than the mass of the Υ'' particle and less than the mass of the Υ''' particle, so that the decay of Υ''' into two B bosons is allowed, while that of Υ'' into two B bosons is forbidden. In this case we find a natural restriction on the mass of the B boson: $M(\Upsilon'') < 2M_B < M(\Upsilon''')$ or (Table 4.10) $10.3 < 2M_B < 10.5\ \mathrm{GeV}/c^2$, i.e., $M_B \approx 5.2\ \mathrm{GeV}/c^2$. It also follows that the mass of the b quark is close to that of the B boson. The properties of the Υ particles and the B bosons are presently being studied intensely at Novosibirsk, Cornell, Hamburg, and Stanford.

This concludes our extremely cursory review of the properties of the new J/ψ and Υ particles and the D, F, and B bosons. Although these particles are produced in electromagnetic interactions, they are of a QCD nature, as we have seen. Research on the properties of the J/ψ and Υ families and the determination of their quantum numbers lean heavily on QED. It has been assumed everywhere that QED is valid for describing these phenomena, and models describing the properties of the new particles have been constructed on the basis of QED with the help of QCD.

Chapter 5
The $\gamma\gamma$ interaction

1. Stages in research on the $\gamma\gamma$ interaction

In quantum electrodynamics, in contrast with classical electrodynamics, it is postulated that an electromagnetic field can interact with an electromagnetic field; i.e., a $\gamma\gamma$ interaction can occur. So far, there has been no successful experiment in which real photons interact with each other.

The phenomena which we will be discussing below can be classified as $\gamma\gamma$ interactions only with some reservations. Nevertheless, the agreement between the experimental data and the theoretical predictions is so impressive that we have every right to assume that we are indeed studying the $\gamma\gamma$ interaction.

We are talking here about the interpretation of processes in which beams of electrons and positrons collide:

$$e^{\pm} + e^{-} \rightarrow e^{\pm} + e^{-} + \gamma^{*} + \gamma^{*} \rightarrow e^{\pm} + e^{-} + X, \tag{5.1}$$

where γ^{*} is a virtual photon, and X is some arbitrary final state allowed by the conservation laws (Fig. 5.1).

We know that the electromagnetic field of a moving charged particle may be represented as a spectrum of photons, whose shape can be calculated accurately in a certain order of a QED perturbation theory or in various approximations. In semiclassical form it was calculated in Refs. 150 and 151 and labeled the "method of equivalent photons." In QED perturbation theory, the interaction of two particles (an electron and a positron) can be thought of as the exchange of two virtual photons (Fig. 5.1) which results in the appearance of state X. In fact, the interaction of high-energy colliding $e^{+}e^{-}$ beams [see (5.1)] is presently regarded as a laboratory for studying the $\gamma\gamma$ interaction.

A $\gamma\gamma$ interaction process can be distinguished from reactions of other types (Fig. 5.2) when one of the photons is real. This is the "Primakoff effect." The final state X may include both leptons and photons, on the one hand, and strongly interacting particles (bosons and baryons), on the other. In the former case we are dealing with "pure QED," while in the second we must incorporate QCD in the theoretical interpretation of phenomena.

Interest in research on the $\gamma\gamma$ interaction intensified after it was shown[152–154] that under certain conditions the effective cross section for process (5.1) could turn out to be larger in absolute value than that for one-photon annihilation.

$$e^{+} + e^{-} \rightarrow \gamma^{*} \rightarrow X, \tag{5.2}$$

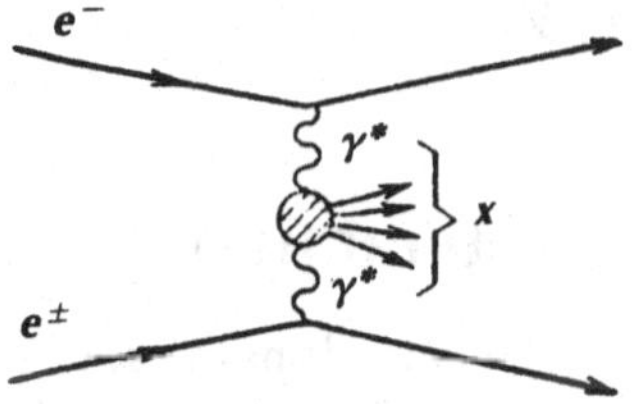

Figure 5.1. Interaction of the two leptons in the approximation of the $\gamma\gamma$ interaction (see the diagrams of class V in Fig. 1.4).

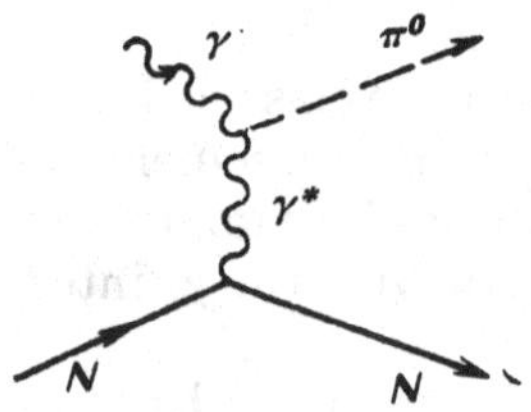

Figure 5.2. Production of a π^0 meson $\gamma + N \to \pi^0 + N$.

even though the cross section for process (5.2) is proportional to α^2 ($\alpha = e^2/\hbar c = 1/137$), while that for process (5.1) is proportional to α^4 and should be four orders of magnitude smaller, by virtue of the small value of α. It turns out that with increasing energy (E) of the colliding e^+e^- beams the cross section for one-photon annihilation (5.2) falls off in proportion to E^{-2}, while that for process (5.1) increases logarithmically: as $\ln^2(E/m_e)\ln^n(E/m_\pi)$, where $n > 1$ is an integer (for the case in which X consists of π mesons exclusively). If the state X is a two-pion state, the cross sections become comparable at a total energy of the colliding beams near 3 GeV. The energies of the colliding beams at the PEP and PETRA accelerators, presently in operation, are high enough for an extensive program of research in the physics of $\gamma\gamma$ interactions. This capability has stimulated theoretical study of $\gamma\gamma$ interactions. This branch of physics is attracting scientists because of its unique possibilities for studying the properties of new particles and the interactions of $\pi\pi$ and $K\bar{K}$ mesons, the possibility of testing the fundamental positions of QCD (e.g., the local nature of the gluon-quark interaction), the search for new particles (Higgs bosons and heavy leptons), etc.

The years 1969–1973 should be identified as the first stage in the active research on the $\gamma\gamma$ interaction, involving the acquisition of experimental data at accelerators at Novosibirsk, Frascati, Orsay, Stanford, and Hamburg. This stage included tests of QED and the equivalent-photon method, measurement of the widths for meson decays of the type $M \to \gamma\gamma$, study of the form factors of mesons and the low-energy $\pi\pi$ and $K\bar{K}$ interactions, a test of the predictions of current algebra, a study of the asymptotic behavior of the vertex functions, a study of the problem of searching for pairs of heavy leptons, and certain other questions.

In the realm of theoretical research this stage was characterized by the extensive use of the apparatus of QED and the ideas of current algebra, the use of the dispersion-relation method to study the $\pi\pi$ and $K\bar{K}$ interactions in

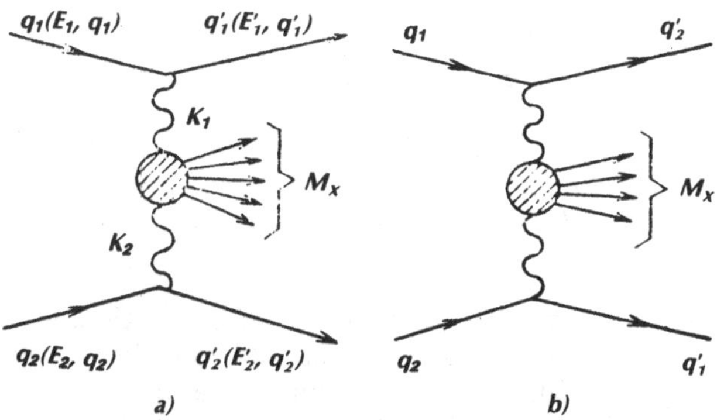

Figure 5.3. (a) Diagram describing the kinematics of deep inelastic scattering of leptons; (b) the antisymmetric diagram.

the processes $\gamma\gamma \to \pi\pi$, $\gamma\gamma \to K\overline{K}$, and the first use of parton models to describe phenomena associated with the $\gamma\gamma$ interaction. The beginning of the second, and present, stage of research on the $\gamma\gamma$ interaction should be placed in 1976–1977, when it became obvious that energies of 20–30 GeV would be reached at e^+e^- colliders within two or three years and that certain aspects of QCD could be tested best in $\gamma\gamma$ interactions in colliding-beam experiments, as we mentioned earlier. This stage is characterized by the appearance of a significant amount of experimental information on the characteristics of the $\gamma\gamma$ interaction; the interpretation of this information has been based exclusively on QCD (under the assumption that QED is valid).

In this chapter we will concentrate on analyzing the progress in the second stage of research on the $\gamma\gamma$ interaction (1976–1980). Earlier results will be discussed here only to the extent necessary for understanding the overall problem.

2. Kinematics of the process $\gamma + \gamma \to X$; the equivalent-photon method

To analyze the kinematics of process (5.1), we introduce the variables (Fig. 5.3a)

$$\left.\begin{aligned} &s=(q_1+q_2)^2;\quad \tilde{s}=(k_1+k_2)^2=M_X^2;\\ &k_i=(q_i-q_i');\quad r_i=(q_i+q_i');\quad \varkappa=(k_1k_2)^2-k_1^2k_1^2;\\ &s=q_1^2+q_2^2+2q_1q_2=2m^2+2E_1E_2-2\mathbf{q}_1\mathbf{q}_2. \end{aligned}\right\} \tag{5.3}$$

In the c.m. frame of the colliding beams we have $\mathbf{q}_1 = -\mathbf{q}_2 = \mathbf{q}$ and $E_1 = E_2 = E$. We can thus write

$$s = 2m^2 + 2q^2 + 2E^2 = 4E^2 .$$

We will also use the variables

$$\omega_1 = E - E_1' = (1/2)(\omega + q); \quad \omega_2 = E - E_2' = (1/2)(\omega - q);$$

$$\tilde{s} = \omega^2 - q^2 = 4\omega_1\omega_2 = 4(E - E_1')(E - E_2'),$$

where ω and q are the energy and momentum of the M_X system.

For definiteness we examine the process

$$e^- + e^- \to e^- + e^- + X, \tag{5.4}$$

where X is a hadronic state, in this subsection.

For e^-e^- collisions we should consider the diagram in Fig. 5.3b [it is, correspondingly, the antisymmetric amplitude for process (5.4)], in addition to the diagram in Fig. 5.3a, in lowest-order perturbation theory.

The cross section for process (5.4), in which a summation is carried out over the final states of the hadrons, X, and over the polarization of the electrons, can be written as follows:

$$\frac{d^4\sigma}{dE_1'\,d\Omega_1'\,dE_2'\,d\Omega_2'} = \left(\frac{\alpha}{8\pi^2}\right)^2 \frac{1}{k_1^2 k_2^2} \frac{E_1'}{E_1} \frac{E_2'}{E_2} \rho_1^{\mu\mu'} \rho_2^{\nu\nu'} W_{\mu\nu\mu'\nu'} \,. \tag{5.5}$$

Here $d\Omega_i = \sin^2\theta_i\, d\theta_i\, d\varphi_i$, and θ_i is the angle through which the ith electron is scattered with respect to the axis of the colliding beams.

After an average is taken over the initial electron states, and a summation over the final electron states, the electron density matrix $\rho_i^{\alpha\beta}$ becomes

$$\rho_i^{\alpha\beta} = (1/k_i^2) \sum \bar{u}(q_i)\gamma_\alpha u(q_i')\bar{u}(q_i')\gamma_\beta u(q_i)$$

$$= g^{\alpha\beta} - q_i^\alpha q_i^\beta / k_i^2 + r_i^\alpha r_i^\beta / k_i^2 \,. \tag{5.6}$$

The tensor $W_{\mu\nu\mu'\nu'}$ is the imaginary part of the amplitude for forward elastic $\gamma\gamma$ scattering. It can be expressed in terms of the amplitudes for the scattering of virtual photons with various polarizations. For this purpose, the amplitude for the scattering

$$\gamma + \gamma \to X \tag{5.7}$$

is examined in its proper c.m. frame through the introduction of helicity amplitudes.

Invariance under time reversal and parity conservation leave only eight invariant helicity amplitudes in the tensor $W_{\mu\nu\mu'\nu'}$. Of these eight, five are associated with transitions from the initial state to the final state which are of such a nature that helicity is conserved for each photon; the three other amplitudes are associated with transitions in which photons exchange helicity, but the total helicity of the state is conserved.

By virtue of the symmetry of the matrix $\rho_i^{\alpha\beta}$ [see (5.6)], only six combinations of the eight amplitudes are physically observable. We introduce the following notation: σ_{tt} is the cross section for the scattering of a photon with a transverse polarization t by a photon with a transverse polarization t; σ_{St} is the cross section for the scattering of a scalar photon S by a photon with a transverse polarization t; and τ_{St} is the cross section for the scattering of a scalar photon S by a photon with a transverse polarization t. The latter cross section corresponds to an amplitude with an exchange of helicities. In terms of these amplitudes, we can write the following final expression for the cross section for process (5.4):

$$\frac{d^4\sigma}{dE_1' d\Omega_1' dE_2' d\Omega_2'} = \left(\frac{\alpha}{4\pi^2}\right)^2 \frac{\sqrt{\mathcal{K}}}{2k_1^2 k_2^2} \frac{E_1' E_2'}{E_1 E_2} (4\rho_1^{11}\rho_2^{11}\sigma_{tt} + 2\rho_1^{11}\rho_2^{00}\sigma_{tS} + 2\rho_1^{00}\rho_2^{11}\sigma_{St} + \rho_1^{00}\rho_2^{00}\sigma_{SS} + 2\,\mathrm{Re}\,\rho_1^{1,-1}\rho_2^{1,-1}\tau_{tt} + 8\,\mathrm{Re}\,\rho_1^{10}\rho_2^{10}\tau_{tS}) . \quad (5.8)$$

The functions σ_{ik} and τ_{ik} can be calculated from models and determined through a comparison with experimental data.

If an experiment is carried out with polarized colliding beams, it becomes possible in principle to distinguish all eight invariant amplitudes. Incorporating the polarization of the electron and positron beams in the initial state adds an antisymmetric part to density matrix $\rho_i^{\alpha\beta}$ [see (5.6)]. However, it is proportional to the electron mass and contributes little to cross section (5.8); we could hardly expect to carry out any experiment with the goal of determining all of the invariant amplitudes in the near future.

For large values of E_i and small values of k_i^2 and E_i', expression (5.8) simplifies substantially, since in it we can ignore the cross sections σ_{tS}, σ_{SS}, σ_{tS}, and τ_{tS} (because they are proportional to k_i^2 and contribute little). Small values of k_i^2 and E_i' correspond to small scattering angles of the colliding particles. In this approximation, the expression for cross section (5.8) becomes

$$\frac{d^4\sigma}{dE_1' d\Omega_1' dE_2' d\Omega_2'} \approx \left(\frac{\alpha}{4\pi^2}\right)^2 \frac{\sqrt{\mathcal{K}}}{k_1^2 k_2^2} \frac{E_1' E_2'}{E_1 E_2} (2\rho_1^{11}\rho_2^{11}\sigma_{tt} + \mathrm{Re}\,\rho_1^{1,-1}\rho_2^{1,-1}\tau_{tt}) , \quad (5.9)$$

where the cross sections σ_{tt} and τ_{tt} can be studied approximately on the mass shell for values $k_1^2 = k_2^2 = 0$. In particular, we have $\sigma_{tt}\ (M_x^2, k_i^2) \to \sigma_{\gamma\gamma}(M_X^2)$; i.e., σ_{tt} becomes the cross secton for the scattering of real photons for reaction (5.7).

If we ignore the dependence of the amplitude $\rho_i^{\alpha\beta}$ on the azimuthal angle, which is defined as the angle between the scattering planes of the electrons (this approach is legitimate in the approximation of small scattering angles θ_1 and θ_2), and if we integrate (5.9) over θ_1 and θ_2 (also setting $E_1 = E_2 = E$), we find the Williams–Weizsäcker approximation:

$$\frac{d^2\sigma}{d\omega_1 d\omega_2} = n(\omega_1)n(\omega_2)\sigma_{\gamma\gamma}(4\omega_1\omega_2) , \quad (5.10)$$

where $\omega_i = E - E_i'$; and $n(\omega_i)$ is the spectrum of equivalent photons of the ith electron.

Expressions for the equivalent-photon spectra have been derived by many investigators. Here are a few of these spectra: the Williams–Weizsäcker spectrum

$$n(\omega) = (2\alpha/\pi)(1/\omega)[\ln(E/\omega) - 0.4];$$

the Kessler–Kessler spectrum[155]

$$n(\omega) = (2\alpha/\pi)(1/\omega)(1 - \omega/E + \omega^2/2E^2)[\ln(E/m) - 0.5]\,; \qquad (5.11)$$

and

$$n(\omega) = \frac{\alpha}{\pi}\frac{1}{\omega}\left[\left(1 - \frac{\omega}{E} + \frac{\omega^2}{2E^2}\right)\ln\frac{k^2_{\max}}{k^2_{\min}} - \left(1 - \frac{\omega}{E}\right)\left(1 - \frac{k^2_{\min}}{k^2_{\max}}\right)\right],$$

where E is the electron energy, $k^2 = \omega^2$, and $k^2 \approx m^2\omega^2/E^2$ (Ref. 156). In the derivation of an equivalent-photon spectrum, the integration over the electron scattering angle θ is carried out in a certain kinematically allowed region. The particular integration limits which are chosen influence the shape of the spectrum. The shape of the spectra in (5.11) may also depend substantially on other approximations which are used in the calculation of $n(\omega)$. The legitimacy of the equivalent-photon approximation is still being debated today. The problem of the applicability of the equivalent-photon method and the associated problem of extracting information on the $\gamma\gamma$ interaction from experimental data obtained in an $ee \to eeX$ reaction through a factorization of the cross section for the process $\gamma\gamma \to X$ (back-factorization) are of foremost importance. They must be solved (a) because of the particular conditions in which experiments are carried out, (b) so that we can improve the accuracy of experiments, and (c) because we wish to verify that the equivalent-photon method does indeed lead to a correct extraction of the $\gamma\gamma$ interaction from the reaction under consideration. In this connection it becomes important (on the one hand) to develop excellent elementary-particle detectors and to increase the luminosity of colliding-beam accelerators and (on the other) to pursue experimentally and theoretically the problem of distinguishing the background in the reaction $ee \to eeM_x$ and of calculating radiative corrections to the processes under study.

Cross section (5.10) can be integrated over one of the variables (ω_1 or ω_2), and a distribution in the square of the invariant mass of the particles produced in the $\gamma\gamma$ collisions, $\tilde{s} = M_X^2$, can be derived. For simplicity we restrict the discussion to the case $E \to \infty$, in which case we need retain only the logarithmic term in spectra (5.11):

$$n(\omega_i) \approx (\alpha/\pi)(1/\omega_i)\ln(k^2_{i\max}/k^2_{i\min})\,. \qquad (5.12)$$

From (5.10) and (5.11) we find

$$\frac{d\sigma}{d\tilde{s}} = \sigma_{\gamma\gamma}(\tilde{s})\int_{\omega_{1\min}}^{\omega_{1\max}} n(\omega_1)\,n\left(\frac{\tilde{s}}{4\omega_1}\right)d\omega_1\,, \qquad (5.13)$$

where $\tilde{s} = 4\omega_1\omega_2$, $\omega_{1\min} = \tilde{s}/4\omega_{2\min}$, and $\omega_{1\max} \to \infty$.

We restrict (5.13) to the leading-log approximation, finding the following cross section for the production of hadrons in terms of the variables $\tilde{s}$:

$$\frac{d\sigma}{d\tilde{s}} = 2\left(\frac{\alpha}{\pi}\right)^2\frac{\sigma_{\gamma\gamma}(\tilde{s})}{\tilde{s}}\ln^2\frac{E}{m}\left[\left(2 + \frac{\tilde{s}}{s}\right)^2\ln\sqrt{\frac{s}{\tilde{s}}} - \left(3 + \frac{\tilde{s}}{s}\right)\left(1 - \frac{\tilde{s}}{s}\right)\right]. \qquad (5.14)$$

It follows from the derivation of (5.14) that it holds in cases in which the scattered electrons are not detected in the final state. If they instead are detected, in some angular interval between $\theta_{\min}$ and $\theta_{\max}$, the result will be the replacement of $\ln(E/m)$ by a factor of $\ln(\theta_{\max}/\theta_{\min})$. The effect will be to significantly reduce the cross section $d\sigma d\tilde{s}$. This result is not surprising, since a small interval $\theta_{\min} \leqslant \theta \leqslant \theta_{\max}$ (in which the scattered electrons are detected) has now been cut out of the entire allowed range of electron scattering angles in (5.14).

Finally, we can integrate over $\tilde{s}$ and find an expression for the total cross section $\sigma(ee \to eeX)$:

$$\sigma(ee \to eeX) \approx 2\left(\frac{\alpha}{\pi}\right)^2 \left(\ln\frac{E}{m}\right)^2 \int \frac{d\tilde{s}}{\tilde{s}}\, \sigma_{\gamma\gamma}(\tilde{s})\, f(\tilde{s}), \tag{5.15}$$

where

$$f(\tilde{s}) \approx \left(2 + \frac{\tilde{s}}{s}\right)^2 \ln\sqrt{\frac{s}{\tilde{s}}} - \left(3 + \frac{\tilde{s}}{s}\right)\left(1 - \frac{\tilde{s}}{s}\right).$$

For large values of $\tilde{s}$, the total cross section for the process $\gamma\gamma \to$ hadrons, which is described by the quantity $\sigma_{\gamma\gamma}(\tilde{s})$ under the assumption of a factorization of the pomeron and leading Regge trajectories, can be written as follows, where we are making use of the asymptotic behavior of the cross sections $\sigma_{\text{tot}}(pp)$ and $\sigma_{\text{tot}}(\gamma p)$:

$$\sigma_{\gamma\gamma}(\tilde{s}) \approx \left(0.24 + \frac{0.27}{\sqrt{\tilde{s}}}\,\text{GeV}\right) \times 10^{-30}\,\text{cm}^2, \tag{5.16}$$

where

$$\sigma_{\gamma\gamma}(\tilde{s}) \to [\sigma_{\text{tot}}(\gamma p)]^2/[\sigma_{\text{tot}}(pp)] \approx 0.24 \times 10^{-30}\,\text{cm}^2.$$

We will analyze expression (5.16) below in the course of a comparison with experimental data on the total cross setion.

The cross section corresponding to the production of a resonance X_R with a mass M_R, a width γ_R, and a spin J_R is given in the equivalent-photon approximation by

$$\sigma(ee \to eeX_R) \approx \frac{(4\alpha)^2}{M_R^3}(2J_R + 1)\Gamma(X_R \to 2\gamma) \ln^2\frac{E}{M_R} f(\tilde{s}_R). \tag{5.17}$$

In the lowest-order perturbation theory, with which we are concerned here, several diagrams other than those in Fig. 5.3, a and b, contribute to the cross section for process (5.4). The diagrams in Fig. 5.4, a–f, are all of the same order in α. Of them, we are interested in only two (a and d); the others are a "theoretical background" for the $\gamma\gamma$ interaction. This background must be calculated and eliminated. In addition to the theoretical background, there is an experimental background, associated with the experimental conditions; it may include errors in the identification of the particles, particle-missing effects, allowance for collisions not associated with ee collisions, etc.

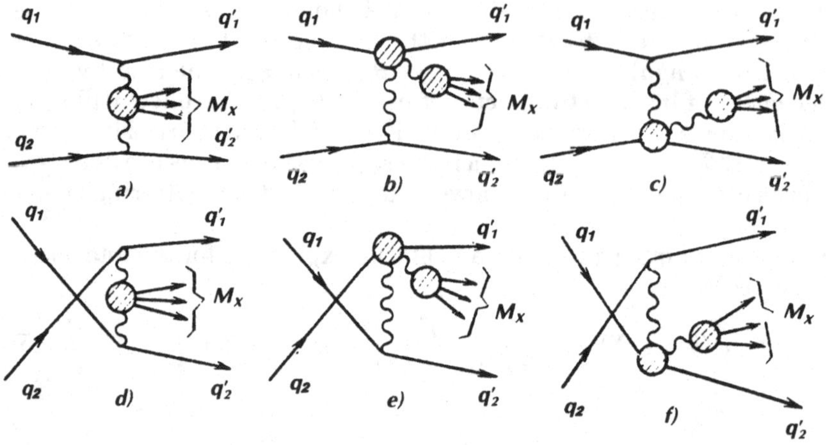

Figure 5.4. (a, d) C-even states M_X in $e^-e^- \to e^-e^-X$; (b, c, e, f) C-odd states.

The states X which are produced in the $\gamma\gamma$ interaction (Fig. 5.4a) are C-even states, in contrast with the states X which appear as a result of the fragmentation of one photon (Fig. 5.4, b, c, e, and f). Because of this dependence of the state X on the charge-conjugation operation, the diagrams in Fig. 5.4, a and d, do not interfere with the four other diagrams. The matrix element corresponding to the diagram in Fig. 5.4a will contain two photon propagators, proportional to $1/k_1^2$ and $1/k_2^2$. If k_1^2 and k_2^2 are sufficiently small, the contribution of this diagram will be large.

The diagrams in Fig. 5.4, d–f, correspond to the backscattering of electrons. As a result, the values of k^2 are large, while the contributions to the cross section are correspondingly small in comparison with that from the diagram in Fig. 5.4a. Finally, the diagrams in Fig. 5.4, b and c, correspond to the situation in which one of the momenta may be small, while the other—at whose vertex the state X appears—cannot be small. Under certain kinematic conditions these diagrams may contribute significantly to the scattering and should be taken into consideration (we will find an estimate below). The diagrams of higher order in α introduce small corrections and have not yet been taken into account.

A corresponding set of diagrams is considered for the collision of oppositely directed e^+e^- beams,

$$e^+ + e^- \to e^+ + e^- + X. \qquad (5.18)$$

In addition to the diagrams in Fig. 5.5, a–f, there is the possibility of the diagrams shown in Fig. 5.6, a–e ("background" diagrams). As we mentioned earlier, the cross section for the process $e^+e^- \to X$ is dominant at colliding-beam energies up to 1.5 GeV, at which point it begins to fall off rapidly with the energy. The radiative corrections to the process $e^+e^- \to X$ fall as $1/E^2$ and can also be distinguished. Two-photon annihilation contributes little to the

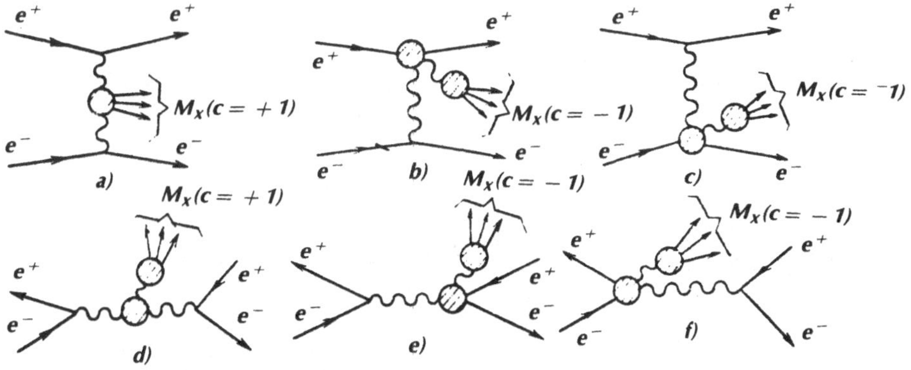

Figure 5.5. (a, d) C-even states M_X in $e^+e^- \to e^+e^-X$; (b, c, e, f) C-odd states.

cross section and can be ignored. The contributions to the $e^+e^- \to e^+e^-X$ cross section made by diagrams of higher order in α are small.

Consequently, as in the case of colliding e^-e^- beams, under certain kinematic conditions it is possible to distinguish the contribution of the diagram in Fig 5.5a to reaction (5.18) and to use the equivalent-photon method to distinguish the cross section for the $\gamma\gamma$ interaction.

To illustrate the estimation of the background, we present the results of one of the early calculations.[157]

At a colliding-beam energy $E = 3$ GeV, the contributions of the diagrams in Figs. 5.4b and 5.5b to cross sections (5.4) and (5.18), respectively, were equal to the contribution of the diagrams in Figs. 5.4a and 5.5a at the following values of θ and M_X:

Process	θ (deg)	M_X (GeV)
$ee \to ee\mu\mu$	36	0.25
$ee \to ee\pi\pi$	40	1
$ee \to eeK\overline{K}$	6	1.02
$ee \to eeM_X$	23	1.40

As the energy of the colliding beams is increased, the situation gets worse. The background must accordingly be taken into account carefully.

Computer programs are now available for calculating the various contributions to $\gamma\gamma$ interactions. In particular, the contributions to the cross section for the reaction $e^+e^- \to e^+e^-\mu^+\mu^-$ from the diagrams in Fig. 5.7, a and b, have been compared with the contribution from the diagrams in Fig. 5.7, c–f.

The kinematics in which both of the scattered leptons (e^+ and e^-) are detected has been studied. For $\theta_e > 2°$, the following values were found for the cross section (for a total energy of the beams of about 30 GeV): $d\sigma/dM_{\mu\mu} = (220 \pm 5)\times 10^{-36}$ cm^2 (the contribution from only the diagrams in Fig. 5.7, a and b, was taken into account; $M_{\mu\mu}$ is the invariant mass of the $\mu^+\mu^-$ pair) and $d\sigma/dM_{\mu\mu} = (249 \pm 5)\times 10^{-36}$ cm^2 (all six of the diagrams in Fig. 5.7 were taken into account; the cross section is exceeded by about 10%).

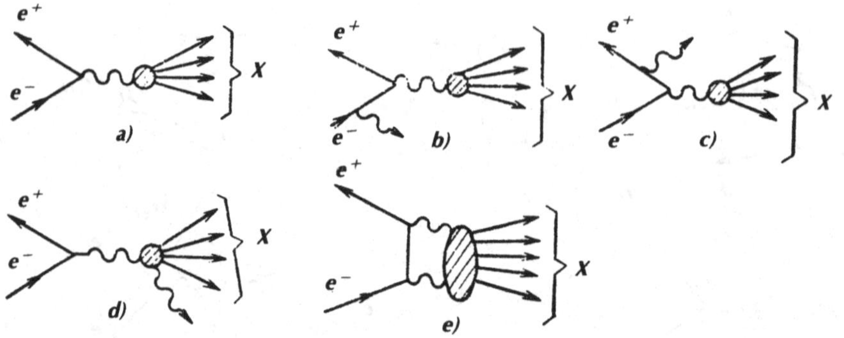

Figure 5.6. (a) One-photon annihilation; (b–d) radiative corrections to that annihilation; (e) two-photon annihilation.

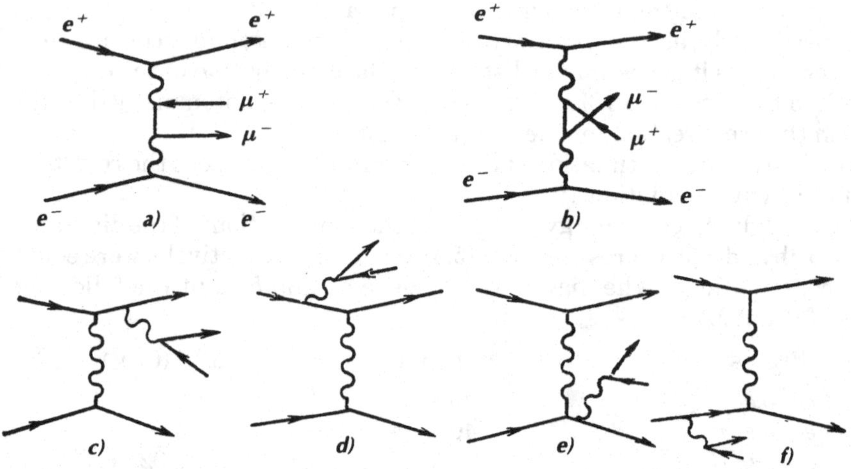

Figure 5.7. Diagrams describing $e^+e^- \to e^+e^-\mu^+\mu^-$ in lowest-order perturbation theory.

The diagrams in Fig. 5.7, c–f, make a noticeable contribution only at values $M_{\mu\mu} \lesssim 1$ GeV. At $M_{\mu\mu} \gtrsim 2$ GeV, the cross section is determined completely by the first two diagrams.

A hard background can result from purely electromagnetic processes (e.g., $ee \to ee_{\gamma\gamma}$ and $ee \to eee^+e^-$). Such processes can be distinguished from the production of hadrons through the detection of at least one hadron in the final state.

In all of the reactions $\gamma\gamma \to X$ for which measurements have been carried out so far, it has been possible to distinguish the background fairly accurately.

3. Tests of quantum electrodynamics

A test of QED in the $\gamma\gamma$ interaction consists of measuring the cross sections for the reactions $ee \to eee^+e^-$ and $ee \to ee\mu^+\mu^-$ and comparing them with the QED predictions. In experiments carried out by the PLUTO group at DESY,[158] data were obtained on the reactions $e^+e^- \to e^+e^- +$ (two charged particles). Included among the charged particles, in addition to e^+e^- and $\mu^+\mu^-$ pairs, were $\pi^+\pi^-$ pairs. Figure 5.8 compares the experimental data with a theoretical curve. Radiative corrections to the 2γ process were ignored. The fact that the experimental points lie above the theoretical curve in the interval $M = 1$–1.5 GeV is attributed to a contribution of an f^0 meson which decays into two π mesons. It can be concluded from Fig. 5.8 that QED holds in the region considered here, $Q^2 \lesssim 0.25\ (\text{GeV}/c)^2$, and at energies of the 2γ system up to 3 GeV in the perturbation theory of lowest order in α (α^4).

4. The $\gamma\gamma$ interaction at low energies

The production of resonances in a state with $C = +1$ should be recognized as one of the interesting effects which occur at low energies. Two virtual spin-1 photons generate states $J^{PC} = 0^{\pm +}, 2^{\pm +}, \ldots$. In $\gamma\gamma$ interactions it is thus possible to study the production of $\pi^0, \eta, \eta', A_2, f, f', \ldots$, mesons and states which are formed by charmed, beautiful, and other (heavier) quarks: η_c, χ_c, η_b, χ_b, etc.

It can be seen from (5.17) that the total cross section for the production of a resonance (or a meson) in the equivalent-photon approximation is proportional to the decay width $\Gamma(R \to \gamma\gamma)$. Experimental values of the decay width $\Gamma(R \to \gamma\gamma)$ have been obtained at the SLAC and DESY accelerators for several mesons:

$$\left.\begin{aligned}
&\Gamma(\pi^0 \to \gamma\gamma) = (7.95 + 0.55)\ \text{eV};\\
&\Gamma(\eta \to \gamma\gamma) = (0.32 \pm 0.05)\ \text{keV};\\
&\Gamma(\eta' \to \gamma\gamma) = (5.9 \pm 1.6)\ \text{keV};\\
&\Gamma(f_0 \to \gamma\gamma) = (2.3 \pm 0.5)\ \text{keV};\\
&\Gamma(f_0 \to \gamma\gamma) = (4.1 \pm 0.4)\ \text{keV};\\
&\Gamma(f_0 \to \gamma\gamma) \leqslant 4.7\ \text{keV}.
\end{aligned}\right\} \tag{5.19}$$

The confidence level for the last value is 95%. Limitations have been found on the values of Γ and $B\Gamma(R \to \gamma\gamma)$ (R is the final state; regarding B, see the note in Table 4.9); the results are shown in Table 5.1.

A study of the production of mesons in $\gamma\gamma$ collisions provides a test of several theoretical hypotheses. Let us consider the decay of a π^0 meson into two photons. The vertex function describing this process is determined from

$$T^{\mu\nu} = \varepsilon^{\mu\nu\rho\sigma} k_{1\rho} k_{2\sigma} F_{\pi^0}(k_1^2, k_2^2, m_\pi^2). \tag{5.20}$$

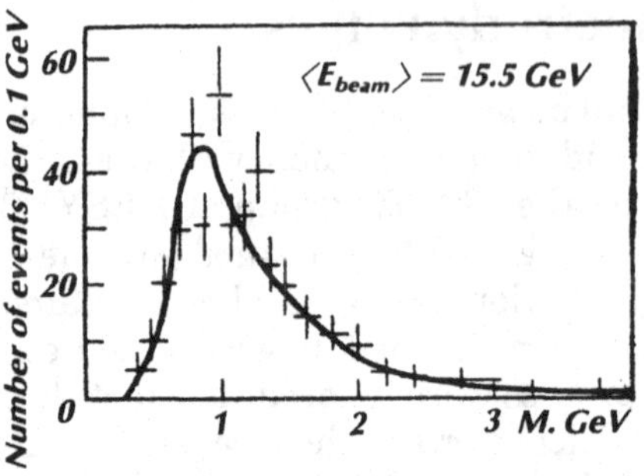

Figure 5.8. Distribution in the invariant mass M of the two charged particles in the reaction $e^+e^- \to$ two charged particles ($\mu^+\mu^-$ or $\pi^+\pi^-$).

Table 5.1. Limitations on the value of $\Gamma(R\to\gamma\gamma)B$.

Final state	Meson	σ, 10^{-33} cm^2	$\Gamma_{\gamma\gamma}$, keV
$\pi^+\pi^-\gamma$	$f(1270)$	$<0.14\ B(f\to\rho\gamma)\sigma$	$<0.8\ B(f\to\rho\gamma)\Gamma_{\gamma\gamma}$
$\rho^\pm\pi^\mp$	$A_2(1310)$	<0.36	<0.25
K^+K^-	$f(1270)$	<4.2	$\leqslant 24$
K^+K^-	$A_2(1310)$	<2.6	$\leqslant 17$
K^+K^-	$f'(1515)$	$<0.052\ B(f'\to K^+K^-)\sigma$	$<0.6\ B(f'\to K^+K^-)\Gamma_{\gamma\gamma}$

The decay width $\Gamma(\pi^0\to\gamma\gamma)$ is expressed in terms of the π^0 form factor F_{π^0}:

$$\Gamma(\pi^0\to\gamma\gamma) = [\,|e^2F_{\pi^0}(0,0,m_{\pi^0}^2)|^2/64\pi\,]\,m_{\pi^0}^3\,. \tag{5.21}$$

Using the hypothesis of partially conserved axial current, and incorporating the anomalous term, we can write the Ward–Takahashi identity as

$$T^{\mu\nu}(k_1,k_2)|_{m_{\pi^0}^2=0} = -(s/2\pi^2 f_\pi)\varepsilon^{\mu\nu\rho\sigma}k_{1\rho}k_{2\sigma} + \text{terms proportional to } k_1^2,k_2^2. \tag{5.22}$$

Here s is a constant. A few words are in order regarding the hypothesis of a partially conserved axial current. The Hamiltonian for the β decay of a neutron is written in the form

$$\mathscr{H} = (G/\sqrt{2})[\,\overline{e}\gamma_\alpha(1+\gamma_5)\nu\,][\,\overline{p}\gamma_\alpha(1+\gamma_5)n\,] + \text{the Hermitian-adjoint term},$$

where the operators (spinors) $\bar{e}$, ν, $\bar{p}$, and n represent the electron field, the neutrino field, the proton field, and the neutron field, respectively. If we take strong interactions into account—if, in terms of Feynman diagrams, we use the generalized diagram in Fig. 2.4—we should write the nucleon current in a more general form, taking account of the form factors (or structure functions):

$$\bar{p}\gamma_\alpha(1+\gamma_5)n = \bar{p}\gamma_\alpha n + \bar{p}\gamma_\alpha\gamma_5 n \to V_\alpha + A_\alpha\,,$$

where V_α and A_α are the generalized vector part and the generalized axial part of the current. We know that the vector current is conserved: $\partial_\alpha V_\alpha = 0$.

As for the axial current, we note that the experimental data agree with the PCAC hypothesis (partial conservation of axial current), which is expressed by the formula

$$\partial_\alpha A_\alpha = f_\pi m_\pi^2 \pi ,$$

where the constant f_π has a value of 95 MeV, $m_\pi^2 \approx 0.02\ \text{GeV}^2$, and the operator π represents the π-meson field. An important role is played in the hypothesis of partial conservation of axial current by a procedure of extrapolating the square mass of the π meson from the point $m_\pi^2 \neq 0$ to the point $m_\pi^2 = 0$ (the approximation of massless π mesons). Going back to (5.22), we note that in the limits $k_1^2 \to 0$ and $k_2^2 \to 0$ with $s = 0$ we would have

$$T^{\mu\nu}(0,0)|_{m_\pi^2} = 0 .$$

When the anomaly ($s \neq 0$) is taken into account, however, we find that the vertex function becomes nonzero:

$$T^{\mu\nu}(0,0,m_\pi^2 \approx T^{\mu\nu}(0,0,0) = -(s/2\pi^2 f_\pi)\varepsilon^{\mu\nu\rho\sigma}k_{1\rho}k_{2\sigma} . \tag{5.23}$$

We can now go back to definition (5.20). The form factor $F_{\pi^0}(k_1^2,k_2^2 m_\pi^2)$ can be normalized either to unity, as was done for the electric form factor of the nucleon (Chap. 2), or to the coupling constant $g_{\pi_0\gamma\gamma}$: $F_{\pi_0}(0,0,m_\pi^2) = g_{\pi_0\gamma\gamma}$. It is a matter of convention. Choosing the normalization to the coupling constant, we find

$$F_{\pi^0}(0,0,0) = -s/2\pi^2 f_\pi = g_{\pi^0\gamma\gamma} , \tag{5.24}$$

where $s = 1/2$ for fractional quark charges (the three-triplet quark model) or $s = 1/6$ for the Gell-Mann–Zweig model. Substituting (5.24) and (5.21), and using $s = 1/2$, we find $\Gamma(\pi^0 \to \gamma\gamma) \approx 7.3$ eV. This value agrees well with the experimental value.

A study of the process $\gamma\gamma \to \pi^0$ thus makes it possible to test the prediction of the hypothesis of partial conservation of axial current and to select one of the quark models.

Corresponding problems are solved in a determination of the decay widths of other mesons (η,η'). For example, the quark model with integer quark charges predicts a value of $\Gamma(\eta' \to \gamma\gamma)$ which differs from that in a quark model with fractional charges.

Some unexpected complications sometimes arise in a study of the properties of low-lying resonance states. For example, if we use the well-tested superconverging sum rules for real photons, and if we saturate the imaginary parts of the integrands in these rules with low-lying pseudoscalar, scalar, and tensor resonances, we find a theoretical estimate of the minimum width for the decay of the f meson into two photons: $\Gamma(f \to \gamma\gamma) \approx 9$ keV. Restricting the contribution to only the pseudoscalar mesons, we then find the value $\Gamma(f \to \gamma\gamma) \gtrsim 3.6$ keV. The experimental value, $\Gamma(f \to \gamma\gamma) = (2.3 \pm 0.5)$ keV [see (5.19)], turns out to be smaller than the minimum theoretical value; this circumstance mandates a further test of this quantity in new experiments. Other data agree better with the theoretical predictions.

Table 5.2. Values of $\Gamma(R\to\gamma\gamma)$ and cross sections for the production of resonances at various energies.

		$\sigma(ee\to ee\,R)$, 10^{-33} cm^2		
Resonance	$\Gamma(R\to\gamma\gamma)$, keV	3 GeV	15 GeV	70 GeV
π^0	7.95×10^{-3}	1.0	2.9	3.7
η	0.324	0.2	0.9	1.7
η'	5.9	0.8	2.6	5.3
A_2	1.8	0.4	1.3	2.9
f	5	1.1	4.2	8.9
f'	0.4	0.04	0.18	0.40
η_c	6.4	6×10^{-3}	5×10^{-2}	0.14
χ_0	1	5×10^{-4}	5×10^{-3}	1.4×10^{-2}
χ_2	4/15	1×10^{-4}	1×10^{-3}	3.2×10^{-3}
η_b	0.4	0	4×10^{-5}	1.8×10^{-4}

It is presently believed that mesons consist of quarks and antiquarks. In $SU(3)$ symmetry, and under the assumption that all the states with given J^P have identical spatial wave functions (nonet symmetry), we find the values of the decay widths for the decays of resonances into two photons[159] (Table 5.2). A test of these values would also be of considerable interest.

Extremely interesting from the theoretical standpoint are the processes $\gamma\gamma\to\gamma\gamma$, $\gamma\gamma\to\pi\pi$, and $\gamma\gamma\to K\bar{K}$, whose study can yield important information on the $\pi\pi$, $\pi\pi\to K\bar{K}$, and $K\bar{K}$ interactions at low energies and also offers the rare possibility of a search for gluonium in isotopically pure states (Sec. 8 of Chap. 5).

Theoretical research on the reaction $\gamma\gamma\to\pi\pi$ began in 1960. At the time, this reaction was of interest for finding an estimate of its effect on Compton scattering by a nucleon.[160] The dispersion-relation method proved convenient for calculating the amplitude for the process $\gamma\gamma\to\pi\pi$. That method led to singular linear integral equations for the partial waves of the process $\gamma\gamma\to\pi\pi$ which turned out to be exactly solvable. The exact solutions depended on the energy dependence of the phase shifts of the partial waves in $\pi\pi$ scattering; that dependence was not known.

In the early 1970s it became feasible to study the process $ee\to ee\pi\pi$ at accelerators with colliding $e^\pm e^-$ beams. A study of the reactions $\gamma\gamma\to\gamma\gamma$, $\gamma\gamma\to\pi\pi$ and $\gamma\gamma\to K\bar{K}$ acquired independent interest. By that time, experimental data had been accumulated on the $\pi\pi\to\pi\pi$, $\pi\pi\to K\bar{K}$, and $K\bar{K}\to K\bar{K}$ interactions, whose amplitudes are a fundamental part of the problem of finding the amplitudes for the processes $\gamma\gamma\to\gamma\gamma$, $\pi\pi$ and $K\bar{K}$. That problem can be solved by the dispersion-relation method. Several theoretical papers[161] appeared on the reactions $\gamma\gamma\to\gamma\gamma$, $\gamma\gamma\to\pi\pi$, and $\gamma\gamma\to K\bar{K}$; the $\gamma\gamma$, $\pi\pi$, and $K\bar{K}$ interactions in the final state were taken into account (strong interactions were taken into account). A comprehensive study of the reactions $ee\to ee\pi\pi$, $ee\to eeK\bar{K}$, and $ee\to ee\gamma\gamma$ was carried out by the dispersion-relation method in Ref. 162. The discussion below basically follows those earlier studies.[162]

Figure 5.9. Diagram describing $\gamma\gamma \rightarrow \pi\pi$.

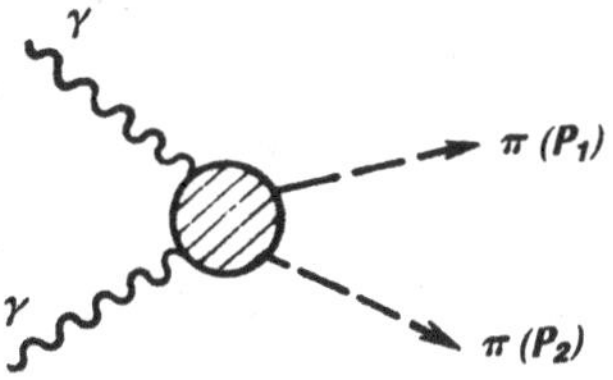

We first consider the reaction $\gamma\gamma \rightarrow \pi\pi$. In the c.m. frame of the virtual photons we introduce the relativistic amplitudes (Fig. 5.9)

$$\langle f|\hat{S}|i\rangle - \frac{i}{(2\pi)^2}\frac{1}{4\omega k_0}[\delta(k_1 + k_2) - p_1 - p_2]\, T^j_{\alpha\beta}(t,\cos\varphi_t)\varepsilon^\alpha \varepsilon'^\beta , \quad (5.25)$$

where

$$k_1^2 = k_2^2 = -m_\gamma^2; \quad p_1^2 = p_2^2 = m_\pi^2; \quad t = 4k_0^2 = 4\omega^2;$$

$\omega_1 = \omega_2 = \omega$ is the energy of the virtual photons; φ_t is the angle through which the π mesons are scattered; α and β take on the three values x, y, and z; and ε is the polarization of the photons. The amplitudes $T^j_{\alpha\beta}$ describe an interaction of two virtual photons with a definite polarization (longitudinal or transverse). The two-pion final state can have total isospin values $T = 0,1,2$. The requirement that the amplitude for the process $\gamma\gamma \rightarrow \pi\pi$ be C invariant eliminates the value $T = 1$. We thus have

$$\left.\begin{aligned} T^{\pi^+\pi^-}_{\alpha\beta}(t,\cos\varphi_t) &= (1/\sqrt{3})T^{(T=0)}_{\alpha\beta} + (1/\sqrt{6})T^{(T=2)}_{\alpha\beta}\,; \\ T^{\pi^0\pi^0}_{\alpha\beta}(t,\cos\varphi_t) &= (2/\sqrt{6})T^{(T=2)}_{\alpha\beta} - (1/\sqrt{3})T^{(T=0)}_{\alpha\beta}\,. \end{aligned}\right\} \quad (5.26)$$

In the direct reaction channel $\gamma\gamma \rightarrow \pi\pi$ the variables t, s, and u take the form

$$t = (k_1 + k_2)^2 = 4k_0^2 = 4\omega^2\,;$$

$$s = (k_1 - p_2)^2 = -m_\gamma^2 + m_\pi^2 - \frac{t}{2} - 2\cos\varphi_t\sqrt{\left(\frac{t}{4} + m_\gamma^2\right)\left(\frac{t}{4} - m_\pi^2\right)}\,;$$

$$u = (k_1 - p_1)^2 = -m_\gamma^2 + m_\pi^2 - \frac{t}{2} + 2\cos\varphi_t\sqrt{\left(\frac{t}{4} + m_\gamma^2\right)\left(\frac{t}{4} - m_\pi^2\right)}\,. \quad (5.27)$$

Dispersion relations can be written first for the isotopic amplitudes $T^T_{\alpha\beta}$. The integrals along the left-hand cuts (Fig. 5.10; over s and u) from the crossing channels, i.e., from the reactions $\gamma + \pi \rightarrow \gamma + \pi$, are taken into account approximately. The contributions from only the ω and ρ resonances in the Born approximation are taken into account.

The amplitudes $T^{(T)}_{\alpha\beta}$ must be normalized to the Thomson limit $(t \rightarrow 0, m_\pi^2 \rightarrow 0)$. This normalization is not an exact operation, since an approximate expression for the $\gamma\gamma \rightarrow \pi\pi$ amplitude is analytically continued to the nonphysical point $t = 0$. However, this is the only way to choose a physical solution from among all possible solutions of the dispersion relations.

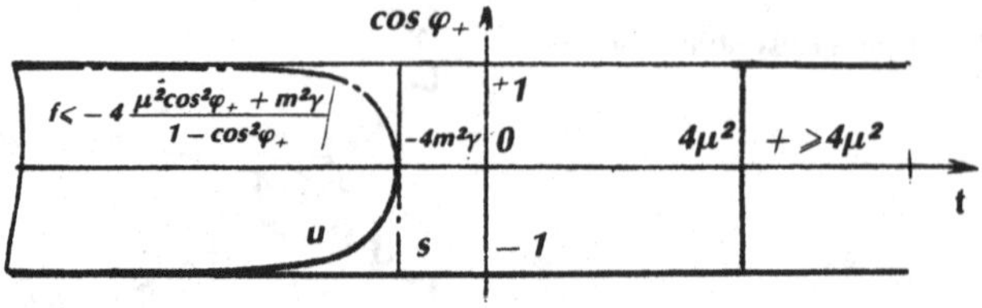

Figure 5.10. Cuts in the $(t, \cos\varphi_t)$ plane.

For charged π mesons, the Thomson limit corresponds to the well-known contact diagram with two photons and two mesons at a vertex, i.e., to the quantity $2e^2$. The $\gamma\gamma\to\pi^0\pi^0$ amplitude must be normalized to a zero charge of the neutral pions. We thus have

$$T_{\alpha\beta}^{(T=0)}(t\to 0, m_\pi^2\to 0) = 4e^2/\sqrt{3};\ T_{\alpha\beta}^{(T=2)}(t\to 0, m_\pi^2\to 0) = 4e^2/\sqrt{6}\,. \quad (5.28)$$

The isotopic amplitudes are normalized to the Thomson limit by means of a single subtraction:

$$\widetilde{T}_{\alpha\beta}^{(T)} = T_{\alpha\beta}^{(T)} - T_{\alpha\beta}^{(T)}(t=0, m_\pi^2=0) + (4e^2/\sqrt{3})(1/\sqrt{2})^{T/2}\,. \quad (5.29)$$

The final expression for the dispersion relation for the isotopic amplitudes is therefore

$$\widetilde{T}_{\alpha\beta}^{(T)}(t, \cos\varphi_t) = \frac{t}{\pi}\int_{4m_\pi^2}^{\infty} \frac{\operatorname{Im}\widetilde{T}_{\alpha\beta}^{(T)}(t', \cos\varphi_t)}{t'(t'-t-i\varepsilon)}\,dt' + \widetilde{B} + \widetilde{P} + \frac{4e^2}{\sqrt{3}}\left(\frac{1}{\sqrt{2}}\right)^{T/2}, \quad (5.30)$$

where $\widetilde{B}$ and $\widetilde{P}$ are the contributions of the Born terms from the ω and ρ resonances, respectively.

A system consisting of two π mesons (the final state of the reaction $\gamma\gamma\to\pi\pi$) has a positive charge parity, so that in the reaction $\gamma\gamma\to\pi\pi$ there are no waves with an odd orbital angular momentum l. In the low-energy region ($E_\gamma \lesssim 1.5$ GeV) in which we are interested here, we can restrict the analysis to the lowest-order (s and d) partial waves. In this approximation the amplitudes $\widetilde{T}_{\alpha\beta}^{(T)}$ can be written

$$\widetilde{T}_{\alpha\beta}^{(T)}(t, \cos\varphi_t) = f_s^{(T)}(t) + 5f_d^{(T)}(t)P_2(\cos\varphi_t)\,, \quad (5.31)$$

where $f_{s,d}^{(T)}$ are the s and d partial waves of the amplitude $\widetilde{T}_{\alpha\beta}^{(T)}(t, \cos\varphi_t)$.

If we choose $\cos\varphi_t = 1/\sqrt{3}$, we find

$$P_2(\cos\varphi_t) = 0 \text{ and } \widetilde{T}_{\alpha\beta}^{(T)}(t, 1/\sqrt{3}) = f_s^{(T)}(t)\,.$$

To distinguish the d wave, we set $\cos\varphi_t = \sqrt{7/15}$ in (5.31), since we have $P_2(\sqrt{7/15}) = 1/5$.

The equations for the s and d waves for the process $\gamma\gamma\to\pi\pi$ are thus written in the form

$$f_s^{(T)}(t) = \frac{t}{\pi}\int_{4m_\pi^2}^{\infty}\frac{\operatorname{Im} f_s^{(T)}(t')}{t'(t'-t)}\,dt' + \widetilde{P}_{\alpha\beta}^{(T)}\left(t,\frac{1}{\sqrt{3}}\right)$$
$$+\widetilde{B}_{\alpha\beta}^{(T)}\left(t,\frac{1}{\sqrt{3}}\right) + \frac{4e^2}{\sqrt{3}}\left(\frac{1}{\sqrt{2}}\right)^{T/2}; \tag{5.32}$$

$$f_d^{(T)}(t) = \frac{t}{\pi}\int_{4m_\pi^2}^{\infty}\frac{\operatorname{Im} f_d^{(T)}(t')}{t'(t'-t)}\,dt' + \widetilde{P}_{\alpha\beta}^{(T)}\left(t,\sqrt{\frac{7}{15}}\right)$$
$$-\widetilde{P}_{\alpha\beta}^{(T)}\left(t,\frac{1}{\sqrt{3}}\right) + \widetilde{B}_{\alpha\beta}^{(T)}\left(t,\sqrt{\frac{7}{15}}\right) - \widetilde{B}_{\alpha\beta}^{(T)}\left(t,\frac{1}{\sqrt{3}}\right). \tag{5.33}$$

The quantity $\operatorname{Im} f_l^T$ ($l = s, d$) in the integral is replaced by the expression which is found for it from the two-particle unitarity condition:

$$\operatorname{Im} f_l^{(T)} = \exp[-i\delta_l^{(T)}(t)]\sin\delta_l^{(T)}(t)\, f_l^{(T)}(t). \tag{5.34}$$

Here $\delta_l^T(t)$ are phase shifts of the corresponding partial amplitudes for $\pi\pi$ scattering with an isospin T.

Substituting (5.34) into (5.33) and (5.32), we find linear singular integral equations for the partial amplitudes $f_s^{(T)}$ and $f_d^{(T)}$.

These equations can be solved exactly through reduction to a Riemann boundary-value problem.[163] The extent to which the solution of the Riemann boundary-value problem is ambiguous is related to the index $\varkappa$, which is defined by

$$\varkappa = \frac{1}{2\pi i}\int_{4m_\pi^2}^{\infty} d\ln\exp[2i\delta_l^T(t)]. \tag{5.35}$$

If the phase shift $\delta_l^{(T)}(t)$ asymptotically vanishes in the limit $t\to\infty$, the solutions of integral equations (5.32) and (5.33) are unique. The total cross section for the process $\gamma\gamma\to\pi\pi$ is expressed as the sum of the s and d partial cross sections:

$$\sigma(t)_{\gamma\gamma\to\pi\pi} = \sigma_s(t) + \sigma_d(t). \tag{5.36}$$

The energy behavior of the s wave of the reaction $\gamma\gamma\to\pi\pi$ depends on the choice of the phase shift $\delta_s^0(t)$ for the $\pi\pi$ scattering. There are two parametrizations for this phase: down and down-up. The down-up phase is determined by the circumstance that it passes through 90°, while the down phase never reaches 90°.

Figure 5.11, a and b, shows results calculated on the s-wave cross sections of the reactions $\gamma\gamma\to\pi\pi$ for transverse photons, $\sigma^{\perp\perp}$. All of the curves shown here, except for the case in which the $\pi\pi$ system has an isospin $T = 2$, exhibit a resonance behavior near the σ meson (a resonance state of the $\pi\pi$ system with quantum numbers $J^{PC} = \sigma^{++}$). The cross sections for the interaction of longitudinal photons are smaller than those for the interaction of transverse photons. The cross sections for the interaction of longitudinal photons enter the total cross section for $\gamma\gamma\to\pi\pi$ with a factor m_γ^2; since m_γ^2 is small, the contribution of these cross sections is extremely small.

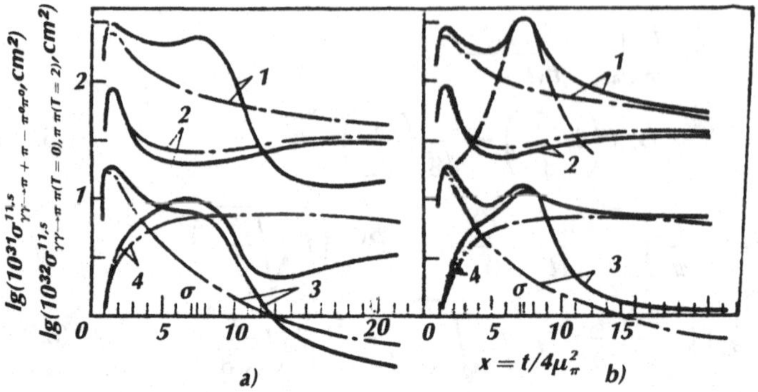

Figure 5.11. *s*-wave cross sections for $\gamma\gamma \to \pi\pi$. (a) Down-parametrization of the phase shift $\delta_s^0(t)$ ($\Gamma\rho \to \pi\gamma \approx .1$ MeV); (b) down-up-parametrization, the phase shift $\delta_s^0(\infty)_{\text{down-up}} = \pi(\Gamma_{\rho \to \pi\gamma} \approx 0.1$ MeV). 1, $\gamma\gamma \to \pi\pi$ ($T = 0$); 2, $\gamma\gamma \to \pi^+\pi^-$ ($T = 2$); 3, $\gamma\gamma \to \pi^+\pi^-$; 4, $\gamma\gamma \to \pi^0\pi^0$ (Dot-dashed lines) Contribution of ω-, ρ-, and π-exchange diagrams; (dashed lines) simplest Breit-Wigner approximation of the σ meson.

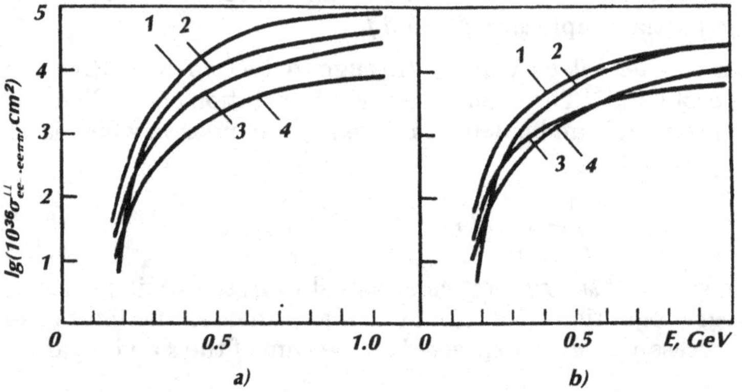

Figure 5.12. Cross section for the interaction of colliding beams in the process $ee \to ee\pi\pi$ as calculated from the *s* wave of $\gamma\gamma \to \pi\pi$. (a) For down-up parametrization; (b) for down-parametrization. 1–4, $\pi^0\pi^0$, $\pi^+\pi^-$, $\pi\pi$ ($T = 0$), and $\pi\pi$ ($T = 2$) two-pion final states.

Figure 5.12, a and b, shows cross sections for the interaction of oppositely directed beams resulting in the production of two π mesons in the process

$$e + e \to e + e + \pi + \pi . \tag{5.37}$$

The cross section for the process $ee \to ee\pi^0\pi^0$ is larger than that for $ee \to ee\pi^+\pi^-$. The cross sections corresponding to different parametrizations of the phase shift $\delta_s^0(t)$ for $\pi\pi$ scattering are so different that it is completely feasible to distinguish them experimentally. There is thus the possibility in principle of distinguishing the down–up solution from the down solution for the phase shift δ_s^0 in $\pi\pi$ scattering. This opportunity is important for a study of the $\pi\pi$ interaction. Only at energies near the threshold does the change of a threshold nature in the phase shift $\delta_s^0(t)$ significantly alter the behavior of the $\gamma\gamma\to\pi\pi$ cross sections; it causes only a slight change in the behavior of the cross section for process (5.37) over the entire range under consideration here. This circumstance must be taken into account in a determination of scattering lengths from an analysis of data on the $\pi\pi$ interaction.

The d-wave cross sections of the reaction $\gamma\gamma\to\pi\pi$ are about two orders of magnitude smaller than the s-wave cross sections and exhibit a resonance behavior near the mass of the f meson. The d-wave component of the total cross section is comparable to the s-wave component only near the f resonance.

The choice of a parametrization for the phase shift in $\pi\pi$ scattering is important. It may be that the down phase shift has a nonresonant form and that in the limit $t\to\infty$ we have $\delta_s^0\to 0$. The down-up phase shift is of resonant form; it passes through the point $\pi/2$, and in the limit $t\to\infty$ it may tend toward either π or 0. In the former case, the index for the down-up phase shift is $\varkappa=1$, and it is a very important matter to consider the general solution of the homogeneous equation in the solution of the dispersion relations.

If $\delta_l^T(\infty)=n\pi$ $(n>0)$, the corresponding partial waves have n resonances, according to Levinson's theorem.[164] To the usual solution of the dispersion relation, which is a particular solution of the inhomogeneous equation, we must add the general solution of the homogeneous equation, which contains an arbitrary polynomial of degree $\varkappa-1$. The solutions corresponding to the indices $\varkappa=0$ and $\varkappa=1$ are thus quite different. If the general solution of the homogeneous equation is ignored, it is not possible to find the expected resonance behavior of the corresponding wave.

It is interesting to compare the curve of the cross section found by the dispersion-relation method with that found through a very simple approximation of the resonance by means of a Breit–Wigner curve (Fig. 5.13). Only near the maximum of the resonance are these cross sections approximately equal. At other energies, they differ substantially.

Let us take a brief look at the results of research on the processes $\gamma\gamma\to K\bar{K}$ and $\gamma\gamma\to\gamma\gamma$ by the dispersion-relation method.[162] A study of the process $\gamma\gamma\to K\bar{K}$ is of independent interest; in addition, once we know its cross section we can estimate its contribution in the $\gamma\gamma\to\gamma\gamma$ cross section (which corresponds to an intermediate $K\bar{K}$ state).

The imaginary part of the partial amplitudes for the process $\gamma\gamma\to K\bar{K}$ contains two-particle π-meson and two-particle K-meson states. Since we have

$$\sigma(K\bar{K}\to S^*\to K\bar{K})/\sigma(\pi\pi\to S^*\to KK)\approx 0.1\,, \tag{5.38}$$

however, we need consider only the two-pion state in the unitarity condition. For $T=0$, for example, this conclusion means that we have

$$\operatorname{Im} f_l\sim(f_{\gamma\gamma\to\pi\pi})_l(f^*_{\pi\pi\to K\bar{K}^*})_l\,. \tag{5.39}$$

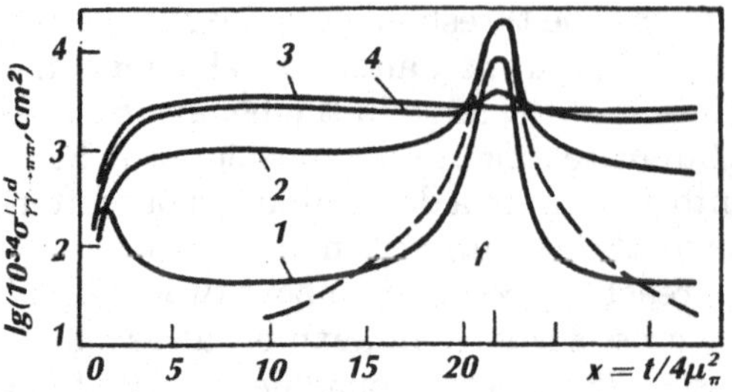

Figure 5.13. *d*-wave cross sections for the reaction $\gamma\gamma\to\pi\pi$. The notation is the same as in Fig. 5.12. (Dashed line) Breit-Wigner curve for the *f* meson.

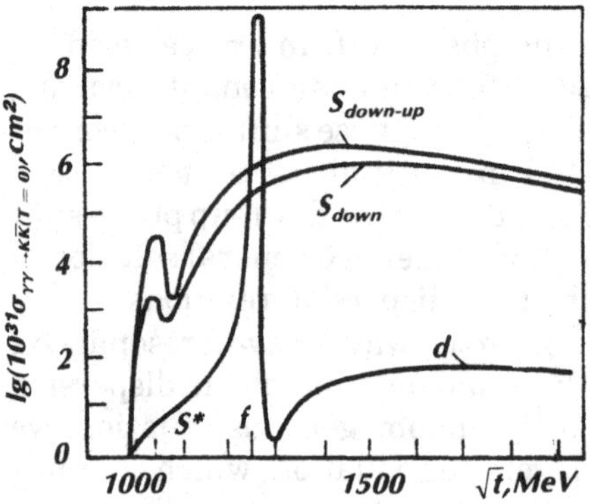

Figure 5.14. Cross section for $\gamma\gamma\to K\bar{K}$ for the *s* and *d* waves.

In other words, in order to use the dispersion-relation method to determine the partial amplitudes for the process $\gamma\gamma\to K\bar{K}$, we need to know the partial amplitudes for the processes $\gamma\gamma\to\pi\pi$ and $\pi\pi\to K\bar{K}$. The former are known, as follows from the discussion above; the latter were calculated in Ref. 165.

As for the process $\gamma\gamma\to\pi\pi$, the $\gamma\gamma\to K\bar{K}$ cross section is calculated for down and down-up solutions for the $\delta_s^{T=0}$ phase shift in $\pi\pi$ scattering. Figure 5.14 shows the behavior of the partial cross sections for $\gamma\gamma\to K\bar{K}$ $(T=0)$; Fig. 5.15 shows the cross section for $ee\to eeK\bar{K}$ $(T=0)$ according to calculations by the equivalent-photon method. This cross section is about two orders of magnitude smaller than the $ee\to ee\pi\pi$ cross section. We can now calculate the cross section for $\gamma\gamma$ scattering, since we know the cross sections for $\gamma\gamma\to\pi\pi$ and $\gamma\gamma\to K\bar{K}$, which appear in the unitarity condition for $\gamma\gamma$ scattering.[13] The values found for the cross sections for $\gamma\gamma$ scattering for the *s* and *d* partial waves for an intermediate $\gamma\gamma\to K\bar{K}$ state are small (Fig. 5.16, a and b) in comparison with the contribution from an intermediate $\pi\pi$ state. It should be noted that the contributions from the intermediate $\pi\pi$ and $K\bar{K}$ states do not interfere with the pole contributions to the process $\gamma\gamma\to\gamma\gamma$ from pseudoscalar mesons (π^0,η^0,χ^0), since they have a positive parity. Figure 5.17 makes a qualitative comparison of the cross sections for $\gamma\gamma$ scattering through intermediate $\pi\pi$ and $K\bar{K}$ states (curves 4 and 5) with the cross section for $\gamma\gamma$ scattering calculated in the approximation of an exchange of only resonance states

Figure 5.15. Cross section for $ee \to eeK\bar{K}$ in the state $T = 0$.

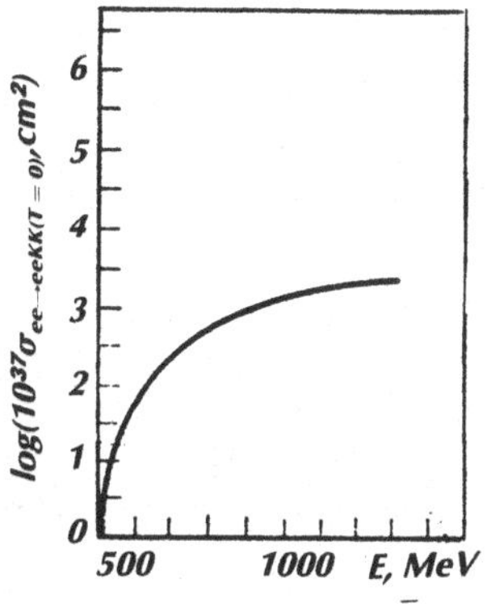

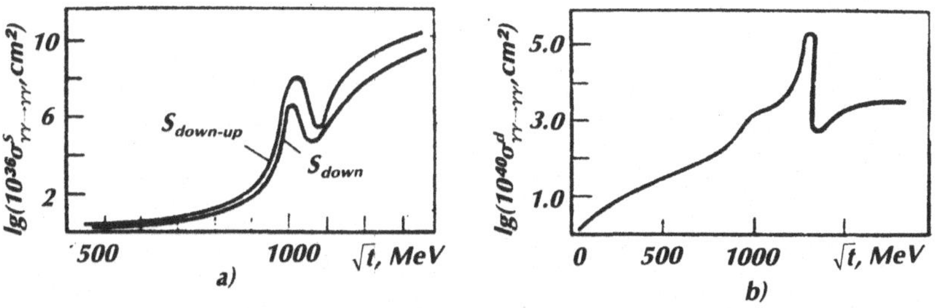

Figure 5.16. Cross section for $\gamma\gamma$ scattering. (a) s wave; (b) d wave.

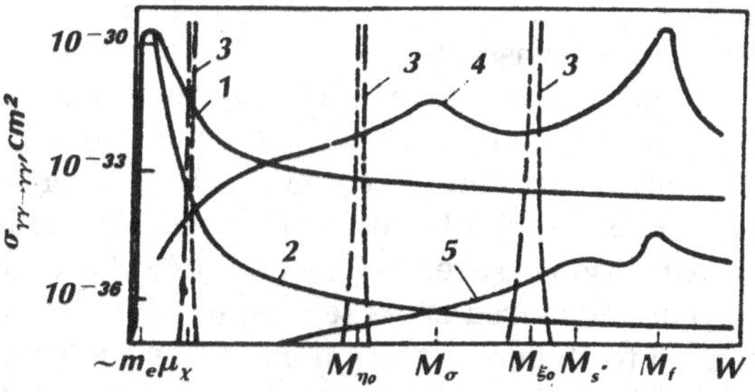

Figure 5.17. Qualitative comparison of the contributions of various intermediate states to $\gamma\gamma \to \gamma\gamma$. 1, $4\pi(d\sigma/d\Omega)$ $(\Theta = 0)$; 2, $4\pi(d\sigma/d\Omega)$ $(\theta = \pi/2)$ (Ref. 168); 3, scattering through intermediate resonance states[167]; 4, scattering through an intermediate two-pion state[162]; 5, scattering through a $K\bar{K}$ intermediate state $(\sigma s_{\text{down-up}} + \sigma^d)$ (Ref. 162).

(curves 3) and with the cross section found in QED (curves 1 and 2). This comparison shows that the cross section for $\gamma\gamma$ scattering in the case in which only e^+e^- pairs are considered in the intermediate state (QED) is substantial only near the threshold for the production of an e^+e^- pair, while the intermediate $\pi\pi$ and $K\bar{K}$ states contribute significantly over a broad energy range, from 300 to 1200 MeV.

Table 5.3. Data on the cross section for $e^+e^- \to e^+e^- - \pi^+\pi^-$.[170,171]

Quantity	Theory (Born approximation)	Experiment
$\sigma_{\mu\mu}$	356×10^{-36} cm^2	$(330\pm45)\times10^{-36}$ cm^2
$\sigma_{\mu\mu}$	30×10^{-36} cm^2	$(70\pm20)\times10^{-36}$ cm^2
$\dfrac{\sigma_{\mu\mu}}{\sigma_{ee}}$	0.55	0.52 ± 0.07
$\dfrac{\sigma_{\pi\pi}}{\sigma_{\mu\mu}+\sigma_{ee}}$	0.030	0.072 ± 0.022

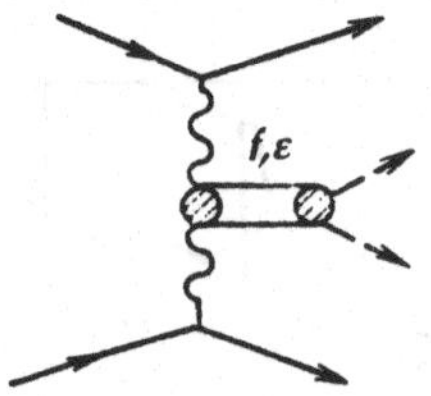

Figure 5.18. Model for $\gamma\gamma$ scattering incorporating the resonances f and ϵ.

Further progress in research on the reactions $\gamma\gamma\to\pi\pi$ and $\gamma\gamma\to K\overline{K}$ will involve taking up a coupled system of equations including equations for the reactions[169] $\gamma\gamma\to\pi\pi$ and $\gamma\gamma\to K\overline{K}$. However, any simplifying assumptions used in approximating the left-hand cut in this case must be carefully justified, for otherwise they may lead to serious errors in the solution.

Experimental data on the production of $\pi^+\pi^-$ pairs in $\gamma\gamma$ collisions (the reaction $e^+e^- + e^+e^-\pi^+\pi^-$) were published in 1979–1980 for beam energies up to 2 GeV. In that experiment, the scattered electrons and positrons were detected in the final state: this circumstance is of particular importance for selecting $\gamma\gamma\to\pi\pi$ events. The data are of a preliminary nature (Table 5.3).

The measured cross section $\sigma_{\mu\mu}$ for the process $e^+e^-\to e^+e^-\mu^+\mu^-$, normalized to the cross section $\sigma_{e^+e^-}$ (for the process $e^+e^-\to e^+e^-e^+e^-$), agrees well with the QED prediction (these calculations were carried out in the equivalent-photon approximation).

However, the experimental values of the cross section for the $\pi^+\pi^-$ interaction are roughly twice as large as that expected theoretically according to calculations in the Born approximation.[154] Evidently, we are already beginning to see the effect of the $\pi\pi$ interaction in the final state; this interaction is dealt with in a systematic way by the dispersion-relation method.[162] In this case we find a good agreement between experimental data and the theoretical predictions. An attempt to describe the reaction $e^+e^- + e^+e^-\pi^+\pi^-$ by a nonrelativistic quark model through the incorporation of only two mesons [f(1270) and ε(1300) resonances] (Fig. 5.18) for $\pi\pi$ energies of about 2 GeV cannot be judged successful, since the model ignores the interaction of π me-

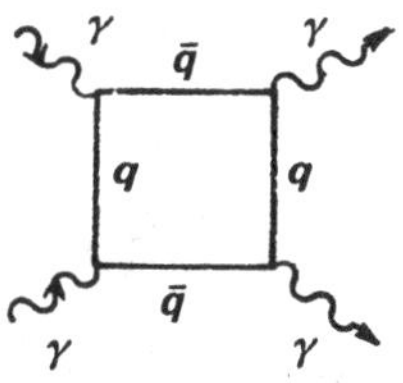

Figure 5.19. Contribution of the $q\bar{q}$ state to $\gamma\gamma$ scattering.

sons in the final state and is capable of describing the reaction $\gamma^*\gamma^* \to \pi\pi$ satisfactorily only in a narrow neighborhood of the f and ε resonances.

5. Multiple production of particles in $\gamma\gamma$ interactions

As the energy of the virtual photons increases, the final state acquires progressively more particles. The cross section for the process $\gamma\gamma \to$ hadrons at high energies can be estimated by the Regge approach [see (5.16)]. However, expression (5.16) does not correctly describe the behavior of the cross section as $\tilde{s}$ is increased. Specifically, if the contribution of the integral

$$\int \frac{d\tilde{s}}{\tilde{s}} \sigma_{\gamma\gamma}(\tilde{s}) f(\tilde{s}) \tag{5.40}$$

to the cross section for $ee \to eeX$ [see (5.15)] is approximated by the contribution to $\sigma_{\gamma\gamma}(\tilde{s})$ from resonance states alone, we find

$$\int \frac{d\tilde{s}}{\tilde{s}} \sigma^R_{\gamma\gamma}(\tilde{s}) f(\tilde{s}) \approx 1.4 \times 10^{-30} \text{ cm}^2 . \tag{5.41}$$

If in the same integral, (5.40), we replace $\sigma_{\gamma\gamma}(\tilde{s})$ by expression (5.16), we find that its value turns out to be 0.4×10^{-30} cm^2, which is smaller by a factor of more than 3 than estimates (5.41), which is clearly on the low side since it ignores the background. Duality (contribution of resonance states $\leftrightarrow$ contribution of Regge trajectories) is thus disrupted. However, a satisfactory agreement with experimental data can be achieved by simply adding to the contributions of the Regge terms in (5.16) a contribution from the quark diagram in Fig. 5.19 (Ref. 172). If we use $m_q \sim 300$ MeV/c^2 for the mass of the quark, we find the following value for the integral:

$$\int \frac{d\tilde{s}}{\tilde{s}} \sigma^{\text{quark}}_{\gamma\gamma}(\tilde{s}) f(\tilde{s}) \approx 0.6 \times 10^{-30} \text{ cm}^2 .$$

Summing this value with 0.4×10^{-30} cm^2, we find a value which is close to the contribution from resonance terms, (5.41). The agreement of the theoretical curve with the experimental data is demonstrated in Fig. 5.20. We see that incorporating the contribution from the diagram in Fig. 5.19 substantially improves the agreement of experimental data with the theoretical predictions. However, we are still a long way from a complete understanding of the

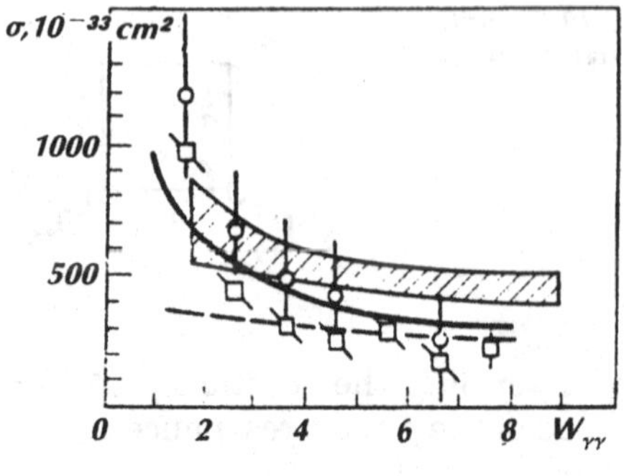

Figure 5.20. Comparison of the measured values of $\sigma_{\gamma\gamma}$ with theoretical predictions. (Hatched band) Data obtained by the TASSO group (in the error corridor); (circles) data obtained by the PLUTO group, extrapolated to the point $Q^2 = 0$, obtained at $\langle Q^2\rangle = 0.1$ (GeV/c)^2; (squares) the same, for $\langle q^2\rangle = 0.25$; (solid line) Ref. 172; (dashed line) prediction of the vector dominance model.

phenomenon. The data obtained by the PLUTO and TASSO groups do not agree completely. The PLUTO group describes its own data (cross section, 10^{-33} cm^2) by the empirical formula

$$1.22(240 + 270/W_{\gamma\gamma}) + 710/W_{\gamma\gamma}^2,$$

while the TASSO group describes its data by the formula

$$380 + 520/W_{\gamma\gamma}.$$

The statistical error is large (~25%).

We have two comments regarding the contribution of the diagram in Fig. 5.19 to the cross section for the $\gamma\gamma$ interaction. First, this cross section is proportional to the fourth power of the charges of the quarks:

$$\sigma_{\gamma\gamma}^{\text{quark}} = \frac{4\pi\alpha^2}{\tilde{s}} \sum_{\substack{\text{color,}\\ \text{flavor}}} e_i^4 \ln\frac{\tilde{s}}{m_q^2},$$

where e_i is the fractional charge of a quark. This expression describes a point-like interaction of photons with quarks; more-accurate experimental data will make it possible to decide whether this interaction is indeed point-like. Second, it is still difficult to say whether the contribution of this diagram alone will be sufficient for explaining the experimental data.

An indication of a point-like nature for the interaction of photons with quarks was found in experiments at the SLAC accelerator. For the ratio

$$R_{\gamma\gamma} = \sigma(\gamma\gamma\to\text{hadrons})/\sigma(\gamma\gamma\to\mu\mu)$$

the value $1.1 \pm 0.3 \pm 0.3$ was found experimentally [measurements were carried out over the interval $0 < M_x^2 \leqslant 6.6$ GeV$^2/c^4$ and $0.7 \leqslant Q^2 \leqslant 0.3$ (GeV/c)^2 for virtual photons]. Theoretical models incorporating a point-like interaction of u, d, and s quarks with photons predict values $R_{\gamma\gamma} = 2/3$ for fractional quark charges and $R_{\gamma\gamma} = 4$ for integer quark charges.[173] Again in this case, however, we can not assert that the point-like interaction has been proved for fractional values of the quark charges, since the experimental data have been obtained in a rather narrow interval of Q^2 values.

With increasing energy of the virtual photons in the reaction $\gamma^* + \gamma^* \to M_x$ (or $e^+e^- \to e^+e^- M_X$), a kinematic analysis of the events becomes more difficult because of the increase in the number of particles in the final state. Events with a large number of particles will constitute most of the events at future high-energy accelerators. Instead of studying in detail the kinematics of each particle separately, one might take the approach of studying global characteristics of the event, by using calorimetric measurement methods: (a) to measure the angular distribution of the energy or "antenna pattern"[14] carried off by hadrons (the hadron energy flux per unit solid angle in the given direction, divided by the energy of the incident radiation per unit area of the target[174]); and (b) to measure the energy correlation in the event, i.e., the correlation between the hadronic energy fluxes through the two calorimeters which detect the state M_x as a function of the angle between the calorimeters.

At this point, these characteristics have been calculated for the process $e^+e^- \to e^+e^- q\bar{q}$; the effects of QCD corrections and bound states on these distributions have been studied; the antenna pattern has been calculated for the production of a pair of heavy quarks ($\gamma^*\gamma^* \to Q\bar{Q}$); and the effect of structure functions of the photon on this pattern has been studied.[175] We can construct a mathematical formulation of the antenna pattern and of energy correlation on the basis of the following definitions. The expression

$$\frac{d\Sigma(\theta)}{d\Omega} = \sum_i \int d\mathbf{q}_i \, \frac{E_i d\sigma_i}{d\mathbf{q}_i} \, d(\Omega_i - \Omega) \tag{5.42}$$

describes the antenna pattern in terms of the inclusive one-particle cross section $d\sigma_i/d\mathbf{q}_i$ for the process $\gamma^* + \gamma^* \to h_i + X_i$, where h_i is the inclusive hadron from the state X. The expression

$$\frac{d^2\Sigma(\Omega,\Omega')}{d\Omega \, d\Omega'} = \sum_{i,j} \int d\mathbf{q}_i d\mathbf{q}_j \delta(\Omega_i - \Omega)\delta(\Omega_j - \Omega) \frac{E_i E_j d^2\sigma_{i,j}}{d\mathbf{q}_i d\mathbf{q}_j} \tag{5.43}$$

describes the energy correlation of the energy with the help of the two-particle inclusive cross section $d^2\sigma_{i,j}/d\mathbf{q}_i \, d\mathbf{q}_j$ for the process $\gamma^* + \gamma^* \to h_i + h_j + X_{ij}$, where h_i and h_j are the two inclusive hadrons from state X. Expression (5.43) is obviously defined in such a way that it should become (5.42) after an integration over one of the solid angles (Ω or Ω').

Definitions (5.42) and (5.43) can be rewritten with the help of expression (5.10):

$$\frac{d\Sigma}{d\Omega} = \int n(\omega_1)d\omega_1 \int n(\omega_2)d\omega_2 \sum_i \int d\mathbf{q}_i \, \frac{d\sigma_i(4\omega_1\omega_2)}{d\mathbf{q}_i} \, \frac{E_i}{\sqrt{\tilde{s}}} \, \delta(\Omega - \Omega_i)\,; \tag{5.44}$$

$$\frac{d^2\Sigma}{d\Omega \, d\Omega'} = \int n(\omega_1)d\omega_1 \int n(\omega_2)d\omega_2 \sum_{i,j} \int d\mathbf{q}_i \int d\mathbf{q}_j \, \frac{d^2\sigma_{ij}(4\omega_1\omega_2)}{d\mathbf{q}_i d\mathbf{q}_j} \times \frac{E_i E_j}{\tilde{s}} \, \delta(\Omega - \Omega_i)\delta(\Omega - \Omega_j)\,. \tag{5.45}$$

Expressions (5.44) and (5.45) are not based on a model of any sort, but they are afflicted with the disadvantages of the equivalent-photon method. It may be that the production of hadronic state X occurs by the scheme $e^+e^- \to e^+e^- q\bar{q}$

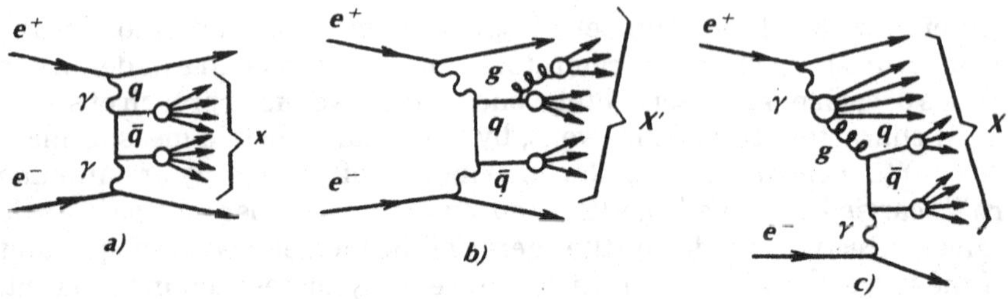

Figure 5.21. Feynman diagrams describing (a) two-jet events and (b, c) three-jet events.

$\to e^+e^-X$, i.e., that state X forms through a fragmentation of the quark q and the antiquark $\bar{q}$ into hadrons (a two-jet scheme) or by the scheme

$$e^+e^-\to e^+e^-q\bar{q}g\to e^+e^-X\,,$$

where state X is formed through the fragmentation of a quark, an antiquark, and a gluon g into hadrons (a three-jet scheme). In expressions (5.44) and (5.45) we would incorporate QCD corrections (at first, in the lowest order in α_s).

For example, to the basic process $e^+e^-\to e^+e^-X$, shown in Fig. 5.21a, we could add diagrams containing QCD corrections for the bremsstrahlung emission of a gluon g (Fig. 5.21b) and for the structure function (Fig. 5.21c). Several other diagrams of this sort could be written out. Direct calculations have shown that all the corrections of the lowest order in α_s make positive contributions to the cross section. Color confinement (gluon confinement) can lead (as can be seen, for example, from the diagram in Fig. 5.21b) to either the production of a third jet of hadrons, if the gluon is hard and carries off a large amount of energy, or the broadening of the quark jet, if it is soft and causes a slight "blurring" of the transverse component of the momentum of the quark (or antiquark). Within the framework of the antenna-pattern method, we could say that the broadening of the angular distribution of jets at high energies is proportional to the quantity $\langle p_\perp \rangle/\tilde{s}$, where $\langle p_\perp \rangle$ is the mean transverse momentum of the hadrons.

The QCD corrections to the energy correlation lead to a symmetry breaking, which should be manifested in the c.m. frame of the colliding virtual photons during the production of a $q\bar{q}$ pair in the process $\gamma^*\gamma^*\to q\bar{q}$ (Fig. 5.21a). For a two-jet event the correction for the emission of a gluon, which causes a broadening (proportional to $\langle p_\perp \rangle/\tilde{s}$) of one of the jets, leads to the same symmetry breaking in the energy correlation.

Consequently, to the problems of the point nature of the interaction of photons with quarks and that of determining the fractional charge of quarks, which we have already discussed, we can add the problems which stem from the QCD predictions regarding the antenna pattern and the energy correlation. These problems will be studied in the multiple production of particles in $\gamma\gamma$ interaction at high energies.

6. Production of jets in $\gamma\gamma$ interactions

Quantum chromodynamics predicts a jet distribution of the hadrons produced in one-photon annihilation (Fig. 5.22). As was mentioned in Chap. 4, a jet distribution of particles in the final state of the one-photon annihilation of colliding e^+e^- beams was found experimentally in a pioneering study by a SLAC–LBL group.[176] In an analysis of this effect in some subsequent studies at the PLUTO installation, at a resultant energy of the colliding beams up to 10 GeV, neutral particles were taken into account. Some interesting characteristics of the behavior of the jets were found.[177] We would expect that in the inelastic e^+e^- scattering $e^+e^- \to e^+e^- + X$, we would also see a manifestation of a jet nature of the particle distribution in the final state as the energy of the colliding beams increases (Fig. 5.21a). It has been suggested that the quark and antiquark may fragment into hadrons, forming jets. Figure 5.21, a–c, shows diagrams for two- and three-jet events. Four-jet events could of course also form, but we will not discuss them because of their complexity.

The production of jets in $\gamma\gamma$ collisions has been studied theoretically in several papers.[178] There is no need to go into detail here. We will examine only the simplest theoretical results, in the expectation of new experimental data. The first experimental results show indications that the number of jets expected theoretically in $\gamma\gamma$ interactions is consistent with the number observed experimentally, and they provide data on the $p_\perp^2$ distribution of hadrons in two-jet events. Let us take a more detailed look at the $p_\perp^2$ distribution of hadrons in two-jet processes. The simplest subprocess which leads to the production of two jets in $\gamma\gamma$ collisions is the production of a quark-antiquark pair (Fig. 5.21a). If we assume that the virtual quark is described by the same propagator as an electron (or μ meson), the cross section for the process $\gamma\gamma \to q\bar{q}$ will be described by an expression similar to that for the process $\gamma\gamma \to e^+e^-$ (or $\gamma\gamma \to \mu^+\mu^-$). Consequently, the cross section for the production of two jets can be written in the form [this is an analog of expression (5.14)]

$$E_{\text{jet}} \frac{d\sigma_{ee \to ee + \text{two jets}}}{dp_{\text{jet}}^2} \sim \left(\frac{\alpha}{2\pi} \ln \eta\right)^2 \frac{\alpha^2}{p_\perp^4} f(\chi_R^{\text{jet}}, \theta_{\text{c.m.}}^{\text{jet}}),$$

where

$$\eta = \begin{cases} \dfrac{s}{4m_e^2} & \text{for electrons which are not detected;} \\ \dfrac{\theta_{\max}^2}{\theta_{\min}^2} & \text{for electrons which are detected;} \end{cases}$$

$$x_R = E_{\text{jet}} / E_{\text{beam}} \ .$$

Figure 5.23 shows preliminary data on the $p_\perp^2$ distribution of charged particles in two-jet events obtained by the PLUTO group at the PETRA accelerator. The solid line shows the QCD predictions in a model with u, d, s, and c quarks with fractional charges. The experimental data are seen to lie about

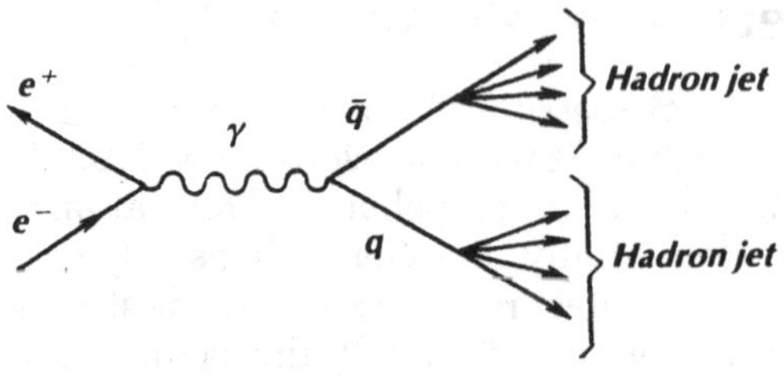

Figure 5.22. Two-jet events for one-photon e^+e^- annihilation.

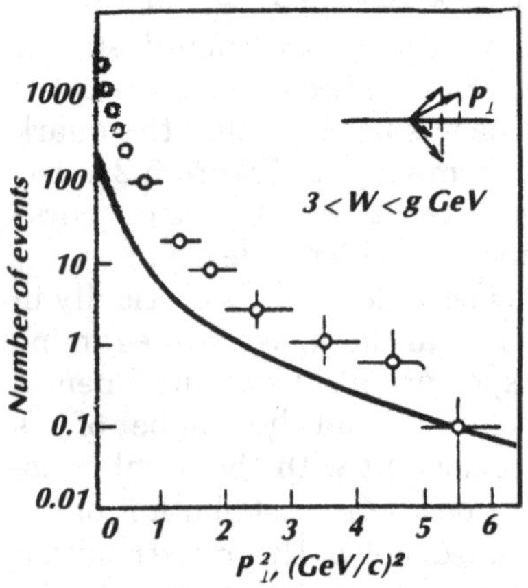

Figure 5.23. Distribution of the transverse momenta of the charged particles with respect to the direction of the colliding e^+e^- beams.

an order of magnitude above the line. At energies up to 10 GeV the experimental cross sections for the process $ee \to ee$ + two jets fall off more rapidly than $1/p_\perp^4$.

It has recently been shown in several studies[179] that if the parameters of the theory are chosen appropriately it is possible to generate a reasonable explanation of the experimental data on deep inelastic scattering of leptons by nucleons, the ratio $R = \sigma(ee \to \text{hadrons})/\sigma(ee \to \mu\mu)$, the width of the f_0 meson, and the cross section for the reaction $\sigma(\gamma p \to \gamma X)$ in a $U(1) \times SU_L(2) \times SU(3)$ theory of strong, electromagnetic, and weak interactions based on quarks with integer charges and charged and neutral gluons. There is the possibility that the behavior of the cross section shown in Fig. 5.23 can also be explained successfully in a model with integer quark charges.

As in the case of the $e^+e^- \to \gamma^* \to$ hadrons, a ratio of cross sections is introduced for the $\gamma\gamma$ interaction:

$$R_{\gamma\gamma} = d\sigma(e^+e^- \to e^+e^- + \text{hadrons})/d\sigma(e^+e^- \to e^+e^- + \mu^+\mu^-),$$

where

$$R_{\gamma\gamma} = 3 \sum e_i^4 [1 + o(\alpha_s/\pi)] .$$

The quantity R in the expression for the cross section for $e^+e^- \to \gamma^* \to$ hadrons contains a sum of the squares of quark charges; a sum of the fourth

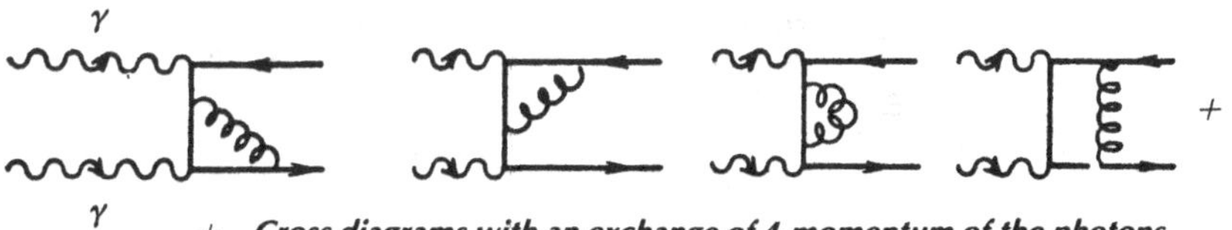

+ *Cross diagrams with an exchange of 4-momentum of the photons*

Figure 5.24. Correction for the emission of a virtual gluon.

+ *Cross diagrams with an exchange of 4-momentum of the photons*

Figure 5.25. Correction for the emission of soft virtual gluons.

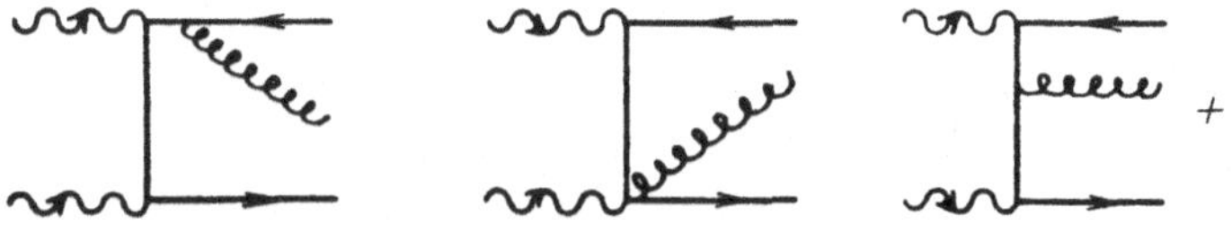

+ *Cross diagrams with an exchange of 4-momentum of the photons*

Figure 5.26. Correction for the emission of hard gluons moving in a collinear fashion.

powers of the quark charges appears in $R_{\gamma\gamma}$. This circumstance is important for selecting among the various quark models. For the model with fractional charges and four quarks(u,d,s,c), the following expression is found for the ratio of cross sections in the leading-log approximation:

$$R_{\gamma\gamma} = 3\times 2[(2/3)^4 + (1/3)^4] = 34/27 \approx 1\,.$$

In a model which considers the mixing of a color singlet component and an octet component below the threshold for the production of color, the value of $R_{\gamma\gamma}$ can reach 10/3.

Corrections of three types on the order of α_s/π to $R_{\gamma\gamma}$ have been considered[180]: a correction for the emission of virtual gluons (Fig. 5.24), one for the emission of soft gluons (Fig. 5.25), and one for the emission of hard, collinearly directed gluons (Fig. 5.26). The calculations show that for the values $\alpha_s \approx 0.3$, $p_\perp^{\min} = 4$ GeV/c, and $\sqrt{s} = 30$ GeV, these corrections reach 11%; i.e., they are small.

A point-like coupling of a photon with a quark should give rise to two noncollinear jets, since some of the energy and momentum will be carried off by the scattered electrons and positrons (without jets). In contrast, jets in the vector-dominance model must be directed along the path of e^+e^- beams (Fig. 5.27). Diagrams of this type contribute little to two-jet events. If two-jet events

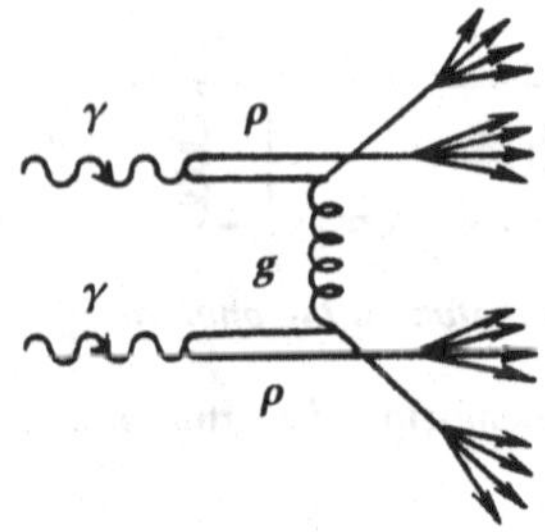

Figure 5.27. Jets produced in the approximation of the vector dominance model.

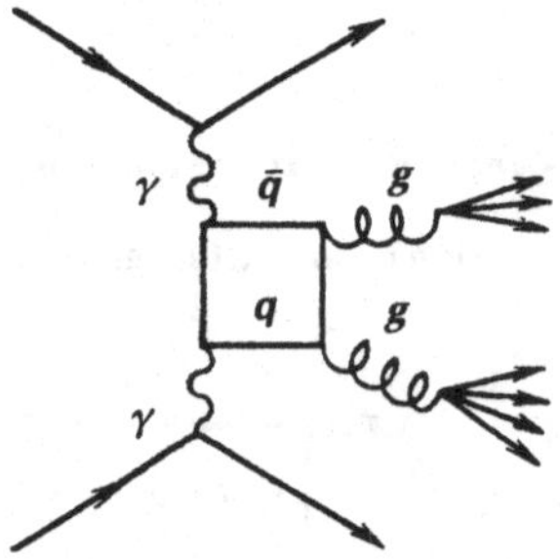

Figure 5.28. Jets formed as a result of the fragmentation of gluons into hadrons.

were not observed experimentally, the implication would be that the perturbation-theory concepts in QCD are not applicable at small distances.

Two-jet events may also arise in the diagram in Fig. 5.28, i.e., through the fragmentation of gluons into hadrons. The presence of a fermion loop leads to the additional factor of $(\alpha_s/\pi)^2$. The contribution of this mechanism to $R_{\gamma\gamma}$ is small in comparison with that from the process $\gamma\gamma\to q\bar{q}$ (less than 0.1). In addition to increasing in the number of two-jet events in $\gamma\gamma$ collisions, this additional mechanism leads to some qualitative differences from the production of jets through a quark-antiquark state: It leads to a broadening of the angular distribution of the hadrons in a jet, and it gives rise to polarization phenomena in the jet (because of the polarization of the gluons).

The polarization property of the jets is of particular interest. In the reaction $e^+e^-\to q\bar{q}g$, there is also a polarization effect, because of the presence of a gluon.[181] In this case, there is an additional but not equivalent opportunity for testing QCD.

The polarization of the jets for a tree diagram (Fig. 5.21a) is zero in the limit of a zero mass of the quarks because of γ_5 invariance.[182] The polarization of the jets which arise from the fragmentation of gluons, $P(\theta_{\rm c.m.})$, turns out to be significant, and it increases with increasing scattering angle $\theta_{\rm c.m.}$ in the c.m. frame of the two photons[183] (Fig. 5.29). The linear polarization of the gluons in the reaction $\gamma\gamma\to gg$ is described by

$$P(\theta_{\rm c.m.}) = (d\sigma_\perp - d\sigma_\parallel)/(d\sigma_\perp + p\sigma_\parallel),$$

where $\perp$ means a polarization perpendicular to the scattering plane of the jets, while $\parallel$ means a linear polarization in the scattering plane. Since the process

Figure 5.29. Polarization of the jets, $P(\theta_{c.m.})$, vs the angle $\theta_{c.m.}$.

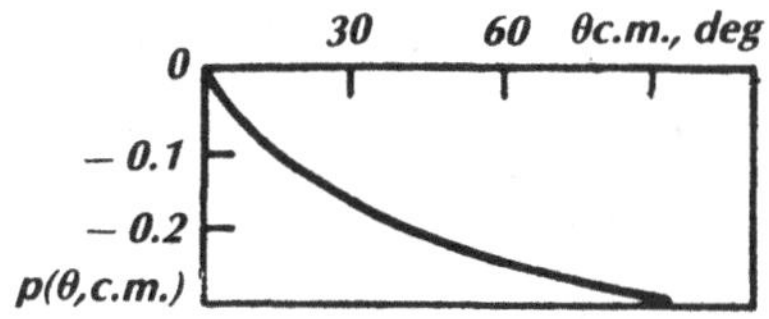

$\gamma\gamma\to q\bar{q}$ does not contribute to the polarization of the jets, the gluon asymmetry can be distinguished without eliminating the background. Neither the imaginary nor real part of the amplitude $\gamma\gamma\to gg$ contains exotic intermediate states in this approximation (see the diagram in Fig. 5.28); if exotic resonances are observed experimentally in the final state in the reaction $\gamma\gamma\to X$, they will find no QCD explanation in terms of a single-loop quark mechanism. On the other hand, they might be interpreted as (for example) multiquark states, gluonium, or glueballs.

7. Deep inelastic $e\gamma$ scattering

By studying the collisions of e^+e^- beams we can carry out an experiment on deep inelastic scattering of electrons by photons and thus learn about the form factors of the photon. For this purpose one chooses the kinematics of the process $e^+ + e^- \to e^+ + e^- + X$ in such a way that one of the leptons (say e^-; Fig. 5.30) is scattered through a large angle and emits a photon with a large 4-momentum q^2, while the other lepton (e^+) is scattered through a very small angle and emits a photon with a nearly zero 4-momentum: $k^2 \approx 0$. In this case the scattering of an electron by a positron may be thought of as the scattering of an electron by a nearly real photon:

$$e^-(p) + \gamma(k) \to e^-(p') + X. \tag{5.46}$$

The kinematic variables are related by the following equalities:

$$(k+p-p')^2 = M_X^2; \quad p=(E,\mathbf{p}); \quad p'=(E',\mathbf{p}');$$
$$k^2 + (p-p')^2 + 2k(p-p') = M_X^2; \quad k=(E_\gamma,\mathbf{k}); \quad |\mathbf{k}| \approx E_\gamma.$$

Here $k^2 \approx 0$ and $(p-p')^2 = -Q^2 = q^2$, so that we have $2kq = M_X^2 + Q^2$. Using these relations, we introduce the customary variables x and y:

$$x = -\frac{q^2}{2kq} = \frac{Q^2}{M_X^2 + Q^2}; \quad y = \frac{kq}{kp} = 1 - \frac{E'}{E}\cos^2\frac{\theta}{2}. \tag{5.47}$$

Using the relations

$$\left.\begin{aligned} &2k(p-p') = 2E_\gamma(E-E') - 2\mathbf{k}(\mathbf{p}-\mathbf{p}'); \\ &|\mathbf{p}| \approx E; \quad |\mathbf{p}'| \approx E', \end{aligned}\right\} \tag{5.48}$$

we can find the energy of the photon by which the electron is scattered:

$$E_\gamma = (Q^2 + M_X^2)/4[E - E'\cos^2(\theta/2)], \tag{5.49}$$

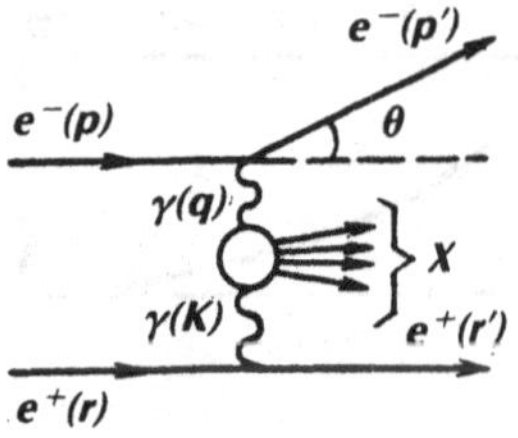

Figure 5.30. Kinematics of deep inelastic scattering of an electron by a photon.

where θ is the angle through which the electron is scattered (Fig. 5.30). If we assume that the invariant mass (M_X^2) of the hadronic state is known, we can determine the energy E_γ unambiguously. In this case, both of the variables (x and Q^2) are known in the form factors of the photon, $F_i^\gamma(x, Q^2)$, so that the functions $F_i^\gamma(x, Q^2)$ can be measured experimentally.

In this case the differential cross section for the process $e^+e^- \to e^+e^- +$ hadrons takes the form[184]

$$\frac{d^4\sigma}{dE_\gamma dy\, dx\, d\Phi} = \frac{2\alpha^2(kp)}{Q^4}[1+(1-y)^2]2xF_T(x,Q^2) + \varepsilon(y)F_L(x,Q^2) + \varepsilon(\xi)\varepsilon(y)\cos 2\Phi F_3(x,Q^2)] , \tag{5.50}$$

where

$$\varepsilon(y) = 2(1-y)/1+(1-y)^2; \quad \xi = E_\gamma/E; \tag{5.51}$$

Φ is the angle between the scattering planes for the scattering of the electron and that of the positron; and F_T, F_L, and F_3 are structure functions of the photon, which are related to the functions of the standard notation, F_1 and F_2, by

$$\left.\begin{aligned} F_1^\gamma(x,Q^2) &= F_T^\gamma(x,Q^2); \\ F_2^\gamma(x,Q^2) &= [F_L^\gamma(x,Q^2) + 2xF_T^\gamma(x,Q^2)] . \end{aligned}\right\} \tag{5.52}$$

The function $F_3(x,Q^2)$ can be measured only if the positron scattering angle is known. If a particle deflected through a small angle (in our case, the positron) is not detected, then only two functions are measured: $F_1(x,Q^2)$ and $F_2(x,Q^2)$.

The form factors $F_T(x,Q^2)$, $F_L(x,Q^2)$, and $F_3(x,Q^2)$ have been calculated in a simpler approximation:

(a) The contribution from the interaction of the photon with the hadrons ρ, ω, and φ is calculated (Fig. 5.31).

(b) The contribution from the interaction of photons with $q\bar{q}$ pairs is calculated (a "box" diagram; Fig. 5.32).

(c) Quantum-chromodynamics corrections to the box diagram are calculated (Fig. 5.33).

Here are explicit expressions for the contributions of the ρ, ω, and φ mesons to the form factors:

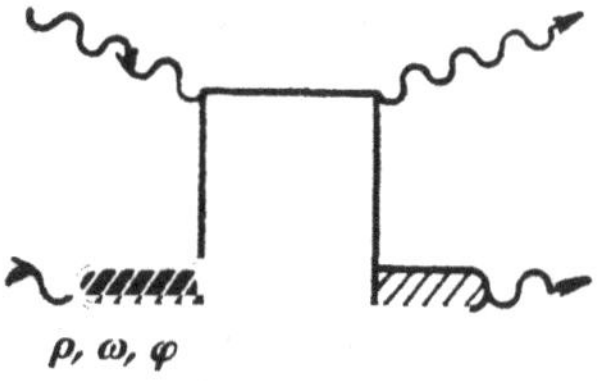

Figure 5.31. Contribution to the form factors of the photon from the interaction of photons with ρ, ω, and φ mesons.

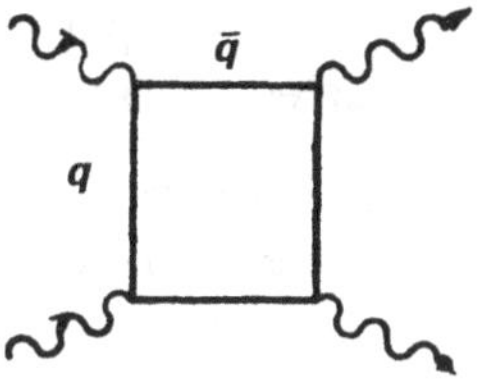

Figure 5.32. Contribution to the photon form factors from a "box" diagram (from the interaction of photons with quarks).

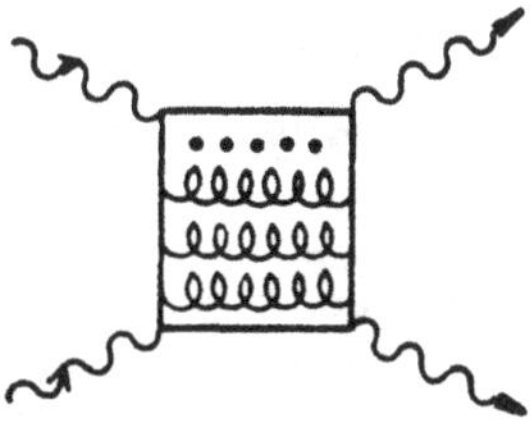

Figure 5.33. QCD corrections to a "box" diagram.

$$F_T^{\gamma}(x,Q^2) \approx \left(\frac{e}{f_\rho}\right)^2 F_T^{\rho^0}(x,Q^2) \approx \frac{1}{2x}\left(\frac{e}{f_\rho}\right)^2 \frac{1}{4}(1-x)\,,$$

$$\frac{f_\rho^2}{4\pi} \approx 2.2;\; F_L^{\gamma}(x,Q^2) \approx o(\langle p_\perp^2\rangle/Q^2) + o(\alpha_s(Q^2))\,; \tag{5.53}$$

$$F_3^{\gamma}(x,Q^2) \approx o(\langle p_\perp^2\rangle/Q^2) + o(\alpha_s(Q^2))\,.$$

Here are explicit expressions for the contributions from the box diagram to the form factors of the photon[185]:

$$F_T^{\gamma}(x,Q^2)|_{\text{box}} = \frac{\alpha}{2\pi}\sum e_i^4 x[x^2 + (1-x)^2]\,\ln\frac{M_X^2}{\Lambda^2}\,;$$

$$F_L^{\gamma}(x,Q^2)|_{\text{box}} = \frac{4\alpha}{\pi}\sum e_i^4 x^2(1-x^2)\,; \tag{5.54}$$

$$F_3^{\gamma}(x,Q^2)|_{\text{box}} = -\frac{\alpha}{\pi}\sum e_i^4 x^3\,.$$

It can be seen from (5.53) and (5.54) that in the limit $x \to 0$ the contribution from the ρ, ω, and φ mesons to $F_T^{\gamma}(x,Q^2)$ is at a maximum, while in the limit $x \to 1$ the meson contribution tends toward zero; the contribution of the box diagram varies as $\ln(Q^2/\Lambda^2)$ with increasing x and Q^2, i.e., becomes the dominant contribution.

The QCD corrections of the next higher-order perturbation theory turn out to be negligible (see also Sec. 6 of Chap. 5). Structure functions for the photon were calculated in the lowest order in the electric charge, but all orders in the strong-interaction constant were taken into account. It turns out that the Q^2 dependence found in approximation (c) (Fig. 5.33) remains the same, but the x dependence of the structure functions changes substantially. The first data on the behavior of the function $F_2(x,Q^2)$ are shown in Fig. 5.34 (Ref. 186).

We should perhaps point out certain differences between the behavior of the nucleon form factors F_i^N and the photon form factors F_i^γ:

(1) It is known that in the limit $x \to 1$ the functions F_i^N vary as $(1-x)^n$, while the functions F_i^γ increase logarithmically.

(2) If the higher twist corrections to the functions $F_i^N(x,Q^2)$ can mask the contributions from the leading logarithms ($1/\ln Q^2$), the contributions of the twist-correction type will be negligibly small for the photon functions $F_i^\gamma(x,Q^2)$.

Finally, we wish to point out that the process

$$e + \gamma \to e + X(q_2 < 0) \tag{5.55}$$

is related to

$$e^+ + e^- \to \gamma^* \to \gamma + X. \tag{5.56}$$

At large q^2 both are described primarily by ladder diagrams, and a switch from one to the other can be made by replacing $q^2 > 0$ by $q^2 < 0$. It becomes possible to directly test the principle of analyticity in q^2 for the amplitudes for processes (5.55) and (5.56) in QCD.

8. Gluonium; Higgs bosons; rare processes

Gluonium is defined as a bound state of two or more gluons. Its production may accompany other production processes in $\gamma\gamma$ collisions (Fig. 5.35). Searching for gluonium in $\gamma\gamma$ interactions may not be the most effective way to find it, but searching for gluonium and studying its properties are important for testing the foundations of QCD. A few words are in order here.

Photons do not interact directly with gluons, so the production of gluonium in $\gamma\gamma$ collisions must occur through intermediate states. The diagram in Fig. 5.36 can serve as an example of the simplest diagrams describing gluonium production. The quark propagator in this diagram might be a b quark or a c quark, for example. In QCD, gluonium must exist. Unfortunately, in QCD itself there are no reference points of any sort for constructing gluonium from gluons, so there are significant variations in the determination of both the limits on the mass of gluonium and its decay width.[187–194] It has been suggested in some studies that the mass of gluonium should be 1–1.5 GeV, while in others it has been suggested that the minimum mass of gluonium is equal to

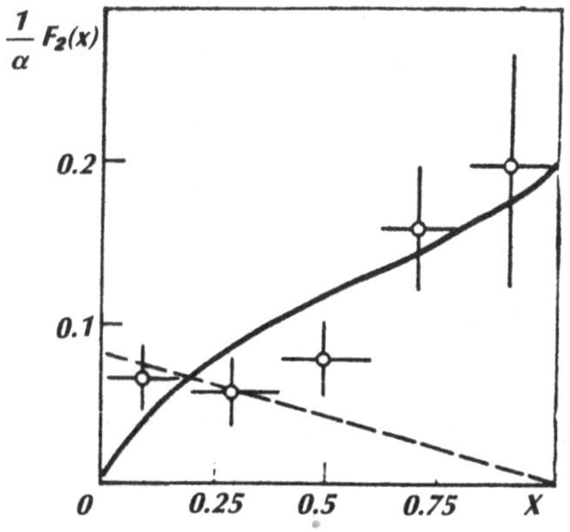

Figure 5.34. Behavior of $F_2(x)$ at $\langle Q^2\rangle = 5$ $(\text{GeV}/c)^2$ (PLUTO data) and theoretical curves of $F_\gamma^\gamma(x, Q^2)$ [solid curve; see (5.54)] and $F_\gamma^\gamma(x, Q^2)$ [dashed curves, calculated in the approximation of contributions from vector mesons; see (5.53)].

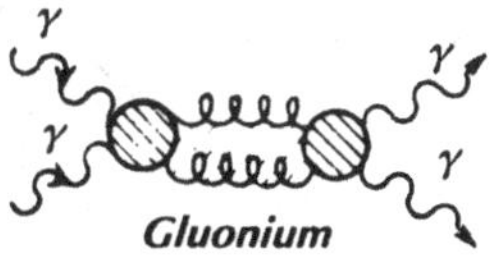

Figure 5.35. Production of gluonium in a $\gamma\gamma$ interaction.

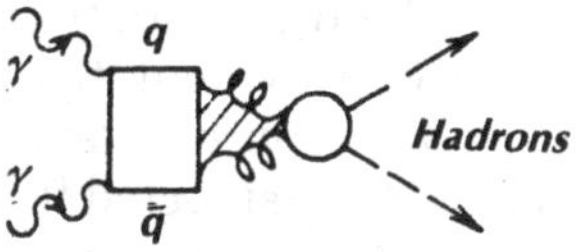

Figure 5.36. Production of gluonium in the reaction $\gamma\gamma\to$hadrons.

the mass of the $X(2.8\ \text{GeV})$ boson. Regarding the decay width of gluonium, the theory has nothing definite to say.

Two-gluon states are the simplest. All of them should be C-even. Two colored "electric" gluons can form $J^P = 0^+$ and $J^P = 2^+$ states, while two colored "magnetic" gluons can form another, hyperfine 0^+ and 2^+ multiplet. Electric and magnetic gluons can form a triplet, with the opposite parity: $0^-,1^+,2^-$ (Ref. 195).

Gluonium can be constructed in a model of bilinear currents.[196] In this case the set of two-gluon states reduces to $0^{++},0^{-+},2^{++},2^{-+}$. There is no state with $J^{PC} = 1^{-+}$ here since the corresponding matrix element for the production of gluonium vanishes. Certain studies, however, postulate the existence of a state with $J^{PC} = 1^{-+}$. Table 5.4 shows the allowed and disallowed states which can be constructed from $q\bar{q}$ states of two or three gluons.

The 1^{-+} state is present as an allowed state here. It can be seen from Table 5.4 that all J^{PC} states are allowed in a $3g$ system.

Opinion is therefore divided even with regard to the classification of the states of gluonium, and the conclusions which are reached depend on the particular models which are used. Equally dependent on the models are the conclusions which are reached to the mass and decay width of gluonium, since

Table 5.4. Allowed (a) and disallowed (d) states.[194]

State	$q\bar{q}$	$2g$	$3g$	State	$q\bar{q}$	$2g$	$3g$
0^{++}	*a*	*a*	*a*	2^{++}	*a*	*a*	*a*
0^{+-}	*d*	*d*	*a*	2^{+-}	*d*	*d*	*a*
0^{-+}	*a*	*a*	*a*	2^{-+}	*a*	*a*	*a*
0^{--}	*d*	*d*	*a*	2^{--}	*a*	*d*	*a*
1^{++}	*a*	*a*	*a*	3^{++}	*a*	*a*	*a*
1^{+-}	*a*	*d*	*a*	3^{+-}	*a*	*d*	*a*
1^{-+}	*d*	*a*	*a*	3^{-+}	*d*	*a*	*a*
1^{--}	*a*	*d*	*a*	3^{--}	*a*	*d*	*a*

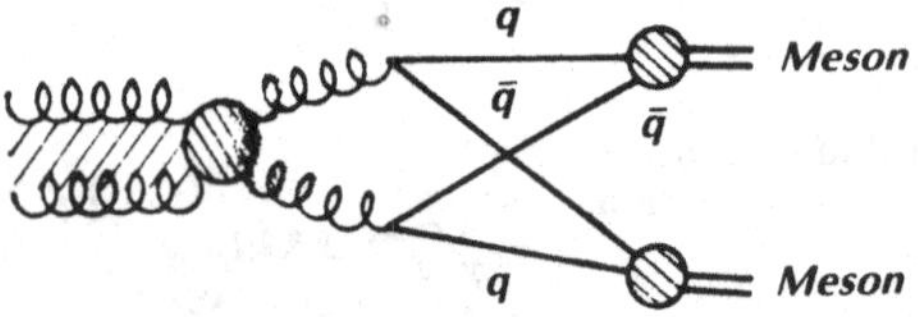

Figure 5.37. Decay of gluonium into a pair of mesons.

(in particular) theoretical predictions of these quantities require knowledge of the gluon distribution function in a hadron, but a function of this sort cannot yet be calculated in QCD, as we discussed in Chap. 3.

It has been suggested that gluonium should decay into ordinary mesons which are constructed from quarks (Fig. 5.37). If the decay width turns out to be small, the decay of gluons will be difficult to distinguish from the decays of ordinary resonance states. If the decay width is instead fairly large, this circumstance will distinguish it from hadronic resonances. Table 5.5 shows some examples of decays of this sort.

We wish to repeat that a phase-shift analysis of the $\pi\pi$ and $K\bar{K}$ final states which are produced in isotopically pure states in the reactions $\gamma\gamma\to\pi\pi$, $K\bar{K}$ (e.g., 0^{++} and 2^{++} in the $\pi\pi$ system; Sec. 4 of Chap. 5) has advantages over other possibilities for searching for low-lying gluonium states.

The representative two-particle decays of gluonium states shown in Table 5.5 indicate that the decays of gluonium will be difficult to distinguish from the decays of known resonances. For this reason, one sometimes sees in the literature attempts to identify certain known resonance states with gluonium states. For example, it has been suggested that the resonance X(2.8 GeV), which decays into two photons, might be a low-lying 0^- state of gluonium.

The $\gamma\gamma$ interaction process can be utilized to solve yet another fundamental problem of elementary-particle physics: proving the existence of Higgs bosons. In the standard Weinberg–Salam theory it is assumed that a single scalar (neutral) boson exists. The scheme for the production of this boson is shown in Fig. 5.38. In the lowest-order perturbation theory (fermion and W^{+-}-boson loops are taken into account) the cross section of the boson is on the order of 10^{-40} cm^2. This value is too small to permit observation of a Higgs

Table 5.5. Two-particle decays of gluonium states [under the assumption of SU(3) symmetry].[195]

J^{PC}		
s wave	*p* wave	Final state *f*
0^{++}	–	$\pi\pi$, $K\overline{K}_{32}\eta\eta$, $\eta'\eta'$
0^{-+}	1^{++}	πS, $K\kappa$, $\eta^{+}\eta'$
2^{-+}	1^{++}, 2^{++}, 3^{++}	πA_2, KK^*, $\eta^{+}\eta'$
2^{++}	1^{++}, 2^{++}, 3^{++}	$\rho\rho$, K^*K^*, $\omega\omega$, $\varphi\varphi$
1^{--}	0^{+-}, 1^{+-}, 2^{+-}	$\rho\delta$, $K^*\kappa$

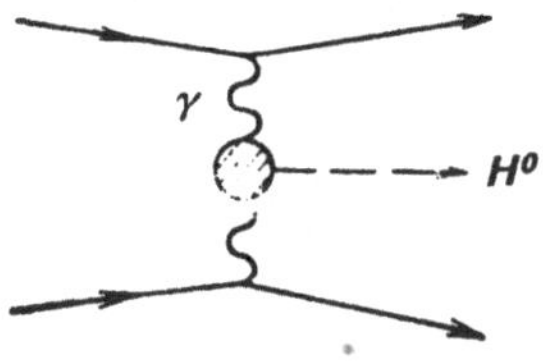

Figure 5.38. Production of a neutral Higgs boson in a $\gamma\gamma$ interaction.

boson. In a model with an expanded Higgs sector, in which it is also assumed that charged Higgs bosons exist, it is possible to find a value on the order of 10^{-36} cm^2 for the cross section. Such cross sections for the production of charged $H^{\pm}$ bosons would be completely accessible to measurements.

Diagrams describing the production of $H^{\pm}$ bosons are shown in Fig. 5.39. Here the cross section for the production of neutral bosons H^0 turns out to be four or five orders of magnitude smaller because of the inclusion of the loop of charged $H^{\pm}$ bosons (Fig. 5.40).

In a model with an expanded Higgs sector, the cross-section values cannot be regarded as anything better than estimates. As for gluonium, QCD gives no hint regarding the masses of Higgs bosons, so that the search for them is seriously complicated.

Finally, in $\gamma\gamma$ collisions one could study the rather interesting couplings of the four-boson type: $\gamma\gamma\to W^+W^-$, $\gamma\gamma\to Z^0Z^0$, etc. These topics are of interest for the theory. It was reported[197] in 1980 that the process $\gamma\gamma\to\rho^0\rho^0$ had been observed in e^+e^- scattering at beam energies of 15 and 18.3 GeV. According to the vector-dominance model, the reaction $\gamma\gamma\to\rho^0\rho^0$ should go primarily through the elastic scattering $\rho^0\rho^0\to\rho^0\rho^0$.

In this case the cross section for the process $\gamma\gamma\to\rho^0\rho^0$ can be written

$$\frac{d\sigma}{dt}(\gamma\gamma\to\rho^0\rho^0)=\left(\frac{\alpha\pi}{\gamma_\rho^2}\right)^2\left(\frac{p^*}{k^*}\right)^2\frac{d\sigma}{dt}(\rho^0\rho^0\to\rho^0\rho^0),$$

where $\gamma_\rho^2/4\pi\approx 0.5$ in the $\gamma\rho$ coupling constant, p^* is the momentum of the ρ^0 meson in the c.m. frame of the two photons, k^* is the momentum of the photon in the same frame, and t is the square of the 4-momentum transfer from the photon to the ρ^0 meson.

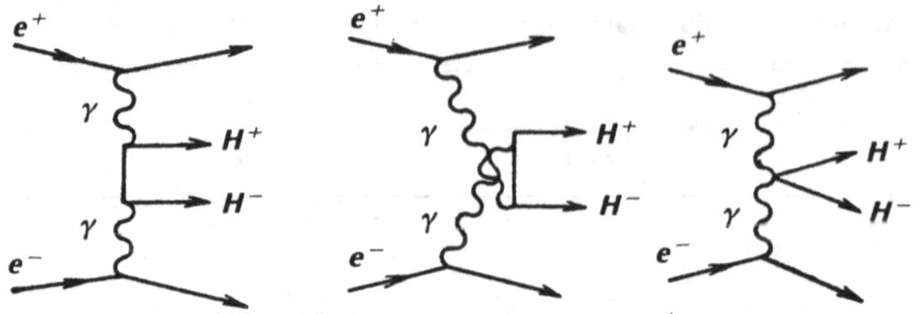

Figure 5.39. Production of charged Higgs bosons in a $\gamma\gamma$ interaction.

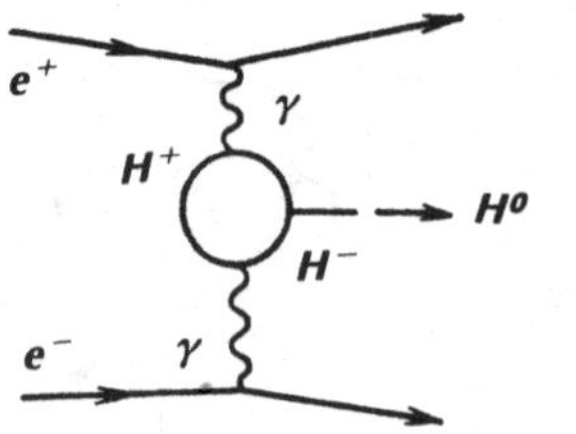

Figure 5.40. Production of a neutral Higgs boson in the reaction $\gamma\gamma \to H^- H^+ \to H^0$.

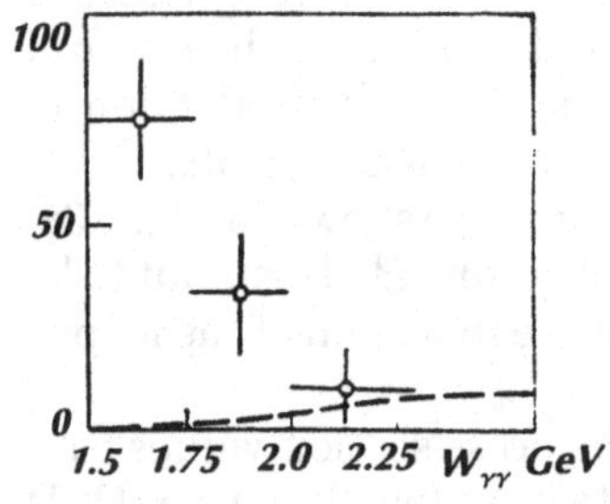

Figure 5.41. Comparison of the measured cross section for $\gamma\gamma \to \rho^0\rho^0$ with the prediction of the vector dominance model.

At high energies it can be assumed within the framework of the quark model that we have $\sigma(\rho^0\rho^0 \to \rho^0\rho^0) \approx (4/9)^2 \sigma(pp \to pp)$. If we also assume $d\sigma/dt \sim \exp(At)$ and $A = 5.6$ GeV, we find (in units of 10^{-33} cm^2/GeV^2)

$$(d\sigma/dt)(\gamma\gamma \to \rho^{00}) \approx (p^+/k^*)^2 100 \exp(5.6t)\,.$$

Integrating this cross section over t, and recalling that we are talking about only the asymptotic contribution to the cross section, we find an expression for the total cross section $\sigma(\gamma\gamma \to \rho^0\rho^0)$ as a function of the invariant energy of the two photons ($W_{\gamma\gamma}$).

The dashed line in Fig. 5.41 shows the energy dependence of the theoretical total cross section. At $W_{\gamma\gamma} < 2$ the experimental points lie far above the predicted course of the cross section. We need to find the meaning of the threshold growth in the cross section for the process $\gamma\gamma^0 \to \rho^0\rho^0$.

Surprises of this sort may await us in a study of any of the $\gamma\gamma$ interaction processes which have not yet been taken up.

* * *

In this chapter we have examined the most important of the current problems regarding the $\gamma\gamma$ interaction. We have listed the most recent experimental results and we have described their present theoretical interpretation. Although significant results have been obtained, questions remain unanswered in each area. For example, the problems of refining the range of applicability of the equivalent-photon method and of expanding its capabilities will remain important as the energy is increased. The functions σ_{St}, σ_{SS}, and τ_{St} must be included in the analysis in order to formulate polarization experiments and experiments in which scattered electrons are detected, in order to achieve a sufficient statistical base, and in order to increase the intensity of the colliding beams. No less important is the problem of accurately calculating corrections of higher orders in the constant α to most of the processes which have been studied.

It is now firmly believed that QED is valid at essentially any value of the square of the 4-momentum transfer, q^2. Nevertheless, a test of the validity of QED will always remain among the most important questions in elementary-particle physics as the energy of the colliding beams and the value of q^2 are increased. Tests of QED in $\gamma\gamma$ collisions should be carried out along with tests in other experiments.

Research on the production of resonances in $\gamma\gamma$ collisions and resonance decays of the type $R\to\gamma\gamma$ is characterized by an unusually broad spectrum of experimental studies and theoretical hypotheses. Testing various symmetries, the point-like nature of the quark-gluon interaction, the hypothesis of a fractional electric charge of quarks, and the consequences of the hypothesis of partial conservation of axial current; incorporating corrections of higher order in the constants α and α_s; and studying the mass spectra of resonance states—these and other possible questions require further research.

Research on the processes $\gamma\gamma\to\gamma\gamma$, $\gamma\gamma\to\pi\pi$, and $\gamma\gamma\to K\bar{K}$ (up to an energy of 2 GeV) should be regarded as lying in the realm of low-energy physics. We have already seen that in these processes it is possible to distinguish isotopically pure states, so that (with a sufficient statistical base) it becomes possible to carry out a qualitative phase-shift analysis of the $\pi\pi$ and $K\bar{K}$ interactions, to find the lowest-lying resonance states in them, and to determine the properties of the resonances. Studies of these reactions may furnish information on the existence of gluonium and Higgs bosons and the masses of these particles. The dispersion-relation method, which gives an excellent description of these processes at low energies, can hardly be replaced by QCD perturbation theory here, since the latter is not applicable at low energies. There is strong interest in searching for new computation methods in QCD at low energies. Perhaps the greatest expectations, however, are pinned on a study of the $\gamma\gamma$ interaction at high energies. The angular distribution of the energy, energy correlation, jets, their polarization, QCD corrections, the structure of the photon and the electron, exotic processes, and the search for gluonium and Higgs bosons—these topics constitute a list of the problems which have been discussed in this chapter and which have been under study for essentially only two or three years. All of these problems might be referred to more accurately as a program for future research. This program is a realistic one, because of the new accelerators which are to come on line in the near future: the LEP (2×50 GeV; 1987, tentatively), the Cornell accelerator (2×50 GeV; 1986 tentatively, if construction is begun in 1983), and the Single Particle Collider (2×50 GeV; 1985, tentatively). In addition, the energy of PETRA is to be increased to 41–46 GeV.

It would be no great surprise if, by the time these accelerators are started up, the list of problems given above will have been extended to include problems involving the asymptotic behavior of the cross sections for $\gamma\gamma$ interactions, electroweak interactions, parity breaking in interactions of polarized particles, and research on the properties of the recently discovered $W^{\pm}$ and Z^0 bosons.

The current decade (through 1990) is thus becoming a stage of development of a powerful new accelerator base for studying phenomena which arise in $\gamma\gamma$ interactions, a stage in which new experimental data on the $\gamma\gamma$ interaction are being accumulated, and a stage of an exceptionally intense study of problems of quantum electrodynamics and quantum chromodynamics at high energies.

Conclusion

Some 15 or 20 years ago, quantum electrodynamics seemed to be a theoretical area in which there was little left to do and which was closed within itself. Today it is a rapidly developing field which promises new discoveries. As the energy of the impinging electrons and photons increases, QED requires corrections of progressively higher order in α and the incorporation of strong and weak interactions in order to explain the experimental data. This situation challenges the theoreticians to carry out calculations on QED effects more accurately and more rapidly. Qualitative changes in computational technology and the development of new computational methods are necessary.

The discovery of the J/ψ and Υ families of particles, of new types of quarks (charmed and beautiful), and of $W^{\pm}$ and Z bosons may be regarded as the beginning of a series of discoveries of unusual particles (and unusual phenomena) which will unfold for physicists after the startup of new accelerators with colliding $e^{+}e^{-}$, $e^{\pm}p$, and $p\bar{p}$ beams.

There is the real possibility that as the energy of the colliding particles increases, QED will be replaced by a new approach to the theory of electromagnetic interactions which will include a universal scale value: a fundamental length.[198]

As the energy increases, and as research continues at a more profound level, any of the phenomena which we have discussed in this book may reveal a new and unexpected side.

Conclusion

References

1. B. Wilk, *Twenty-first Scottish University Summer School in Physics* (Hamilton Hall, Scottish University, 1980).
2. K. Ringhofer and H. Saleker, in *Proceedings of the 1975 International Symposium on Lepton and Photon Interactions at High Energies* (Stanford University, 1975) (Stanford University, Stanford, Calif., 1975), p. 1017.
3. R. N. Faustov, Fiz. Elem. Chastits At. Yadra **3**, 238 (1972) [Sov. J. Part. Nucl.; *Svyazannaya sistema dvukh chastits v kvantovoĭ elektrodinamike (Bound Two-Particle System in Quantum Electrodynamics)* (Joint Institute for Nuclear Research, Dubna, 1974).
4. B. E. Lautrup, A. Peterman, and E. de Rafael, Phys. Rep. **3C**, 193–259 (1972).
5. P. T. Olsen and E. R. Williams, Preprint AMCO-5, Paris, 1975; Phys. Rev. Lett. **42**, 1575–1579 (1979).
6. P. S. Isaev, Fiz. Elem. Chastits At. Yadra **2**, 67–104 (1971) [Sov. J. Part. Nucl. **2**, 45 (1971)].
7. T. Kinoshita, in *Proceedings of the Nineteenth International Conference on High Energy Physics* (Tokyo, 1978) (Phys. Soc., Tokyo, 1979), pp. 571–581.
8. R. S. Van Dyck, Jr., P. B. Schwinberg, and H. G. Dehmelt, Phys. Rev. Lett. **38**, 310–314 (1977).
9. R. S. Van Dyck, Jr., Bull. Am. Phys. Soc. **24**, 758 (1979).
10. T. Kinoshita and W. R. Lindquist, Preprint CLNS-424, April 1979; CLNS-426, June 1979.
11. J. Bailey, K. Borer, F. Combley, *et al.*, Phys. Lett. **68B**, 191–196 (1977).
12. J. Calmet, S. Narison, M. Perrottet, and E. de Rafael, Rev. Mod. Phys. **49**, 21–29 (1977); C. Chlouber and M. A. Samuel, Phys. Rev. D **16**, 3596–3601 (1977).
13. W. E. Lamb, Jr., and R. C. Retherford, Phys. Rev. **72**, 241–243 (1947); **79**, 549–572 (1950); **81**, 222–232 (1951); **86**, 1014–1022 (1952); W. E. Lamb, Jr., Phys. Rev. **85**, 259–276 (1952).
14. H. A. Bethe, Phys. Rev. **72**, 339–341 (1947).
15. S. R. Lundeen and F. M. Pipkin, Phys. Rev. Lett. **34**, 1368–1370 (1975); D. S. Andrews and G. Newton, Phys. Rev. Lett. **37**, 1254–1257 (1976).
16. G. W. Erickson, Phys. Rev. Lett. **27**, 780–783 (1971).
17. P. J. Mohr, Phys. Rev. Lett. **34**, 1050–1052 (1975).
18. V. A. Petrun'kin, Fiz. Elem. Chastits At. Yadra **12**, 692–753 (1981) [Sov. J. Part. Nucl. **12**, 378 (1981)].
19. D. E. Casperson, T. W. Crane, A. B. Denison, *et al.*, Phys. Rev. Lett. **38**, 956–959, 1504 (1977).
20. V. W. Hughes, in *Proceedings of the Nineteenth International Conference on High Energy Physics* (Tokyo, 1978) (Phys. Soc., Tokyo, 1979), pp. 286–291.
21. E. Picasso, Nova Acta Leopold. (Deutsche Akademie der Naturforscher Leopoldina), Suppl. No. 8, Bd. 44, 159–201 (1976).
22. B. N. Taylor, *et al.*, *Fundamental Constants and Quantum Electrodynamics* (Academic, New York, 1969) (Russian translation: Atomizdat, Moscow, 1972).
23. A. A. Logunov and A. N. Tavkhelidze, Nuovo Cimento **29**, 380–399 (1963).
24. S. J. Brodsky and S. D. Drell, Phys. Rev. D **22**, 2236–2243 (1980).

25. R. Gatto, in *Electromagnetic Interactions and Structure of Elementary Particles*, edited by A. M. Baldin (Russian translation: Mir, Moscow, 1969), pp. 42–122.
26. R. Hofstadter, Rev. Mod. Phys. **28**, 214–254 (1956).
27. B. Richter, Phys. Rev. Lett. **1**, 114–116 (1958).
28. J. D. Bjorken and S. D. Drell, Phys. Rev. **114**, 1368–1374 (1959).
29. Chen Mao-chao, Phys. Rev. **127**, 1844–1846 (1962).
30. P. S. Isaev and I. S. Zlatev, Nuovo Cimento **13**, 1–11 (1959); Nucl. Phys. **16**, 608–618 (1960).
31. P. S. Isaev and I. S. Zlatev, Zh. Eksp. Teor. Fiz. **35**, 309–311 (1958) [Sov. Phys. JETP **8**, 213 (1958)]; **37**, 728–734, 1161–1162 (1960) [Sov. Phys. JETP **10**, 519, 826 (1960)].
32. A. A. Logunov and V. S. Vladimirov, Izv. Akad. Nauk SSSR Ser. Mat. Mekh. **23**, 661–676 (1959); A. A. Logunov and P. S. Isaev, Nuovo Cimento **10**, 917–942 (1958).
33. N. N. Bogolyubov and D. V. Shirkov, *Vvedenie v teoriyu kvantovannykh poleĭ* (Nauka, Moscow, 1973) [*Introduction to the Theory of Quantized Fields* (Wiley-Interscience, New York, 1980)].
34. F. A. Berends and R. Kleiss, Nucl. Phys. B **177**, 237–262 (1981).
35. V. Alles-Borelli, M. Bernardini, D. Bollini, *et al.*, Nuovo Cimento **7A**, 345–362 (1972).
36. N. Newman, R. Averill, J. Eshelman, *et al.*, Phys. Rev. Lett. **32**, 483–485 (1974).
37. J.-E. Augustin, A. M. Boyarski, H. Breidenbach, *et al.*, Phys. Rev. Lett. **34**, 233–236 (1975).
38. L. H. O'Neil, B. L. Beron, R. L. Carrington, *et al.*, Phys. Rev. Lett. **37**, 395–398 (1976).
39. E. Hilger, B. L. Beron, R. L. Carrington, *et al.*, Phys. Rev. D **15**, 1809–1814 (1977).
40. B. L. Beron, T. F. Crawford, R. L. Ford, *et al.*, Phys. Rev. D **17**, 2187–2199 (1978).
41. M. L. Perl, G. S. Abrams, A. M. Boyarski, *et al.*, Phys. Rev. Lett. **35**, 1489 (1975); G. J. Feldman, F. Bulos, D. Lüke, *et al.*, Phys. Rev. Lett. **38**, 117–120 (1976).
42. J. Burmester, L. Criegee, H. C. Dehne, *et al.*, Phys. Lett. **68B**, 297–300, 301–304 (1977).
43. S. M. Bilen'kiĭ, *Lektsii po fizike neĭtrinnykh i lepton-nuklonnykh protsessov (Lectures on the Physics of Neutrino and Lepton-Nucleon Processes)* (Energoizdat, Moscow, 1981).
44. G. Passarino and M. Veltman, Nucl. Phys. B **160**, 151–207 (1979).
45. B. H. Wiik, in *Proceedings of the Twentieth International Conference on High Energy Physics* (University of Wisconsin, 1980) (American Institute of Physics, New York, 1980), pp. 1379–1429.
46. M. Greco, G. Pancheri-Srivastava, and Y. Srivastava, Nucl. Phys. B **171**, 118–140 (1980).
47. M. Consoli, Nucl. Phys. B **160**, 208–252 (1979).
48. R. Budny and A. McDonald, Phys. Lett. **48B**, 423–426 (1974); R. Budny, Phys. Lett. **55B**, 227–231 (1975).
49. J. G. Branson, in *Proceedings of the 1981 International Conference on Lepton and Photon Interactions at High Energy* (Bonn, 1981) (University of Bonn, 1981), pp. 279–300.
50. E. H. De Groot, G. J. Gounaris, and D. Schildknech, Phys. Lett. **90B**, 427–430 (1980).
51. V. Barger, W. Y. Keung, and E. Ma, Phys. Rev. Lett. **44**, 1169–1172 (1980).

52. P. Barber, U. Becker, H. Benda, *et al.*, Phys. Rev. Lett. **43**, 1915–1918 (1979).

53. Ch. Berger, H. Genzel, R. Griguel, *et al.* (PLUTO Collaboration), Phys. Lett. **99B**, 489–494 (1981).

54. N. N. Bogolyubov, *Report to the International Congress of Theoretical Physicists* (Seattle, 1956) (see Ref. 33); N. J. Bremmermann, R. Oehme, and J. G. Taylor, Phys. Rev. **109**, 2178–2190 (1958).

55. V. A. Matveev, R. M. Muradyan, and A. N. Tavkhelidze, Lett. Nuovo Cimento **7**, 719–723 (1973); S. J. Brodsky and G. R. Farrar, Phys. Rev. Lett. **31**, 1153–1156 (1973).

56. R. Hofstadter, *Nuclear and Nucleon Structure. A Collection of Reprints with an Introduction* (Benjamin, New York, 1963); S. D. Drell and F. Zachariasen, *Electromagnetic Structure of Nucleons* (Oxford Univ. Press, London, 1961) (Russian translation: Izd-vo Inostr. Lit, Moscow, 1962).

57. M. Gourdin, *Herzeg Novi Lectures 1961. Vol. 2. Lectures of High Energy Physics. (Sixth Summer Meeting, Herzeg Novi)* (Gordon and Breach, New York, 1965).

58. L. N. Hand, D. G. Miller, and R. Wilson, Rev. Mod. Phys. **35**, 335–349 (1963).

59. W. Pauli, Rev. Mod. Phys. **13**, 203–232 (1941).

60. Y. Nambu, Nuovo Cimento **6**, 1064–1083 (1957).

61. N. I. Muskhelishvili, *Singulyarnye integral'nye uravneniya (Singular Integral Equations)*, 2nd ed. (Fizmatgiz, Moscow, 1962); R. Omnes, Nuovo Cimento **8**, 316–326 (1958).

62. P. S. Isaev and V. A. Meshcheryakov, Zh. Eksp. Teor. Fiz. **43**, 1339–1348 (1962) [Sov. Phys. JETP **16**, 951 (1963)].

63. P. S. Isaev, V. I. Lend'el, and V. A. Meshcheryakov, Zh. Eksp. Teor. Fiz. **45**, 294–302 (1963) [Sov. Phys. JETP **18**, 205 (1964)].

64. V. G. Grishin, É. P. Kistenev, and Mu Tszyun', Yad. Fiz. **2**, 886–891 (1965) [Sov. J. Nucl. Phys. **2**, 632 (1966)].

65. C. W. Akerlov, W. W. Ash, K. Berkelman, C. A. Lichtenstein, A. Ramanauskas, and R. H. Siemann, Phys. Rev. **163**, 1482–1497 (1967).

66. R. Marschall, in *Proceedings of the 1977 International Symposium on Lepton and Photon Interactions at High Energy* (Hamburg, 1977) (DESY, Hamburg, 1977), pp. 489–554.

67. G. T. Adylov, F. K. Aliev, D. Yu. Bardin, *et al.*, Phys. Lett. **51B**, 402–406 (1974).

68. E. B. Dally, J. Hauptman, C. May, *et al.*, in *Tr. 18-ĭ Mezhdunar. Konf. po fizike vysokikh energiĭ (Proceedings of the Eighteenth International Conference on High-Energy Physics)* (Tbilisi, 1976) (Joint Institute for Nuclear Research, Dubna, 1976), Vol. 1, D1, 2-1040, pp. A7–A8.

69. V. L. Auslender, G. I. Budker, Yu. Pestov, *et al.*, Phys. Lett. **25B**, 433–435 (1967); V. L. Auslender, G. I. Budker, E. V. Pakhtusova *et al.*, Yad. Fiz. **9**, 114–115 (1969) [Sov. J. Nucl. Phys. **9**, 69 (1969)]; J. Augustin, J. Bizot, J. Buon, *et al.*, Phys. Lett. **28B**, 508–512 (1969).

70. R. Perez-y-Jorba, in *Proceedings of the Fourth International Symposium on Electron and Photon Interactions at High Energy* (Liverpool, 1969) (Sci. Research Council, Liverpool, 1969), pp. 213–227.

71. V. A. Meshcheryakov, S. Dubnička, and I. Furdik, Preprint IC/76/102, Trieste, 1976; S. Dubnička, V. A. Meshcheryakov, and J. Milko, J. Phys. G **7**, 605–612 (1981).

72. E. B. Dally, J. M. Hauptman, J. Kibic, *et al.*, Phys. Rev. Lett. **45**, 232–235 (1980).

73. W. R. Molzon, J. Hoffnagle, J. Rochring, *et al.*, Phys. Rev. Lett. **41**, 1213–1216 (1978).

74. S. B. Gerasimov, Zh. Eksp. Teor. Fiz. **23**, 1559–1564 (1966) [Sov. Phys. JETP **50**, 1040 (1966)].

75. E. Clementel and C. Villi, Nuovo Cimento **4**, 1207–1211 (1956); S. Bergia, A. Stanghellini, S. Fubini, and C. Villi, Phys. Rev. Lett. **6**, 367–371 (1961).

76. R. Hofstadter and R. Herman, Phys. Rev. Lett. **6**, 293–296 (1961).

77. M. N. Rosenbluth, Phys. Rev. **79**, 615–619 (1950).

78. J. G. Rutherglen, in *Proceedings of the Fourth International Symposium on Electron and Photon Interactions at High Energy* (Liverpool, 1969) (Sci. Research Council, Liverpool, 1969), pp. 163–176.

79. P. S. Isaev, in *Mezhdunar. shkola molodykh uchenykh po fizike vysokikh energiĭ (International School of Young Scientists on High-Energy Physics)* (Gomel', 1973), R 1,2-7642 (Joint Institute for Nuclear Research, Dubna, 1973), pp. 36–81.

80. W. Bartel, F.-W. Bčiszer, W.-R. Dix, *et al.*, Phys. Lett. **33B**, 245–248 (1970).

81. L. H. Chan, K. W. Chen, J. R. Dunning, Jr., N. F. Ramsey, J. K. Walker, and R. Wilson, Phys. Rev. **141**, 1298–1307 (1966).

82. G. Hohler, E. Pietarinen, I. Sabba-Stefanescu *et al.*, Nucl. Phys. B **114**, 505–539 (1976).

83. V. Silvestrini, in *Proceedings of the Sixteenth International Conference on High Energy Physics* (National Accelerator Laboratory, Brookhaven, 1972), Vol. 4, pp. 1–40; G. Bossompierre, G. Binder, P. Dalpiaz, *et al.*, Phys. Lett. **68B**, 477–479 (1977); B. Delcourt, I. Derado, J. L. Bertrand, *et al.*, Phys. Lett. **86B**, 395–398 (1979).

84. T. G. Körner and M. Kuroda, Phys. Rev. D **16**, 2165–2175 (1977).

85. V. A. Viktorov, S. V. Golovkin, R. I. Dzhelyadin, *et al.*, Yad. Fiz. **32**, 998–1001, 1005–1008 (1980) [Sov. J. Nucl. Phys. **32**, 516, 520 (1980)].

86. R. P. Feynman, *Photon–Hadron Interactions* (Benjamin, New York, 1972) (Russian translation: Mir, Moscow, 1975); J. Bjorken and E. Paschos, Phys. Rev. **185**, 1975–1982 (1969).

87. H. Politzer, Phys. Rev. Lett. **30**, 1346–1349 (1973); D. Gross and F. Wilczek, Phys. Rev. Lett. **30**, 1343–1346 (1973).

88. M. A. Markov, *Neĭtrino (Neutrinos)* (Nauka, Moscow, 1964); Preprint E2-4370, Joint Institute for Nuclear Research, Dubna, 1969.

89. J. Kuti and V. F. Weisskopf, Phys. Rev. D **4**, 3418–3439 (1971).

90. J. Bernstein, *Elementary Particles and Their Currents* (Freeman, San Francisco, 1968) (Russian translation: Mir, Moscow, 1970).

91. R. E. Taylor, in *Proceedings of the Fourth International Symposium on Electron and Photon Interactions at High Energies* (Liverpool, 1969) (Sci. Research Council, Liverpool, 1969), pp. 251–260.

92. C. G. Callan and D. J. Gross, Phys. Rev. Lett. **22**, 156–159 (1969).

93. O. Nachtman, Nucl. Phys. B **38**, 397–417 (1972).

94. L. I. Sedov, *Metody podobiya i razmernosti v mekhanike (Similarity and Dimensionality Methods in Mechanics)* (Gostekhizdat, Moscow, 1957); K. P. Stanyukovich, *Neustanovivshiesya dvizheniya sploshnoĭ sredy (Time-varying Motions of a Continuous Medium)* (Gostekhizdat, Moscow, 1958).

95. V. A. Matveev, R. M. Muradyan, and A. N. Tavkhelidze, Preprint R2-4578, Joint Institute for Nuclear Research, Dubna, 1969.

96. V. A. Matveev, R. M. Muradyan, and A. N. Tavkhelidze, Fiz. Elem. Chastits At. Yadra **2**, 5 (1971) [Sov. J. Part. Nucl. **2**, 1 (1972)]; T. D. Lee, Fiz. Elem. Chastits At. Yadra **4**, 689 (1973).

97. N. N. Bogolyubov, V. S. Vladimirov, and A. N. Tavkhelidze, Teor. Mat. Fiz. **12**, 305–330 (1972).

98. K. Wilson, Phys. Rev. **179**, 1499–1512 (1969).

99. L. W. Hand, in *Proceedings of the Seventeenth International Conference on High Energy Physics* (London, 1974), pp. IV-61–IV-65.

100. R. E. Taylor, in *Proceedings of the 1975 International Symposium on Lepton and Photon Interactions at High Energy* (Stanford University, 1975) (Stanford University, Stanford, Calif., 1975), pp. 679–708.

101. S. D. Drell and Tung-Mow Yan, Phys. Rev. Lett. **24**, 181–186 (1970); G. B. West, Phys. Rev. Lett. **24**, 1206–1209 (1970).

102. L. W. Mo, in *Proceedings of the 1975 International Symposium on Lepton and Photon Interactions at High Energy* (Stanford University, 1975) (Stanford University, Stanford, Calif., 1975), pp. 651–678.

103. H. L. Anderson, V. K. Bharadwaj, and N. E. Booth, in *Tr. 18-ĭ Mezhdunar. Konf. po fizike vysokikh energiĭ (Proceedings of the Eighteenth International Conference on High-Energy Physics)* (Tbilisi, 1976) (Joint Institute for Nuclear Research, Dubna, 1977), Vol. 2, D1, 2-10400, pp. B53–B55.

104. G. Smadja, in *Proceedings of the 1981 International Symposium on Lepton and Photon Interactions at High Energy* (Bonn, 1981) (University of Bonn, 1981), pp. 444–460; J. Dress, *ibid.*, pp. 474–507.

105. L. N. Hand, in *Proceedings of the 1977 International Symposium on Lepton and Photon Interactions at High Energy* (Hamburg, 1977) (DESY, Hamburg, 1977), pp. 417–466.

106. N. N. Bogolyubov, B. V. Struminskiĭ, and A. N. Tavkhelidze, Preprint D-1968, Joint Institute for Nuclear Research, Dubna, 1965; M. Y. Han and Y. Nambu, Phys. Rev. **B139**, 1006–1011 (1965).

107. M. Gell-Mann, Acta Phys. Austriaca Suppl. IX, 733–761 (1972) (Schladming Lectures, 1972).

108. A. J. Buras, Rev. Mod. Phys. **52**, 199–276 (1980).

109. L. N. Lipatov, Yad. Fiz. **20**, 181–198 (1974) [Sov. J. Nucl. Phys. **20**, 94 (1974)]; G. Altarelli and G. Parisi, Nucl. Phys. B **126**, 298–319 (1977).

110. Yu. L. Dokshitser, D. I. D'yakonov, and S. I. Troyan, *Fizika elementarnykh chastits. Mat. 13-ĭ zimneĭ shkoly LIYaF (Elementary Particle Physics. Proceedings of the Thirteenth Winter School of the Leningrad Institute of Nuclear Physics)* (Leningrad Institute of Nuclear Physics, Leningrad, 1978).

111. A. V. Efremov and I. F. Ginzburg, Fortschr. Phys. **22**, 575–610 (1974); A. V. Efremov and A. V. Radyushkin, Teor. Mat. Fiz. **44**, 17–33, 157–171, 327–342 (1980).

112. P. S. Isaev and S. G. Kovalenko, Yad. Fiz. **32**, 756–764 (1980) [Sov. J. Nucl. Phys. **32**, 390 (1980)]; P. S. Isaev and S. G. Kovalenko, Hadronic J. **3**, 919–940 (1980); I. S. Zlatev, Yu. P. Ivanov, P. S. Isaev, and S. G. Kovalenko, Yad. Fiz. **35**, 454–463 (1982) [Sov. J. Nucl. Phys. **35**, 260 (1982)]; V. A. Bednyakov, I. S. Zlatev, P. S. Isaev, and S. G. Kovalenko, Yad. Fiz. **36**, 745–757 [Sov. J. Nucl. Phys. **36**, 436 (1982)].

113. A. A. Logunov, M. A. Mestvirishvili, and Nguyen Van Hieu, Phys. Lett. **25B**, 611–614 (1967).

114. C. N. Yang, in *Proceedings of the Third International Conference on High Energy Collisions of Hadrons* (Stony Brook, New York, 1969), pp. 509–517; J. Benecke, P. T. Chou, C. N. Yang, and E. Yen, Phys. Rev. **188**, 2159–2169 (1969).

115. K. P. Scruler, in *Proceedings of the Twentieth International Conference on High Energy Physics* (University of Wisconsin, 1980) (American Institute of Physics, New York, 1980), pp. 781–783.

116. J. D. Bjorken, Phys. Rev. D **1**, 1376–1379 (1970).

117. F. E. Close, Nucl. Phys. B **80**, 269–299 (1974).

118. J. Kour, Nucl. Phys. B **128**, 219–252 (1977).

119. J. Schwinger, Nucl. Phys. B **123**, 223–240 (1977).

120. M. Gourdin, Nucl. Phys. B **38**, 418–429 (1972).

121. R. McElhaney and S. P. Tuan, Nucl. Phys. B **72**, 487–502 (1974).

122. A. I. Akhiezer and M. P. Rekalo, Fiz. Elem. Chastits At. Yadra **4**, 662–688 (1973) [Sov. J. Part. Nucl. **4**, 277 (1974)].

123. Yu. V. Bushutin, R. I. Dzhelyadin, A. F. Dunaĭtsev, *et al.*, in *Tr. 18-ĭ Mekhdunar. Konf. po fizike vysokikh energiĭ (Proceedings of the Eighteenth International Conference on High-Energy Physics)*, (Tbilisi, 1976) (Joint Institute for Nuclear Research, Dubna, 1977), Vol. 2, D1, 2-10400, p. B60.

124. C. Y. Prescott, W. B. Atwood, H. Cottreu, *et al.*, Phys. Lett. **77B**, 347–352 (1978).

125. E. Derman, Phys. Rev. D **7**, 2755–2775 (1973).

126. C. Bernardini, G. F. Corazza, G. Ghigo, and B. Touschek, Nuovo Cimento **18**, 1293–1295 (1960).

127. K. G. Chetyrkin, A. L. Kataev, and F. V. Tkachev, Phys. Lett. **85B**, 277–279 (1979); M. Dine and J. Sapirstein, Phys. Rev. Lett. **43**, 668–671 (1979); W. Celmaster and R. J. Gonsalves, Phys. Rev. D **21**, 3112–3128 (1980).

128. D. Flugge, International Report, DESY PLUTO-80/01, Hamburg, 1980.

129. A. H. Mueller, Preprint CU-TP-192, Columbia University, 1980.

130. G. Hanson, G. S. Abrams, A. M. Boyarski, *et al.*, Phys. Rev. Lett. **35**, 1609–1612 (1975).

131. R. Brandelik, W. Braunschweig, K. Gather, *et al.*, Phys. Lett. **89B**, 418–422 (1980).

132. R. D. Field and R. P. Feynman, Phys. Rev. D **15**, 2590–2616 (1977); Nucl. Phys. B **136**, 1–76 (1978).

133. J. Ellis, K. Gaillard Mari, and G. G. Ross, Nucl. Phys. B **111**, 253–271 (1976); J. Ellis and I. Karliner, Nucl. Phys. B **148**, 141–147 (1979).

134. R. Brandelik, R. W. Braunschweig, K. Gather, *et al.* (TASSO Collaboration), Phys. Lett. **97B**, 453–458 (1980).

135. Ch. Berger, H. Genzel, R. Grigull, *et al.* (PLUTO Collaboration), Phys. Lett. **97B**, 459–464 (1980).

136. G. J. Feldman and M. C. Perl, Phys. Rep. **33**, 285–365 (1977).

137. G. Wolf, Preprint DESY 80/13, Hamburg, 1980.

138. Ya. I. Azimov, Yu. L. Dokshitser, and V. A. Khoze, Usp. Fiz. Nauk **132**, 443–475 (1980) [Sov. Phys. Usp. **23**, 732 (1980)].

139. P. N. Bogolyubov, N. V. Krasnikov, V. A. Kuz'min, and K. G. Chetyrkin, Fiz. Elem. Chastits At. Yadra **7**, 816–868 (1976) [Sov. J. Part. Nucl. **7**, 325 (1976)].

140. A. B. Govorkov, Fiz. Elem. Chastits At. Yadra **8**, 1056–1105 (1977) [Sov. J. Part. Nucl. **8**, 431 (1977)].

141. J. J. Aubert, U. Becker, P. J. Biggs, *et al.*, Phys. Rev. Lett. **33**, 1404–1406 (1974).

142. J.-E. Augustin, A. M. Boyarski, and H. Breidenbach, Phys. Rev. Lett. **33**, 1406–1408 (1974).

143. G. S. Abrams, D. Briggs, W. Chinowsky, *et al.*, Phys. Rev. Lett. **33**, 1453–1455 (1974).

144. D. L. Scharre, in *Proceedings of the 1981 International Symposium on Lepton and Photon Interactions at High Energy* (Bonn, 1981) (University of Bonn, 1981), pp. 163–189.

145. S. W. Herb, D. C. Hom, L. M. Lederman, *et al.*, Phys. Rev. Lett. **39**, 252–255 (1977); W. R. Innes, J. A. Appel, B. C. Brown, *et al.*, Phys. Rev. Lett. **39**, 1240–1242 (1977); L. M. Lederman, in *Proceedings of the Nineteenth International Conference on High Energy Physics* (Tokyo, 1978) (Phys. Soc., Tokyo, 1979), pp. 706–724.

146. K. Ueno, B. C. Brown, C. N. Brown, *et al.*, Phys. Rev. Lett. **42**, 486–489 (1979).

147. S. L. Glashow, J. Iliopoulos, and L. Maiani, Phys. Rev. D **2**, 1285–1292 (1970).

148. G. Altarelli, N. Gabibbo, and L. Maiani, Nucl. Phys. B **88**, 285–289 (1975); M. B. Einhorn and C. Quigg, Phys. Rev. D **12**, 2015–2030 (1975).

149. R. Brandelick, W. Braunshweig, H.-U. Martyn, *et al.* (DASP Collaboration), Phys. Lett. **70B**, 132–136 (1977); **80B**, 412–417 (1979).

150. E. J. Williams, Proc. R. Soc. London A **139**, 163–186 (1933); Mat. Fyz. Medellingen **13**, 4 (1935).

151. C. F. von Weizsäcker, Z. Phys. **88**, 612–625 (1934).

152. N. Arteaga-Romero, A. Jaccarini, and P. Kessler, C. R. Acad. Sci. B **269**, 153–159, 1129–1133 (1969).

153. V. E. Balakin, V. M. Budnev, and I. F. Ginzburg, Pis'ma Zh. Eksp. Teor. Fiz. **11**, 559–562 (1970) [JETP Lett. **11**, 388 (1970)].

154. S. J. Brodsky, T. Kinoshita, and H. Terazawa, Phys. Rev. Lett. **25**, 972–975 (1970).

155. D. Kessler and P. Kessler, C. R. Acad. Sci. **242**, 3045–3047 (1956); **244**, 1896–1898 (1957); Nuovo Cimento **4**, 601 (1956).

156. V. M. Budnev, I. F. Ginzburg, G. V. Meledin, and V. G. Serbo, Fiz. Elem. Chastits At. Yadra **4**, 239–285 (1973) [Sov. J. Part. Nucl. **4**, 100 (1973)].

157. J. Parisi, J. Phys. (Paris) **35**, C-2, Suppl. to N3, C2-51–C2-61 (1974).

158. W. Wagner, in *Proceedings of the 1979 International Conference in High Energy Physics* (University of Wisconsin, 1980) (American Institute of Physics, New York, 1981), pp. 576–585.

159. F. J. Gilman, in *Proceedings of the 1979 International Conference on Two-Photon Interactions* (Lake Tahoe, California, 1979) (University of California, 1979), pp. 215–233.

160. M. Gourdin and A. Martin, Nuovo Cimento **17**, 224–243 (1960).

161. D. H. Lyth, Nucl. Phys. B **30**, 195–212 (1971); C. J. Brown and D. H. Lyth, Nucl. Phys. B **53**, 323–349 (1973); H. Cheng and T. T. Wu, Nucl. Phys. B **32**, 461–485 (1971); F. J. Yndurain, Nuovo Cimento **7A**, 687–694 (1972).

162. P. S. Isaev and V. I. Khleskov, Pis'ma Zh. Eksp. Teor. Fiz. **16**, 190–193 (1972) [JETP Lett. **16**, 134 (1972)]; Yad. Fiz. **17**, 368–373 (1973) [Sov. J. Nucl. Phys. **17**, 188 (1974)]; Yad. Fiz. **19**, 121–126 (1973) [Sov. J. Nucl. Phys. **19**, 63 (1974)].

163. F. D. Gakhov, *Kraevye zadachi (Boundary-Value Problems)* (Fizmatgiz, Moscow, 1958).

164. R. J. Eden, *High Energy Collisions of Elementary Particles* (Cambridge Univ. Press, New Rochelle, 1968) (Russian translation: Nauka, Moscow, 1970).

165. P. S. Isaev and V. A. Matveev, Yad. Fiz. **4**, 198–204 (1966) [Sov. J. Nucl. Phys. **4**, 144 (1967)].

166. B. De Tollis, Nuovo Cimento **32**, 757–768 (1964); **35**, 1182–1190 (1965).

167. Z. Kunst, R. M. Muradyan, and V. M. Ter-Antonyan, Preprint E2-5424, Joint Institute for Nuclear Research, Dubna, 1980.

168. R. Karplus and H. Neuman, Phys. Rev. **80**, 380–385 (1950); **83**, 776–780 (1951).

169. O. Babelon, J. L. Basdevant, D. Gaillerie, and M. Gourdin, Nucl. Phys. B **114**, 252–270 (1976).

170. M. Brossard, A. Falvard, and J. Jousset, in *Proceedings of the Invited Talk at the International Workshop on $\gamma\gamma$-Interactions* (Amiens, April 1980) (Amiens, 1980), pp. 212–237.

171. A. Courau, in *Proceedings of the Invited Talk at the 1979 International Conference on Two-Photon Interactions* (Lake Tahoe, California, 1979) (University of California, 1979), pp. 174–197; in *Proceedings of the Twentieth International Conference on High Energy Physics* (University of Wisconsin, 1980) (American Institute of Physics, New York, 1981), p. 1612.

172. M. Greco and Y. Srivastava, Nuovo Cimento **43A**, 88–105 (1978).

173. W. R. Frazer and J. F. Gunion, Phys. Rev. D **20**, 147–165 (1979); H. J. Lipkin, Nucl. Phys. B **115**, 104–114 (1979).

174. C. L. Basham, S. L. Brown, S. D. Ellis, and S. T. Love, Phys. Rev. D **17**, 2298–2306 (1978).

175. M. Abud, R. Gatto, and C. A. Savoy, Phys. Rev. D **20**, 2224–2247 (1979); Phys. Lett. **84B**, 229–233 (1979).

176. R. F. Schwitters, in *Proceedings of the 1975 International Symposium on Lepton and Photon Interactions at High Energy* (Stanford University, 1975) (Stanford University, Stanford, Calif., 1975), pp. 5–24.

177. Ch. Berger, W. Lackas, F. Raupach, *et al.* (PLUTO Collaboration), Phys. Lett. **78B**, 176–182 (1978).

178. S. J. Brodsky, T. de Grand, J. F. Gunion, and I. Weis, Phys. Rev. Lett. **41**, 672–676 (1978); Phys. Rev. D **19**, 1418–1443 (1979); C. H. Llewellyn Smith, Phys. Lett. **79B**, 83–87 (1978); S. Berman, J. Bjorken, and J. Kogut, Phys. Rev. D **4**, 3388–3418 (1971).

179. A. N. Tavkhelidze, in *XIV Mezhdunar. shkola molodykh uchenykh po fizike vysokikh energiĭ (Proceedings of the Fourteenth International School of Young Scientists on High-Energy Physics)* (Joint Institute for Nuclear Research, Dubna, 1981), D2-81-158, pp. 9–35; G. M. Vereshkov, S. A. Zharinov, S. V. Ivanov, and S. V. Mikhaĭlov, Yad. Fiz. **32**, 227–235 (1980) [Sov. J. Nucl. Phys. **32**, 117 (1980)]; A. V. Efremov, S. V. Ivanov, and S. V. Mikhaĭlov, Pis'ma Zh. Eksp. Teor. Fiz. **32**, 669–673 (1980) [JETP Lett. **32**, 656 (1980)].

180. F. Berends, Z. Kunszt, and R. Gastmans, Phys. Lett. **92B**, 186–188 (1980).

181. H. A. Olsen, P. Osland, and I. Overbo, Phys. Lett. **89B**, 221–224 (1980).

182. G. L. Kane, J. Pumplin, and W. Repko, Phys. Rev. Lett. **41**, 1689–1691 (1978).

183. A. De Voto, J. Pumplin, W. Repko, and G. L. Kane, Phys. Lett. **90B**, 436–438 (1980).

184. T. F. Walsh, in *Proceedings of the Invited Talk at the International Workshop on $\gamma\gamma$-Interactions* (Amiens, April 1980) (Amiens, 1980), pp. 346–356.

185. E. Witten, Nucl. Phys. B **120**, 189–202 (1977).

186. B. Wiik, in *Proceedings of the Twenty-first Scottish University Summer School in Physics* (Hamilton Hall, Scottish University, 1980).

187. H. Fritzs and P. Minkowski, Nuovo Cimento **30A**, 393–430 (1975).

188. P. G. O. Freund and Y. Nambu, Phys. Rev. Lett. **34**, 1645–1649 (1975).

189. R. L. Jaffe and K. Johnson, Phys. Lett. **60B**, 201–204 (1976).

190. D. Robson, Nucl. Phys. B **130**, 328–348 (1977).

191. P. Roy and T. F. Walsh, Phys. Lett. **78B**, 62–66 (1978).

192. K. Koller and T. F. Walsh, Nucl. Phys. B **140**, 449–467 (1978).

193. K. Ishikawa, Phys. Rev. D **20**, 731–737 (1979).

194. J. J. Coyne, P. M. Fishbane, and S. Meshkov, Phys. Lett. **91B**, 259 (1980).

195. J. D. Bjorken, in *Proceedings of the Summer Institute on Particle Physics* (Stanford University, 1979), SLAC Report No. 224 (Stanford, 1980), pp. 219–289.

196. A. I. Vaĭnshteĭn, V. I. Zakharov, V. A. Novikov, and M. A. Shifman, Fiz. Elem. Chastits At. Yadra **13**, No. 3, 542–612 (1982) [Sov. J. Part. Nucl. **13**, 224 (1982)].

197. R. Brandelik, W. Braunschweig, K. Cather, *et al.* (TASSO Collaboration), Phys. Lett. **97B**, 448–452 (1980).

198. V. G. Kadyshevskiĭ, Fiz. Elem. Chastits At. Yadra **11**, No. 1, 5–39 (1980) [Sov. J. Part. Nucl. **11**, 1 (1980)].

Subject Index

Printed in the United States
By Bookmasters